ANALYTICAL CHEMISTRY SYMPOSIA SERIES — volume 25

electrochemistry, sensors and analysis

Proceedings of the International Conference "Electroanalysis na h'Éireann", Dublin, Ireland, June 10–12, 1986

ANALYTICAL CHEMISTRY SYMPOSIA SERIES — volume 25

electrochemistry, sensors and analysis

Proceedings of the International Conference "Electroanalysis na h'Éireann", Dublin, Ireland, June 10–12, 1986

edited by

Malcolm R. Smyth and Johannes G. Vos

School of Chemical Sciences, National Institute for Higher Education Dublin, Glasnevin, Dublin 9, Ireland

ELSEVIER
Amsterdam — Oxford — New York — Tokyo   1986

0343990₂

ELSEVIER SCIENCE PUBLISHERS B.V.   **CHEMISTRY**
Sara Burgerhartstraat 25
P.O. Box 211, 1000 AE Amsterdam, The Netherlands

Distributors for the United States and Canada:

ELSEVIER SCIENCE PUBLISHING COMPANY INC.
52, Vanderbilt Avenue
New York, NY 10017, U.S.A.

Library of Congress Cataloging-in-Publication Data

International Conference "Electroanalysis na h'Éireann"
    (1986 : Dublin, Ireland)
    Electrochemistry, sensors, and analysis.

    (Analytical chemistry symposia series ; v. 25)
    1. Electrochemical analysis--Congresses. 2. Electro-
chemical analysis--Instruments--Congresses. I. Smyth,
Malcolm R. II. Vos, Johannes G. III. Title. IV. Series.
QD115.I559 1986   543'.0871   86-24070
ISBN 0-444-42719-8

ISBN 0-444-42719-8 (Vol. 25)
ISBN 0-444-41786-9 (Series)

Printed in The Netherlands

CONTENTS

ANALYTICAL VOLTAMMETRY

ANALYTICAL POTENTIOMETRY

BIOELECTROCHEMISTRY

MODIFIED ELECTRODES AND SENSORS

VIII

ELECTROANALYTICAL METHODS IN CLINICAL AND PHARMACEUTICAL CHEMISTRY

ANALYTICAL CHEMISTRY SYMPOSIA SERIES

Volume 1   Recent Developments in Chromatography and Electrophoresis. Proceedings of the 9th
International Symposium on Chromatography and Electrophoresis, Riva del Garda,
May 15—17, 1978
edited by A. Frigerio and L. Renoz

Volume 2   Electroanalysis in Hygiene, Environmental, Clinical and Pharmaceutical Chemistry.
Proceedings of a Conference, organised by the Electroanalytical Group of the Chemical
Society, London, held at Chelsea College, University of London, April 17—20, 1979
edited by W.F. Smyth

Volume 3   Recent Developments in Chromatography and Electrophoresis, 10. Proceedings of the
10th International Symposium on Chromatography and Electrophoresis, Venice, June
19—20, 1979
edited by A. Frigerio and M. McCamish

Volume 4   Recent Developments in Mass Spectrometry in Biochemistry and Medicine, 6. Proceedings of the 6th International Symposium on Mass Spectrometry in Biochemistry and
Medicine, Venice, June 21—22, 1979
edited by A. Frigerio and M. McCamish

Volume 5   Biochemical and Biological Applications of Isotachophoresis. Proceedings of the First
International Symposium, Baconfoy, May 4—5, 1979
edited by A. Adam and C. Schots

Volume 6   Analytical Isotachophoresis. Proceedings of the 2nd International Symposium on Isotachophoresis, Eindhoven, September 9—11, 1980
edited by F.M. Everaerts

Volume 7   Recent Developments in Mass Spectrometry in Biochemistry, Medicine and Environmental Research, 7. Proceedings of the 7th International Symposium on Mass Spectrometry in Biochemistry, Medicine and Environmental Research, Milan, June 16—18, 1980
edited by A. Frigerio

Volume 8   Ion-selective Electrodes, 3. Proceedings of the Third Symposium, Mátrafüred, Hungary,
October 13—15, 1980
edited by E. Pungor

Volume 9   Affinity Chromatography and Related Techniques. Theoretical Aspects/Industrial and
Biomedical Applications. Proceedings of the 4th International Symposium, Veldhoven,
The Netherlands, June 22—26, 1981
edited by T.C.J. Gribnau, J. Visser and R.J.F. Nivard

Volume 10   Advances in Steroid Analysis. Proceedings of the Symposium on the Analysis of Steroids,
Eger, Hungary, May 20—22, 1981
edited by S. Görög

Volume 11   Stable Isotopes. Proceedings of the 4th International Conference, Jülich, March 23—26,
1981
edited by H.-L. Schmidt, H. Förstel and K. Heinzinger

INTRODUCTION TO THE CONFERENCE

The international conference "Electroanalysis na h'Éireann" was
held at the National Institute for Higher Education, Dublin
(NIHED) on June 10-12, 1986. This conference was organised with
the specific aim of bringing together scientists who work in the
fields of electrochemistry, analytical science or biochemistry
and who use electroanalytical techniques or are involved in the
study of electron transport mechanisms.

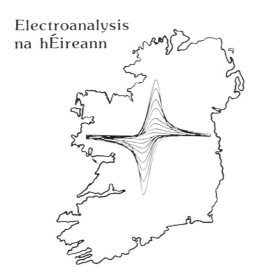

Electroanalysis
na hÉireann

The main theme of the conference was analytical in nature, in
that most papers dealt either with new developments in
electrochemical instrumentation/sensor design or with analytical
applications of electrochemical techniques, especially in the
biomedical sciences. Some papers, however, dealt with
fundamental aspects of electron transport in biological systems
and at electrode surfaces, and highlighted the need for
understanding electron transfer mechanisms prior to the
development of analytical methodology. This interdisciplinarian
approach touched a concordant note with the participants to the
conference, and it is hoped that the sounds will resonate again
at the next meeting of this series in Finland in 1988.

            ..."a terrible beauty is born"...

                              Malcolm R. Smyth
                              Johannes G. Vos
                              Organising Committee

in memory

Professor Dr. Hans Wolfgang Nürnberg 1930-1985

ACKNOWLEDGEMENTS

The Organising Committee would like to thank the many people and organisations whose help, both moral and financial, created the atmosphere for the ultimate success of "Electroanalysis na h'Éireann". In particular we would like to thank Dr. D. O'Hare, the President of the National Institute for Higher Education, Dublin, for opening the conference and for his financial support of this Proceedings Volume. In addition we are deeply indebted to Ms. Vicky La Touche-Price and Mrs. Emer Smyth for secretarial assistance and to all the students and staff of the Institute who gave of their time and patience.

"Go raibh míle maith agaibh"

ANALYTICAL VOLTAMMETRY

M.R. Smyth and J.G. Vos
Electrochemistry, Sensors and Analysis
© Elsevier Science Publishers B.V., Amsterdam — Printed in The Netherlands

ANALYTICAL PULSE VOLTAMMETRY

JANET OSTERYOUNG AND ROBERT OSTERYOUNG
Chemistry Department, State University of New York at Buffalo, Buffalo, NY
14214, (U.S.A.)

SUMMARY
    The most common pulse techniques, normal pulse, differential pulse,
staircase, and square-wave voltammetry, are described, examples are given of
their analytical application, and their strengths and weaknesses are
compared.

INTRODUCTION
    Pulse voltammetry comprises a suite of techniques based on repetitive
application of step-wise changes in potential accompanied by a coordinated
scheme of sampling the measured current. These techniques conceptually are
an elaboration of chronoamperometry, but the earliest qualitative
implementation of these ideas was carried out by Kemula(1). The most
important of these techniques are described in Figure 1.

TECHNIQUES
Normal Pulse Voltammetry
    Of these the simplest is normal pulse voltammetry, which was first
developed by Barker(2). The scheme of potential application in normal pulse
voltammetry comprises a sequence of pulses of successively increasing
amplitude, each of width t_p, separated by a relatively long delay time,
$\tau-t_p$, between pulses. This scheme renders the experiment simple from the
instrumental point of view, because the cycle time, τ, is on the order of
seconds. Because $\tau/t_p \gg 1$, or, as in polarography, the diffusion layer is
renewed by convection between cycles, each cycle constitutes an independent
experiment, so the simple and well-known theories for single potential step

4

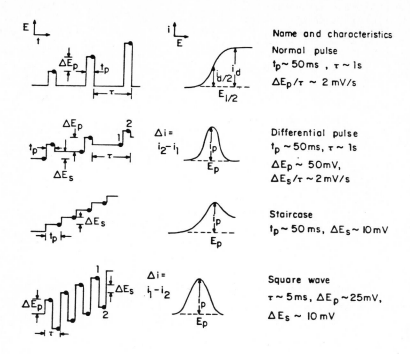

Name and characteristics

Normal pulse
$t_p \sim 50\,ms$, $\tau \sim 1s$
$\Delta E_p / \tau \sim 2\,mV/s$

Differential pulse
$t_p \sim 50\,ms$, $\tau \sim 1s$
$\Delta E_p \sim 50\,mV$,
$\Delta E_s / \tau \sim 2\,mV/s$

Staircase
$t_p \sim 50\,ms$, $\Delta E_s \sim 10\,mV$

Square wave
$\tau \sim 5\,ms$, $\Delta E_p \sim 25\,mV$,
$\Delta E_s \sim 10\,mV$

Fig. 1. Commonly used pulse voltammetric techniques.

chronoamperometry can be applied. The current-potential response is the "normal" sigmoid-shaped wave, hence the name of the technique.

The anodic oxidation of mercury in the presence of a base presents an interesting general scheme for determination of bases. As a specific example, consider the determination of hydroxide by measuring the normal pulse limiting current for formation of $Hg(OH)_2$(aq), which is diffusion-controlled in hydroxide(3). The intrinsic solubility of $Hg(OH)_2$ is ca 0.2 mM. Above the equivalent concentration of hydroxide, 0.4 mM, there are severe irregularities in the limiting current because of precipitation of $Hg(OH)_2$(s) near the electrode. Below this value the waves are well-shaped with flat limiting current plateaus which can be used for direct measurement of hydroxide concentration or indirectly to detect the endpoint in an amperometric titration. Both techniques work well into the low micromolar range of concentration. They provide rapid and accurate analysis in unbuffered solutions under conditions where pH measurements are insufficiently accurate and potentiometric titrations are either very difficult or impossible.

This general concept can be applied to a wide range of bases, some further examples of which will be described below. In addition, substances which on reduction produce a base in the diffusion layer around the electrode can be determined in a similar fashion. Consider the reduction of hydrogen peroxide in unbuffered media(4). This is an irreversible two-electron process which produces hydroxide:

$$H_2O_2 + 2e^- \rightarrow 2OH^- \hspace{4cm} (1)$$

The hydroxide thus produced can be determined as described above. A typical example is shown in Figure 2.

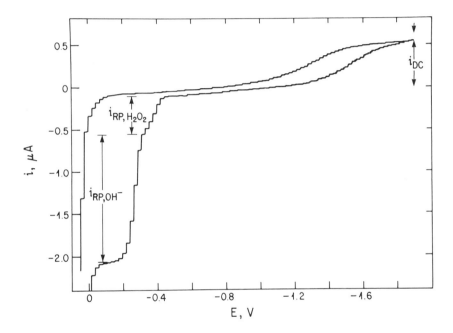

Fig. 2. DC and Reverse Pulse Polarograms of 93.7 μM H_2O_2 in 0.1M $NaClO_4$. τ = 1.1s, t_p = 11.1ms. (Reprinted with permission from Ref. (4). Copyright American Chemical Society, 1983.)

The background curve in Figure 2 is the DC polarogram which gives rise to a poorly defined diffusion-controlled DC current, i_{DC}, from the peroxide reduction. By fixing the initial potential of the normal pulse waveform on this limiting-current plateau, one can operate in the reverse pulse mode, in which the pulse current is used to determine the products formed during the time between pulses. The resulting reverse pulse polarogram has a cathodic branch due to reaction(1). It is displaced to more negative potentials with respect to the DC wave because it appears less reversible on the shorter pulse time scale. There are two anodic waves. The first arises from the reaction

$$H_2O_2 + 2OH^- \rightarrow O_2 + 2H_2O + 2e^- \qquad (2)$$

and is equal in height to the cathodic wave ($i_{RP,H_2O_2} = i_{DC}$). The second is due to oxidation of mercury, and has the same $1OH^-/1e^-$ stoichometry. Thus the total anodic wave height, $i_{RP,OH} + i_{RP,H_2O_2}$, is strictly proportional to the bulk concentration of H_2O_2 and depends otherwise only on the time parameters of the experiment. This wave height may be used directly to determine H_2O_2 or it may be used to detect the endpoint in a diffusion layer titration with a strong acid.

Differential Pulse Voltammetry

Differential pulse voltammetry, also developed by Barker[5] and described in Figure 1, employs the same timing sequence as normal pulse voltammetry, but in this case pulses of constant amplitude are superimposed on a successively increasing (either stepwise or as a slowly-changing ramp) base potential. Differential response is obtained by sampling current both before and after pulse application, giving rise to a peak-shaped response which is approximately the derivative of the normal pulse voltammogram. In contrast with normal pulse voltammetry, the base potential moves through the wave, and hence artifacts common to DC voltammetry associated with electrolysis of the equivalent of many monolayers of material tend also to manifest themselves in differential pulse voltammetry.

The virtues of differential pulse voltammetry, good signal-to-background ratio and adequate sensitivity, are equalled or surpassed by square-wave voltammetry, which is discussed below. Normal and differential pulse are both rather slow techniques, because of the practical requirement that $\tau \gg t_p$. Square-wave (and staircase) voltammetry, in contrast, are fast techniques. Thus square-wave voltammetry is preferred over differential pulse voltammetry. The many examples in the literature of successful application of differential pulse voltammetry[6] should be transferable to square-wave voltammetry with only minor modification.

Staircase Voltammetry

Staircase voltammetry (Figure 1), also promoted by Barker[7], can be viewed as a surrogate for linear scan voltammetry, in that as the step height approaches zero at constant effective scan rate ($\Delta E_s/t_p$) the faradaic response approaches (from below) that expected for linear scan voltammetry at that scan rate. The main virtue of staircase voltammetry in contrast with linear scan voltammetry is that the stepwise scheme of potential application and current sampling discriminates against capacitive current.

Thus, although the sensitivity of staircase voltammetry is less than that of linear scan voltammetry (at the same scan rate) the signal-to-noise ratio is usually greater and hence the detection limit less.

Linear scan voltammetry is widely used and has a well-developed theory. However modern trends in instrumentation are replacing the customary analog linear scan wave form with its digital staircase equivalent. Therefore the uncritical user may be unwittingly carrying out a staircase experiment and analyzing the results with the theory for linear scan. Thus it is practically important to know exactly what the correspondence is between these two techniques. By using a Walsh series approximation for the linear potential profile, it can be shown that the approximate expression for the current in linear scan voltammetry for a reversible reaction, which can be made arbitrarily exact, is identical to the exact expression for the current in staircase voltammetry(7). The practical consequence is that the linear scan and staircase voltammograms at a given scan rate are identical regardless of the time at which current is sampled on each pulse as long as the step height in staircase voltammetry is less than $0.3/n$ mV, where n is the number of electrons. Furthermore, if the current is sampled at one-quarter of the pulse width, the identity holds for $n\Delta E_s \leq 8$ mV(8). The quantitative correspondence between these techniques should make it possible to replace linear scan voltammetry with staircase voltammetry to achieve simplicity of digital design and improved performance of the pulse technique without time-consuming and expensive elaboration of a special theory for staircase voltammetry.

Square Wave Voltammetry

Finally, square-wave voltammetry, also shown in Figure 1, is a modification of a steady-state technique first proposed by Barker(9) and by Kambara(10) and Fujinaga(11). This technique has advantages of precision, flexibility, and speed over those mentioned above(12). Because it is differential, it also gives a peak-shaped response and rejects charging currents well.

To compare differential pulse and square-wave polarography directly we return to the example of anodic oxidation of mercury in the presence of base, in this case the chelons EDTA, EGTA, DCTA, and NTA(13). Figure 3 compares directly voltammograms obtained in the same solution using the same dropping mercury electrode. The square-wave peak currents are about five times larger than the differential pulse peak currents and the background current is diminished in the square-wave response, providing much better resolution of the peak due to NTA. The differential pulse experiment at a

8

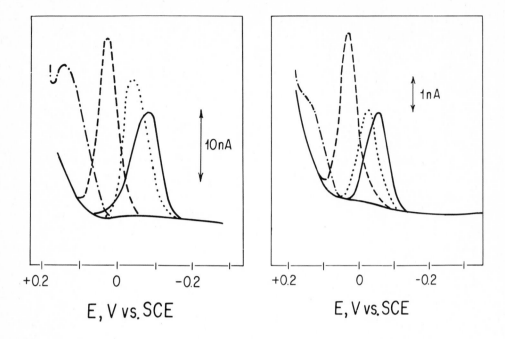

Fig. 3. Oxidation of mercury at the DME in the presence of chelons. m = 0.81 mg/s; 0.1 M NaOAc, pH 5.8; [Y] = 5 x 10^{-7} F; EDTA (-), DCTA (•••), EGTA (---), NTA (-•-). Right: Differential pulse, drop time, 2 s; scan rate 0.5mV/s, pulse amplitude 25mV, pulse width 48s; Left: square-wave, delay time 2s, step height 5 mV, square-wave amplitude 25 mV, frequency 30 Hz. (Reprinted with permission from Ref. (13). Copyright American Chemical Society, 1981.)

scan rate of 0.5 mV/s requires about 17 minutes and 500 drops of mercury, whereas the square-wave scan is carried out on a single drop in 5s.

Square-wave voltammetry also has the interesting property that for reversible reactions the response is invariant in shape and position on the potential scale over a wide range of diffusion conditions(14,15). Thus non-planar or restricted diffusion or convection influences only the magnitude of the response. The reason for this is that the forward and reverse currents are affected in about the same way by these special diffusion conditions, so the net current response is largely unperturbed. This is illustrated for a variety of cases in Figure 4, which shows the forward and reverse currents as well as the net current.

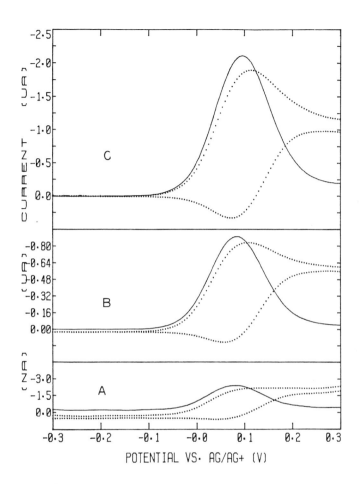

Fig. 4. Square wave voltammograms obtained with various microelectrode geometries. (A) graphite microdisk (8 μm diameter, 50 Hz, $r/\sqrt{DT} = 0.8$); (B) exposed graphite fiber (8 μm diameter, 900 μm length, 50 Hz, $r/\sqrt{Dt} = 0.8$); (C) exposed gold wire (50 μm diameter, 1800 μm length, 60 Hz, $r/\sqrt{Dt} = 5.6$). Step height 5 mV, square-wave amplitude 50 mV. Reference electrode Ag/0.01M AgNO$_3$, 0.1M LiClO$_4$ in acetonitrile. (-) net current, (•••) forward and reverse currents. (Reprinted with permission from Ref. 14.) (Copyright American Chemical Society, 1985.)

Because square-wave voltammetry is a rapid scanning technique, it can be used for voltammetric (as opposed to amperometric) detection in flowing streams. This permits one to characterize as well as to determine the

amount of substances eluted after separation by high-pressure liquid
chromatography(16) or in a process stream. An example of the former is
shown in Figure 5.

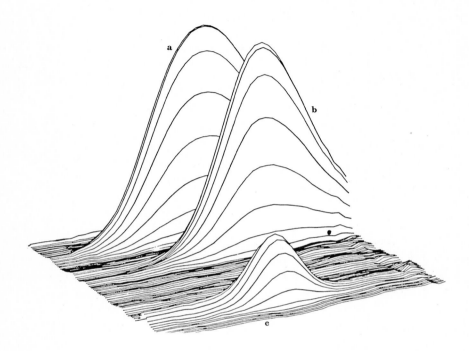

Fig. 5. Chromatopolarogram of N-nitrosodiethanolamine (first peak) and N-
nitrosoproline (second-peak). The small peak is an artifact associated with
injection. Time axis: s; potential axis: V vs SCE; current axis: μA.
ΔE_s = 25 mV, f = 100 Hz. Mobile phase: 1% phosphate (pH 3.5); flow rate:
2 mL/min. [Used with permission. Copyright John O'Dea 1984.]

CONCLUSIONS

It is useful to summarize this brief discussion by classifying the
pulse voltammetric techniques qualitatively according to criteria of time
scale (pulse width), speed (scan rate), sensitivity (slope of calibration

curve, m), rejection of background current, detection limit ($3s_I\sqrt{n}/m$, where
s_I is the standard deviation of the intercept of a calibration curve based
on n measurements in a concentration range near the detection limit),
experimental flexibility (range of parameters accessible), adequacy of
theory, and ease of interpreting experimental results. The classification
is presented in Table 1. Normal pulse shines, despite its rather high

TABLE 1   Comparison of the Pulse Voltammetric Techniques of Figure 1.[a]

	NP	DP	SC	SW
time scale	++++	+++	+++	+++
speed	−	−	+++	+++
sensitivity	++++	+++	+++	++++
background	+	+++	+	++++
detection limit	++	+++	++	++++
flexibility	++++	+	+++	++++
theory	++++	−	+++	+++
interpretation	++++	−	++	+++

[a]See text for discussion; (++++)-excellent, (+++)-very good, (++)-good, (+)-fair, (-)-poor.

background currents which cause detection limits to be rather high (0.1-1µM), because of its experimental and theoretical simplicity. Differential pulse rejects background currents much better, so that although lower in sensitivity it gives improved detection limits (10-100nM). However it is theoretically so complex, especially if employed at a stationary electrode, that the results are practically uninterpretable except as a measure of concentration. Both of these techniques are rather slow. The weak points of staircase voltammetry are rather modest detection limits and the complex shape of the response, which makes interpretation somewhat involved. Square-wave voltammetry tends to combine the best features of the other three techniques and has no weak points by comparison. Particularly notable are speed, excellent detection limits (1-10 nM), and flexibility.

ACKNOWLEDGEMENTS

The work described here was carried out under National Science Foundation grant number CHE 8305748 and previous NSF support as well as support by the Office of Naval Research. J.O. gratefully acknowledges support from the Guggenheim Foundation.

REFERENCES

1   W. Kemula, Collect. Czech. Chem. Commun., 2 (1930) 502-519.
2   G.C. Barker, Z. Anal. Chem., 173 (1960) 79.
3   E. Kirowa-Eisner and Janet Osteryoung, Anal. Chem., 50 (1978) 1062-1066.
4   A. Brestovisky, E. Kirowa-Eisner and Janet Osteryoung, Anal. Chem., 55 (1983) 2063-2066
5   G. C. Barker and A. W. Gardner, AERE C/R 2297, H. M. Stationary Office, London, 1958.
6   G. C. Barker, in "Advances in Polarography," I. S. Longmuir, ed.; Pergamon, New York, 1960, p. 144.

7   M. Seralathen, Robert Osteryoung and Janet Osteryoung, J. Electroanal.
     Chem. in press.
8   R. Bilewicz, R. A. Osteryoung and Janet Osteryoung, Anal. Chem., in
     press.
9   G. C. Barker and I. L. Jenkins, Analyst, 77 (1952) 685-696.
10   T. Kambara, Bull. Chem. Soc. Japan, 27 (1954) 529-534.
11   M. Ishibushi and T. Fujinaga, Bull. Chem. Soc. Japan, 23 (1950) 261.
12   Janet Osteryoung and John O'Dea, Square-wave Voltammetry, in:
     A. J. Bard (Ed.), Electroanal. Chem., Vol. 14, Marcel Dekker, New York,
     1986.
13   Z. Stojek and Janet Osteryoung, Anal. Chem., 53 (1981) 847-851.
14   J. J. O'Dea, M. Wojciechowski, Janet Osteryoung and K. Aoki, Anal.
     Chem., 57 (1985) 954-955.
15   K. Aoki, K. Tokuda, H. Matsuda and Janet Osteryoung, J. Electroanal.
     Chem., in press.
16   R. Samuelson, John O'Dea and Janet Osteryoung, Anal. Chem., 50 (1980)
     2125-2216.

M.R. Smyth and J.G. Vos
Electrochemistry, Sensors and Analysis
© Elsevier Science Publishers B.V., Amsterdam — Printed in The Netherlands

POTENTIALITIES AND APPLICATIONS OF VOLTAMMETRY IN THE ANALYSIS OF THE MARINE ENVIRONMENT

P. VALENTA, L. MART and H.W. NÜRNBERG*

Institute of Applied Physical Chemistry, Nuclear Research Center (KFA) Jülich, P.O.B. 1913, D-5170 Jülich, Federal Republic of Germany

INTRODUCTION

The sea covers 70 % of the globe, yet only recently a number of reliable data on the trace levels of certain heavy metals, e.g. Cd, Cu, Pb and Zn, have been obtained. Although the amount of information is limited, important findings have emerged about the behaviour of these heavy metals and their fate as functions of oceanographic parameters, biological productivity and pollution input.

Since the mid-seventies, many trace element levels have been shown to be orders of magnitude lower than previously reported (refs. 1-7). Drastic examples are given by the ubiquitous heavy metals Pb (Fig. 1) and Zn whose mean levels decreased by a factor of about 200 or more during 40 years. This was mainly result of the improved contamination control during sampling, sample pre-treatment and the determination step, based on the pioneering work of researchers like Patterson, Schaule, Bruland, Boyle and, since 1975, also Mart.

As a consequence of its inherently low accuracy risks, voltammetry, mainly in the mode of differential pulse stripping voltammetry, constitutes a particularly attractive approach, especially for the determination of trace metals in the dissolved state. These concentrations are extremely low in nonpolluted marine regions (refs. 8-10). High ionic strength of sea water of about 0.7 M makes the use of an externally added indifferent electrolyte unnecessary. In situ preconcentration of the heavy metals directly in the voltammetric cell enables the analysis of sea water

*deceased on 12th May 1985

14

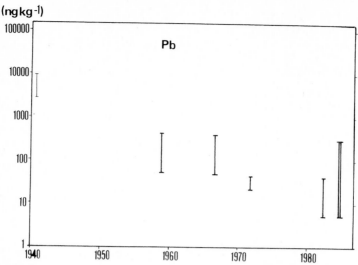

Fig.1. Summary of "relevant" Pb levels in surface waters over the last 4 decades. The drastic decrease is due to an improved contamination control. The last range (double bar) gives realistic Pb levels in surface waters, including rather unpolluted coastal waters.

samples directly without previous chemical preparation or preconcentration steps. Good selectivity of voltammetry allows the simultaneous determination of several metals (up to 5) in one run. Sensitivity to chemical species makes voltammetry attractive for speciation studies (ref. 11). Low costs for instrumentation and operation of about 10^4 $ gives the possibility to use voltammetry widely for the monitoring of heavy metals in the aquatic environment. Disadvantages are the vulnerability of the working electrodes, i.e. the hanging mercury drop electrode (HMDE) or the rotating mercury film electrode (MFE) and the prerequisite of a good electrochemical background.

Risks by dust contamination

Ultimate factors limiting accuracy of the heavy metal determinaton in the marine environment consist mainly of blanks from inadequate sampling methods or airborne contamination during preparative steps, sampling and determination. To overcome these difficulties it is necessary to work either in clean areas or in closed systems (Fig. 2). Bottles from high-pressure polyethylene used for sampling and storage are subjected to scrupulous cleaning procedures in order to leach pollutant trace metals. The last

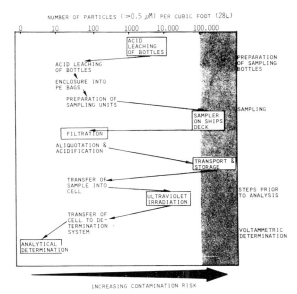

NUMBER OF PARTICLES (>0.5 μM) PER CUBIC FOOT (28L)

Fig.2. Dust contamination during preparative steps, sampling and analysis. Critical operations in highly contaminated surroundings are carried out in quasi-closed systems (Taken from ref. 16).

leaching procedure is performed with pure acids, e.g. Supra-pur^R hydrochloric acid, under a clean bench (ref. 12). Samples are stored deep-frozen at -20^OC to eliminate leaching of trace metals from the container walls and also losses by adsorption. In general, sea-water samples with large amounts of alkali and alkaline earth metal ions competing for adsorption sites, are less affected by losses through adsorption than fresh water samples. Mercury is very sensitive to adsorption onto polyethylene, and therefore subsamples for the determination of mercury have to be stored in glass or quartz bottles at low temperature.

Collection of water samples

Collection of pure surface water directly from a large ship is practically impossible and the research vessel has to be left with a rubber boat, avoiding the area contaminated by the ship. From a distance of at least 500 m, surface water samples are collected from the bow of the boat progressing slowly upwind, by immersing a sample bottle under the water surface (ref. 13). From smaller boats like yachts or fishing boats, the water surface can be reached by extending a telescope pole to appropriate length (up to 4 m). At the end of this pole, a holder for insertion of a sampling bottle is attached. Sampling can be performed from the bow of the ship during slow progression upwind.

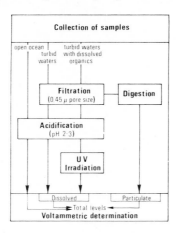

Fig. 3. Flow chart of the procedures from sampling to analysis

Schaule and Patterson have proposed a dynamic sampling of deep sea water samples, i.e. collecting water with the sampler being continually lowered into virgin water during sampling, opened only at the sampling depth and coming back to the surface completely closed. We have adopted the concept of this sampler and have contructed a device for sampling at three different depths in one haul (ref. 14). The sampling units can be prepared in a clean room and will be only inserted, completely closed, into the automatic main frame. Collection of the water samples will be triggered automatically by pressure at the present depth. Intermediate waters (down to 200 m) can be obtained with a modified, commercially available, Go-Flo sampler from General Oceanics, Miami. In coastal and estuarine waters containing appreciable amounts of particulate matter, fractionation is achieved by filtration through a membrane filter (0.45 μm pore size) according to a common convention in aquatic chemistry. A closed filtration pressurized with inert nitrogen is used to avoid contamination of the sample by airborne particles. Samples with a large content of dissolved organic matter (DOM) have to be subjected to UV-irradiation to decompose organic substances, which bind trace metals as inert complex species and are maskered in the voltammetric determination. Particulate matter collected on filters is decom-

posed with a particularly low contamination risk by low-temperature ashing (at 150°C), avoiding losses by evaporation of volati-
le metal compounds. The residue is dissolved in pure acid and pH
is adjusted by addition of alkali.

Airborne contamination has to be avoided, using a laminar flow
clean bench, for all sample operations that are not performed in
closed systems (ref. 15). This results in a drastic diminution of
the level of small particles (0.5 - 5 µm) to below 10 particles
per cubic foot (28 l). In field missions a clean-room container
is installed. Use of protective clean-room overalls and armsheets
together with polyethylene gloves for manipulations at the clean
bench and use of high purity water and reagents is necessary to
achieve low blanks.

Voltammetric determination

Among the heavy metals, Cd, Co, Cr, Cu, Pb, Ni and Zn can be
advantageously determined by voltammetry using various approaches. Amalgam-forming heavy metals such as Cd, Cu, Pb and Zn are
frequently determined by differential-pulse anodic stripping
voltammetry (DPASV) at the hanging (HMDE) or stationary (SMDE)
mercury drop electrode (Fig.3).

These four metals can be determined simultaneously after an
in-situ preconcentration step by cathodic deposition at a potential of -1.2 V vs. Ag/AgCl reference electrode for several minutes with stirring of the solution (900 rpm) at pH 2. The HMDE and
SMDE can be applied for heavy metal concentrations typically down
to 0.1 µg/l or even less. In the ultra-trace range, however, down
to 0.001 µg/l or even lower, the mercury film electrode (MFE)
formed in situ during the plating stage on a specially activated
glassy carbon substrate electrode, is required. During the plating step the electrode is rotating at 2000 rpm to speed up mass
transfer. This very effective mass transfer and the extemely low
thickness of the plated Hg layer of about 300 Å enables to achieve a large ratio (up to 10^5) of the concentrations of the metal
in the amalgam to that in the bulk solution, and thus an excep-
tional sensitivity. In general, the MFE yields currents about a
factor 70x higher, compared to a HMDE, as can be seen in Fig. 4.
Usually, precise determination requires evaluation of the metal
concentrations by standard additions for all types of natural
waters. Precleaned glass or quartz cells are needed because glass,
and to a lower extent, quartz, can release minute amounts of the

18

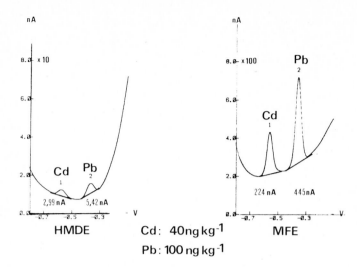

Fig. 4. Comparison of results with a hanging mercury drop electrode (HMDE) and a rotating mercury film electrode (MFE). Surface sea water sample, plating time 10 min. The different current scales should be taken into consideration.

heavy metals to be determined. Usually, two rotating electrode systems are working in parallel in order to speed up the analysis rate. One is used for the preparation (deaeration, film formation), while in the other a determination is performed. The overall RSD, including the replication of sampling, filtration, acidification, UV-irradiation, low-temperature ashing of particulate matter and voltammetric determination, is better than 10 %; the RSD of the voltammetric procedure alone is about 5 % or lower.

Ni and Co do not form amalgams but can be preconcentrated as dimethylglyoxime complexes at pH 9.2 (ammoniacal buffer) on the electrode surface, using a HMDE or SMDE (ref.16). An optimal adsorption potential is -0.7 V (Ag/AgCl), i.e. close to the zero charge potential. A voltammogram is given in fig. 5. The amount of metals adsorbed remains proportional to the overall bulk concentration of the metal, provided that full coverage of the electrode surface is avoided. Analogously, Cr can be determined after adsorption preconcentration as a Cr(III)-DTPA complex at pH 6.2 (acetate buffer with 0.5 M $NaNO_3$ at -1.0 V (Ag/AgCl) (ref. 17). As Cr(VI) is reduced at that potential to Cr(III), followed by complex formation, the total Cr is determined.

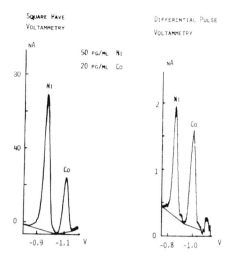

Fig. 5. Comparative Ni-Co analysis with square wave and differential pulse voltammetry at a HMDE (Taken from ref. 18)

For the determination of Co, Cr and Ni by adsorption voltammetry, the square-wave mode is superior to the differential pulse mode. The square-wave voltammetry (SWV), being a modification of the a.c. voltammetry, is especially sensitive to the adsorbed species at the electrode. The other advantage of the SWV is a noticeable increase of the analysis rate. For a simultaneous determination of the Cd, Cu, Pb and Zn having bulk concentrations about 1 µg/l, voltammetry can compete with the electrothermal AAS (Table 1).

Accuracy test

Considering the possible risks of the ultratrace determination of heavy metals in aquatic systems, not only the precision but also the accuracy of the results has to be established. The most efficient way is an interlaboratory comparison of typical sea water samples, using several trace analytical techniques. In one of these campaigns organised by Patterson in 1976, a Pacific water sample has been analyzed for Cd, Pb and Cu by 3 Institutes, applying different determination methods. The results are listed in Table 2a. The agreement between all methods was excellent and proof was given that there exists no methodic bias, provided that contamination has been avoided. A certain advantage of DPASV in

TABLE 1

Time comparison for the determination of heavy metals with voltammetry and AAS (ref. 18)

Determination method	Number of determined elements	Analysis time
Square wave voltammetry $(v = 200 \text{ mVs}^{-1})$	4	20 min
Differential pulse voltammetry $(v = 5 \text{ mVs}^{-1})$	4	40 min
Electrothermal AAS	1	20 min

Remark: The time was calculated for two standard additions

comparison with IDMS in the determination of Pb was its distinctly shorter analysis time (factor 10). The accuracy of the voltammetric method was further proved by another intercomparison (Table 2b). One of the most comprehensive intercomparisons was carried out by Bruland, yielding practically identical results for the different methods applied (ref. 19). Some of his results are given in Table 2c.

Thus, reliable values of the surface water and deep water concentrations of heavy metals are available at present, based on the systematic studies of all possible sources of error discussed above.

TABLE 2a

Accuracy test by interlaboratory comparison of a deep sea water sample. Sample taken 120 km off the Californian coast from 1000 m depth with CIT sampler (Table taken from ref. 19)

	(1)	(2)	(3)
Cd (ng/kg)		105	101
Pb	3,3	–	2,9
Cu	–	115	110

(1) C.C.Patterson, California Institute of Technology, Pasadena CH_3Cl-dithizone extraction and isotopic dilution mass-spectroscopy.
(2) K.W. Bruland, University of California, Santa Cruz, APDC-DDDC-extraction and electrothermal AAS.
(3) L. Mart, Nuclear Research Center, Jülich, DPASV

TABLE 2b

Interlaboratory comparison of sea water. Cadmium in ng/kg

	L.G. Danielsson* APDC-DDTC- extraction GFAAS	L. Mart** DPASV/MFE
Baltic, October 1981	27	29
	28	30
	31	29
Atlantic, October 1981		
surface	24	24
20 m	21	20
100 m	18	18
200 m	18	18
500 m	23	25
1000 m	22	23
2000 m	21	22
2800 m	21	20

* University of Göteborg, Sweden
** Nuclear Research Center, Jülich, FRG

TABLE 2c

California Current intercalibration sea water sample (ng/kg).
Collected at $37^O6'N$, $123^O14'W$. The samples for DPASV were UV
oxidized prior to analysis

	DPASV KFA*	Org. Ext./GFAAS UCSC**	MLML***
Cd	12.1 ± 1.3 (n = 7)	12.3 ± 0.2 (n = 6)	12.7 ± 1.0
Pb	8.6 ± 0.7	≤ 10	~ 6
Cu	72 ± 2 (n = 7)	70 ± 2 (n = 6)	66 ± 3

* Nuclear Research Center Jülich (analyzed by K.W. Bruland during his stay in Jülich)
** Inst. Marine Sciences, Univ. of California, Santa Cruz
*** Moss Landing Marine Laboratory

Constancy of the "background" heavy metal concentration

The background levels of heavy metals in oceans vary according
to the metal type and are functions of oceanographic parameters,
biological productivity and pollution input. These levels are generally constant over a time scale of many years. This is also
true to a limited extent for other natural waters, e.g. lakes,
and even coastal waters. Thus, a mapping of the heavy metal data
in the oceans is possible in order to show their distribution ho-
rizontally, e.g. due to long-scale water movements, and vertical-

ly, i.e. result of the biogeochemical cycle. Voltammetry has contributed substantially (ref. 6) to this task, and at present reliable heavy metal data in several parts of the world oceans are available, most of them from the North East and Central Pacific as well as for the North Atlantic, the North Sea and the Western Mediterranean, including certain coastal zones of both seas, and the Baltic.

RESULTS

A brief summary of the actual knowledge about some trace metal ranges in sea water is given in Table 3. It should be taken into consideration that the ranges given for certain metals like Cd are rather large, due to three extreme situations: first, the Cd level can be extremely low in so-called depleted waters, as the central gyre of the North Pacific; second, it may be distinctly elevated in upwelling areas, where depth waters with elevated Cd levels are carried to the surface; third, the highest Cd levels occur in the deep waters, especially of the Pacific.

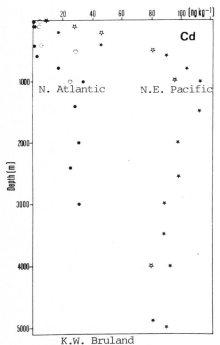

Fig. 6. Comparison of deep sea profiles of Cd in the North Atlantic and the north-east Pacific

• K.W. Bruland
•* K.W. Bruland and R.P. Franks (1983)
o✷ our data (1983)

TABLE 3

Ranges of trace metal concentrations in ocean waters (surface and depth, taken from ref. 20)

Trace metal	Range (ng/kg)
Cd	0.1 - 120
Pb	1 - 40
Cu	30 - 380
Ni	120 - 700
Co	0.6 - 6
Mn	10 - 165

Consistent depth profiles for at least 20 trace metals have been established in various oceanic regions (ref. 20). An example of a profile is given in Fig. 6. The nutrient-like metals as Cd, As, Ni and also Cu are taken up by phytoplankton and carried to various depths with the debris and fecal pellets, where they undergo redissolution from the oxidatively decomposed organic carriers. They are then advected again to the euphotic zone, together with the released nutrients. This cycle can occur many times and is a major source of nutrient-like elements in the surface zone. In contrast to these metals, Pb cannot be correlated with the nutrients. It has a pronounced tendency to be bound to all kinds of suspended matter. Thus, Pb levels will be more elevated at the water surface, due to the major source, aeolian input. A depth profile yields decreasing levels below 100 to 300 m, as a consequence of a rather effective transport to the bottom.

Distribution of heavy metal concentrations between the dissolved phase and the particulate matter

In surface and deep sea water, the concentration of the particulate matter remains negligible so that mainly all the metal ions are present in the dissolved state, defined conventionally for a particle size < 0.45 μm. Thus, in the so-called "dissolved state", small particles, colloids etc., are also present. In coastal and estuarine waters the contribution of metals adsorbed at the suspended matter cannot be neglected. Our investigations of coastal waters and estuaries of large rivers going to the North Sea and to the Mediterranean Sea have revealed that the distribution of trace metals between the dissolved phase and particulate matter varies with the metal type and the water type (refs. 21-23). Pb is mostly bound to particulate matter (up to 99 %) and is sedimented to a large extent within the estuary. Cd prevails

TABLE 4

Ranges of the distribution ratio K_d in the Scheldt estuary, in
rivers and in the oceans (ref. 23)

Trace metal distribution ratio K_d

	Scheldt estuary	Rivers	Oceans
Cd	$3 - 13 \times 10^4$	5×10^4	$4 - 8 \times 10^4$
Pb	$1.3 - 3 \times 10^6$	1×10^6	$2 - 3 \times 10^6$
Cu	$0.5 - 1.2 \times 10^5$	0.7×10^5	$1 - 4 \times 10^6$

mostly in the dissolved state and can cross the sedimentation
belt and becomes dispersed into the open ocean. Ni follows the
same trend as Co; Cu behaves differently according to the water
type. This distribution can be quantitatively described by the
distribution quotient K_d, defined as the ratio of the concentration of the respective metal in the suspended matter to the concentration in the dissolved state (ref. 23). Although the distribution quotient does not represent an equilibrium distribution
constant, it is a useful quantity to describe the complex partition of metals in natural waters. It has been found that the distribution quotient for Cd and Pb has the same range in various
water types (lake, river, sea water), whereas that for Cu is nearly an order of magnitude lower in estuarine and river waters
(Table 4). This is obviously caused by the increased concentration of complexing components of the dissolved organic matter
that make very stable complexes with Cu. Another important finding is that the distribution quotient for Cd is strongly seaso-
nally dependent and rises to high values in estuarine waters during the phytoplankton bloom period. This is a strong indication
for the binding of Cd by living organic matter (ref. 23).

Speciation studies

   The uptake and release of heavy metals between various components of the marine system depends not only on the total heavy
metal concentrations in various phases, but also on their chemical form (speciation). Thus, the transport of heavy metals
through the membrane of a living cell is only possible for the
free ionic form of the metal. Therefore, new analytical procedures are needed to determine the distribution of the overall concentrations of a given metal trace among various chemical spe-

cies. The capability of voltammetry in this respect stems from the fact that it is bassically a species-sensitive method, in contrast to the purely element-sensitive methods, such as atomic absorption spectroscopy, emission spectroscopy, X-ray fluorescence or neutron activation analysis. For certain toxic metals or metalloids, voltammetry provides a particular specificity for certain oxidation states, e.g. As(III), Se(IV), Cr(III) etc. (ref. 24).

Using the classical polarographic approach of the half-wave potential shift as a function of the ligand concentration, stoichiometric complexity constants and ligand numbers can be determined directly in sea water and at realistic trace levels. Thus, the "neovoltammogram" approach allows to work at overall heavy metal concentration of 10^{-9}M.

The complexity constants for $CdCl^+$ and $CdCl_2$ complexes have been determined directly in the sea water medium. Analogously the complexity constants of the Pb-carbonato-complexes have been determined using a neopolarogramm aproach. With these values, species distribution for Cd and Pb in sea water, with a correction for ion pairing between major salinity components, has been evaluated (ref. 25). With respect to the almost constant composition of sea water and the very large inertia of water masses, this distribution is stable and corresponds to a true equilibrium state.

On the other hand the distribution between the non-labile complexes with organic ligands of monomeric and polymeric nature, originating from DOM-components, is more of a dynamic nature. As in this case (in general) two voltammetric signals are obtained, one for the noncomplexed and the other for the complexed form of metal, it is possible to determine the ligand concentration required to achieve a certain complexation degree as function of the major chemical properties, e.g. salinity, pH of the water type etc. Specific side-effects due to salinity components in sea water can be elucidated and prognostic estimations of the modification of the basic metal speciation pattern by a given organic ligand can be made. Moreover, voltammetric studies can give insight into the typical kinetics and the mechanism of the formation of stable complexes with DOM components acting as ligands. By voltammetric titration with a pilot ion, e.g. Cu, Pb, Zn, the sum of the concentrations of all ligands forming non-labile complexes with the pilot ion (complexation capacity) can be determined (ref. 26).

CONCLUSIONS

Numerous applications of voltammetry over the last decade have revealed that this method is advantageous in the direct determination of a number of trace metals, especially in sea water medium. Up to 5 metals can be simultaneously determined in their natural concentration levels. Moreover, conclusions can be done about their binding state (speciation) as well as their partition between the dissolved state and the particulate matter. Interlaboratory comparisons have proved that there exists no methodic bias between the different determination methods, provided that all factors influencing the accuracy have been taken into account.

REFERENCES

1 E. Boyle, S. Huested and S. Jones, On the distribution of copper, nickel and cadmium in the surface waters of the North Atlantic and North Pacific Ocean, J. Geophys. Res. 86 (1981) 8048-8066
2 E. Boyle, F. Sclater and J.M. Edmond, On the marine geochemistry of cadmium, Nature 23 (1976) 42-44
3 K.W. Bruland, Oceanographic distributions of cadmium, zinc, nickel and copper in the North Pacific, Earth Planet. Sci. Lett. 47 (1980) 176-198
4 K.W. Bruland, G.A. Knauer and J.H. Martin, Cadmium in northeast Pacific waters, Limnol. Oceanogr. 23 (1978) 618-625
5 K.W. Bruland, R.P. Franks, G.A. Knauer and J.H. Martin, Sampling and analytical methods for the determination of copper, cadmium, zinc and nickel at the nanogram per liter level in sea water, Anal. Chim. Acta 105 (1979) 233-245
6 L. Mart, H. Rützel, P. Klahre, L. Sipos, U. Platzek, P. Valenta and H.W. Nürnberg, Comparative studies on the distribution of heavy metals in the oceans and coastal waters, Sci. Tot. Environm. 26 (1982) 1-17
7 B. Schaule and C.C. Patterson, The occurrence of lead in the Northeast Pacific and effects of anthropogenic input, in: Lead in the marine environment, Eds. M. Branica, Z. Konrad, Pergamon Press, Oxford 1980, pp. 31-43
8 H.W. Nürnberg, P.Valenta, L. Mart, B. Raspor, L. Sipos, Applications of polarography and voltammetry to marine and aquatic chemistry. II. The polarographic approach to the determination and speciation of toxic metals in the marine environment. Fresenius Z. Anal. Chem. 282 (1976) 357-367
9 H.W. Nürnberg, Voltammetric trace analysis in ecological chemistry of toxic metals, Pure Appl. Chem. 54 (1982) 853-878

10 H.W. Nürnberg, The voltammetric approach in trace chemistry of natural waters and atmospheric precipitation, Anal. Chim. Acta 164 (1984) 1-21

11 P. Valenta, Voltammetric studies on trace metal speciation in natural waters. Part I. Methods, in: G.G. Leppard, ed., Trace element speciation in surface waters and its ecological implications, Plenum Publ. Corp., New York-London 1983, pp. 46-69

12 L. Mart, Prevention of contamination and other accuracy risks in voltammetric trace metal analysis of natural waters. I. Preparatory steps, filtration and storage of water samples. Fresenius Z. Anal. Chem. 296 (1979) 350-357

13 L. Mart, Prevention of contamination and other accuracy risks in voltammetric trace metal analysis of natural waters. II. Collection of surface water samples. Fresenius Z. Anal. Chem. 299 (1979) 97-102

14 L. Mart, H.W. Nürnberg and D. Dyrssen, Low level determination of trace metals in Arctic sea water and snow by differential pulse anodic stripping voltammetry. in: C.S. Wong, E. Boyle, K.W. Bruland, J.D. Burton and E.D. Goldberg (eds.), Trace metals in sea water. Plenum Press, New York 1983, pp. 113-130

15 L. Mart, Minimization of accuracy risks in voltammetric ultratrace determination of heavy metals in natural waters, Talanta 29 (1982) 1035-1040

16 B. Pihlar, P. Valenta and H.W. Nürnberg, New high performance analytical procedure for the voltammetric determination of Ni in routine analysis of waters, biological materials and food, Fresenius Z. Anal. Chem. 307 (1981) 337-346

17 J. Golimowski, P. Valenta and H.W. Nürnberg, Trace determination of chromium in various water types by adsorption differential pulse voltammetry, Fresenius Z. Anal. Chem. 322 (1985) 315-322

18 P. Ostapczuk, P. Valenta and H.W. Nürnberg, Square wave voltammetry – a rapid and reliable determination method of Zn, Cd, Pb, Cu, Ni and Co in biological and environmental samples, Sci. Tot. Environm., in press

19 K.W. Bruland, K.H. Coale and L. Mart, Analysis of sea water for dissolved cadmium, copper and lead, an intercomparison of voltammetric and atomic adsorption methods, Mar. Chem. 17 (1985) 285-300

20 K.W. Bruland, Trace elements in sea water, Chapt. 45, Chemical Oceanography, vol. 8, J.P. Riley and R. Chester, eds., Academic Press, London 1983

21 W. Dorten, Ermittlung der aktuellen Schwermetallbelastungen in den Mittelmeerästuaren von Rhone, Ebro, Po und Arno, Ph. Dr. Thesis, Univ. Bonn, June 1986

22 L. Mart, H.W. Nürnberg, Cd, Pb, Cu, Ni and Co distribution in the German Bight, Mar. Chem. 18 (1986) 197

23 P. Valenta, E.K. Duursma, A.G.A. Merks, H. Rützel, H.W. Nürnberg, Distribution of Cd, Pb and Cu between the dissolved and particulate phase in the Eastern Scheldt and Western Scheldt estuary, Sci. Tot. Environm. 51 (1986) in press

24 H.W. Nürnberg, Investigation on heavy metal speciation in natural waters by voltammetric procedures, Fresenius Z. Anal. Chem. 316 (1983) 557-565

25 L. Sipos, B. Raspor, H.W. Nürnberg, R.M. Pytkowicz, Interaction of metal complexes with coulombic ion-pairs in aqueous media of high salinity, Mar. Chem. 9 (1980) 37-47

26 H.W. Nürnberg, B. Raspor, Applications of voltammetry in studies of the speciation of heavy metals by organic chelators in sea water, Environm. Technol. Lett. 2 (1981) 457-482

M.R. Smyth and J.G. Vos
Electrochemistry, Sensors and Analysis
© Elsevier Science Publishers B.V., Amsterdam — Printed in The Netherlands

30

ADSORPTIVE STRIPPING VOLTAMMETRY OF SELECTED MOLECULES

W. FRANKLIN SMYTH

Pharmacy Department, The Queen's University of Belfast, Belfast BT9 7BL,
Northern Ireland.

SUMMARY
     The principles of the technique of adsorptive stripping voltammetry, as it
applies to the quantitation of selected organic molecules and metal complexes
at mercury and carbon indicator electrodes, are reviewed. The resulting
determination of a wide range of molecules of biological significance such as
drugs, pesticides, naturally occurring compounds and metal ions such as Mo, Al,
V etc., is discussed with particular reference to sensitivity, selectivity
and application to on-line methods of analysis.

INTRODUCTION
     The techniques of stripping voltammetry for the trace analysis of heavy
metal ions and amenable organic molecules have been receiving considerable
attention in recent years. In anodic and cathodic stripping voltammetry,
the species of interest is preconcentrated into or onto a stationary indicator
electrode by electrolytic deposition under controlled hydrodynamic conditions
The measurement step consists of electrolytically stripping the deposited
species back into solution by imposition of a potential step as a d.c. or
pulse voltage ramp. The 1980's have seen an increasing interest in stripping
measurements of organic and inorganic substances that cannot be accumulated by
electrolysis. Alternative preconcentration schemes, based primarly on
adsorptive accumulation, have been used in the deposition step. Adsorptive
stripping voltammetry, AdSV, refers to an electroanalytical technique in
which the analyte or a derivative of the analyte is preconcentrated onto the
indicator electrode by adsorption, followed by voltammetric measurement of the
surface-active species. This results in the quantitation of many organic
molecules and metal complexes, possessing surface activity, at the nanomolar
concentration level. Detection limits can descend to values as low as
10^{-11}M, and are thus comparable to those obtained in conventional stripping
measurements of heavy metal ions. Recent publications, reviewed in this text,
have shown the sensitivity of the technique in trace measurements of drugs
(e.g. tranquillisers, antibiotics, antiulcer and cardiac agents), pesticides
(e.g. thiourea, nitro- and triazine-containing substances) and naturally
occurring compounds (e.g. DNA, progesterone, testosterone, dopamine,

soflavin), with selectivity shown in those direct measurements in complex matrices such as blood, urine, cattle feed etc. The selectivity of the technique has also been exploited in certain flow injection analyses which utilise adsorptive stripping voltammetric detection. In addition to determination of organic molecules, the technique has been successfully applied to measurement of metal ions following the formation and deposition of surface active complexes e.g. Al and V with the Solochrome Violet RS ligand. Low detection limits of the order of 10^{-10}M can be obtained for those metal ions that cannot be conveniently measured by conventional anodic stripping voltammetry due to low solubility in mercury, irreversible redox behaviour or formation of intermetallic compounds.

DISCUSSION

Principles

In AdSV, the actual adsorption step is rapid and the overall process is governed by diffusion of the species to the electrode surface. Koryta (ref. 1) has proposed the equation (1) for these electrochemical conditions.

$$M_{max} = 7.36.10^{-4} \ D^{\frac{1}{2}} \ t^{\frac{1}{2}} \ C \tag{1}$$

where C is the concentration of the surface active compound, of which M_{max} moles cm^{-2} fully cover the electrode surface in time t. D is the diffusion coefficient. The theory for ASV, as reviewed by Bond (ref. 2), puts forward equation (2) for the concentration of reduced metal in the mercury, $C_{M(Hg)}$, using a hanging mercury drop electrode.

$$C_{M(Hg)} = k_2 \ m \ t \ r^{-1} C_{M}n+ \tag{2}$$

where m is the mass transfer coefficient (which involves the diffusion coefficient of M^{n+}, rate of solution stirring and kinematic visosity of the solution), t is the time of deposition, r is radius of the mercury drop and $C_{M}n+$ is the concentration of the metal ion. AdSV and ASV are therefore similar in that adsorption of the species of interest and dissolution of the metal ion in the mercury respectively both depend on their concentration in solution, time of adsorption/deposition and the mass transport process to the electrode. The resulting stripping processes in AdSV and ASV reflect these dependencies e.g. the stripping peak height in DCASV for a reversible process at the hanging mercury drop electrode is given below (ref. 2) by equation (3)

$$i_p = -k_4 \ m \ n^{\frac{3}{2}} \ D_M^{\frac{1}{2}} \ r \ v^{\frac{1}{2}} \ C_{M}n+ \ t \tag{3}$$

where D_M is the diffusion coefficient of the metal in the mercury and v is
the scan rate of the DC potential. Time of adsorption/accumulation time
is critical in AdSV in that the stripping signal does not increase linearly
with concentration beyond an electrode saturation value. Typical times range
from 1-5 min at the 10^{-7} M level to 10-20 min at the 10^{-9} M level. For
most analytes, the rate of accumulation is governed by mass transport. Thus
as in ASV, various forms of forced convection (stirring, rotation, flow etc)
are employed during the preconcentration step. The convective transport is
then stopped and, after a 15s rest period, the voltammetric measurement is
performed in quiescent solution.

The amount of surface active compound accumulated on the electrode surface,
in addition to the variables of solution concentration, time of adsorption
and diffusion coefficient stated in equation (1), is also affected by other
variables such as accumulation potential, electrolyte composition (ionic
strength, pH, solvent), electrode material, and temperature. Optimum
conditions for maximum accumulation of strongly adsorbing species in order
to achieve maximum sensitivity in the subsequent voltammetric stripping
response are usually found by examining the peak current enhancement (at
a given accumulation time as compared to that without accumulation) using a
10^{-6} - 10^{-7} M solution. Attention must also be paid to reproducibility and
compromise may have to be made it and sensitivity.

Accumulation potential and electrolyte composition and their effect on the
adsorptive stripping voltammetric response have been evaluated by Batycka
et al. (ref. 3) and Wang et al. (ref. 4). Preconcentration should proceed
at the potential of maximum adsorption. For neutrally charged compounds, a
potential close to the electrocapillary zero value should be used. The
accumulation potential can be used to improve the selectivity of the
technique by minimising the effect of interferences e.g. in the determination
of thiourea in urine samples (ref. 5). Similarly, specific adsorption of
halide ions or electrolytic deposition of metal ions can also be minimised.
With regard to electrolyte composition, it is possible to use with differential
pulse stripping diluted supporting electrolytes which, in some cases, give
rise to enhanced stripping currents. pH adjustments are used for acidic
and basic analytes or when the stripping current occurs at particularly
high potentials. In addition measurement of metal ions based on the
adsorption of their surface active complexes, requires optimisation of
complexing ligand concentrations.

The hanging mercury drop is widely used for measuring reducible surface
active species, while carbon paste, wax-impregnated graphite and platinum
electrodes are used when oxidisable analytes are concerned. The mercury

electrode, especially with the static mercury drop design, offers the advantages of self-cleaning properties, reproducible surface area and automatic control. A cleaning step can be required with solid electrodes when the species of interest is not desorbed during the voltammetric scan. Using carbon paste electrodes, the adsorptive accumulation of various compounds is accompanied by extraction into the body of the electrode (ref. 6).

The stripping step in AdSV can be performed using a variety of voltammetric waveforms including linear scan (ref. 5), differential pulse (ref. 7), square wave (ref. 8), staircase (ref. 8), alternating current and subtractive modes. The differential pulse mode has been widely used because of its correction for the charging current and commercial availability. When the differential pulse mode offers little improvement in sensitivity over the linear scan mode, then the latter is preferred due to its speed of operation. Brown and Anson (ref. 9) have produced equations for the peak current and potential in differential pulse and linear scan measurements of surface bound species. The stripping step can also be carried out in a simple electrolyte solution as opposed to the original complex matrix in which the preconcentration was carried out (ref. 10), thus improving the selectivity of the technique by elimination of electroactive interferences in the complex matrix. This is also accomplished by the use of flow injection analysis together with adsorptive stripping voltammetric detection (ref. 11). In general, linear calibration plots are observed in the 10^{-7} - 10^{-10}M region with deviations observed at higher concentration due to electrode saturation. Because of lower background currents, mercury electrodes offer lower detection limits (10^{-10} - 10^{-11}M) as compared to solid electrodes (10^{-8} - 10^{-9}M). Relative standard deviations for replicate measurements at mercury electrodes range from 2-6% and from 5-12% at solid electrodes.

Applications

(i) <u>Tranquillisers</u>. Kalvoda (ref. 12) has studied the adsorptive stripping voltammetric behaviour of the 1,4-benzodiazepines, diazepam and nitrazepam and found that they each gave linear calibration plots in the range 10^{-9} - 10^{-6}M, using an accumulation potential of -0.50V (vs. SCE) in an acetate buffer pH 4.6. Using an accumulation potential of -0.40 V (vs. SCE) with a stripping peak potential of -0.72 V (vs. SCE) in 0.2 M NaOH, nitrazepam could be determined in the presence of a 10 fold excess concentration of diazepam, which peaks at -1.25 V (vs SCE) in this supporting electrolyte. The influence of other surface active compounds was also studied in this paper e.g., it was found that the peak of 5 x 10^{-8} M nitrazepam in 0.2 M NaOH at an accumulation deposition time (t_{acc}) of 15s was unaffected by up to

5 mgl^{-1} gelatine but with a t_{acc} value of 60s the peak was totally suppressed. This suggested the use of small t_{acc} values in stirred solutions or a separation technique such as molecular exclusion chromatography (Sephadex G-25) in order to maintain the sensitivity of the AdSV technique.

Jarbawi and Heineman (ref. 13) have shown the major tranquilliser chlorpromazine to be extracted from aqueous samples solutions into a wax-impregnated graphite electrode. Chronocoulometry gave a charge-time curve with an intercept which indicated adsorption of this molecule at the inter-face and a time dependence consistent with chlorpromazine extraction into the body of the electrode. This extraction served as a preconcentration step which lowered the detection limit to 5 x 10^{-9} M when determined by differential pulse voltammetry. An investigation of the extractive/adsorptive accumulative behaviour of various phenothiazine drugs at the carbon paste electrode has recently been presented (ref. 14). The compounds, depending on the structure of the side chains in the molecule, exhibit different degrees of accumulation. A detection limit of 1.5 x 10^{-9} M perphenazine was obtained with 10 min preconcentration. Wang and Freiha (ref. 11) have then used adsorptive stripping voltammetric detection in combination with flow injection analysis for the particularly selective determination of chlorpromazine in 10^2-fold excesses of non-adsorbable solution species with similar redox potentials. Enhanced sensitivity is obtained as a result of the preconcentration step at +0.3 V (vs. Ag/AgCl) in the flowing phosphate buffer stream, which also functions as medium exchange, allowing for reproducible quantitation of chlorpromazine in urine with no sample treatment. At a flow rate of 0.3 ml min^{-1}, injection rates of 24 samples hr^{-1} and detection limits of a few ng were obtained.

(ii) Antibiotics. Preconcentration and quantitative determination of the anti-cancer chemotherapy agent adriamycin have been accomplished by adsorption of the molecule at a carbon paste electrode and differential pulse voltammetry of the resulting surface (ref. 14). By immersion of the electrode in the adriamycin - containing sample for a 3 min period, followed by rinsing of the electrode and placement in a pH 4.5 buffer solution, limits of detection of the order of 10^{-8} M have been achieved with linear calibration in the range 10^{-5} - 10^{-7} M and relative standard deviation of approximately 10%. One of the drug's principal metabolites, adriamycin aglycone is also strongly adsorbed and preconcentrated as well; however, other common chemotherapy agents including cisplatin, mitoxanthrone, mitomycin C and vincristine do not adsorb strongly. The adsorption is sufficiently selective that accurate, reproducible preconcentration of adriamycin species directly from urine samples

possible without any preliminary pretreatment of the sample. The total
analysis time was found to be less than 10 min.

Chaney and Baldwin (ref. 16) have taken this approach a stage further
by adapting the adsorption/preconcentration approach discussed in ref. 15
for use in flow injection analysis of doxorubicin (adriamycin). Not only
have they found ease of operation inherent in flow injection methods but
also the limit of detection has been improved upon by one order of magnitude
down to 10^{-9}M.

(iii)   <u>Other drugs</u>. Webber et al. (ref. 17) have studied the electrochemical reduction and determination of the anti-ulcer drug cimetidine in
the nM-μM concentration range. The high sensitivity of the resulting
electroanalytical method was found due to reactant adsorption and major
metabolites of cimetidine could also be determined. Kalvoda (ref. 18)
has used the static mercury drop electrode to determine codeine, cocaine
and papaverine down to a detection limit of 10^{-8} M. The cardiovascular
drugs, digoxin and digitoxin contain the basic steroidal nucleus and
polarographic reduction occurs at the C=C bond in the five membered ring
containing a conjugated carbonyl group. Digoxin and digitoxin are reduced
at high negative potentials with limits of detection of the order of
10^{-6} M using differential pulse polarography. Wang et al. (ref. 19)
have improved upon this sensitivity significantly using AdSV and achieved
detection limits of 2.3×10^{-10} M and 8×10^{-10} M respectively. These
limits were based on 15 min accumulation at the static mercury drop.

(iv)   <u>Pesticides</u>. Stara and Kopanica (ref. 5) have discussed the
adsorptive stripping voltammetry of thiourea, α-naphthylthiourea and
diphenylthiourea. In perchlorate supporting electrolyte, these compounds were
found to adsorb at the hanging mercury drop electrode at positive potentials
or on open circuit and can be stripped in a cathodic scan. Detection limits
of conventional cathodic stripping voltammetry procedures can be dramatically
reduced to 2.5 ng 1^{-1} (thiourea), 30 ng 1^{-1} (α-naphthylthiourea) and
50 ng 1^{-1} (diphenylthiourea) The technique has been applied to the
:tle feed and its direct analysis in urine.
have applied the technique to nitro- and
ides with a detection limit of 5×10^{-10} M
. Trichlorobiphenyl is electrochemically
but can be determined by adsorptive stripping
ɔuffer pH 6.5 using an accumulation potential

(v) Naturally-occurring compounds . Controlled adsorptive accumulation of riboflavin, flavin mononucleotide (FMN) and alloxazine on the static mercury drop electrode provides the basis for the direct stripping measurement of these compounds at nM concentration (ref. 23). Differential pulse voltammetry, following a 30 min preconcentration yielded a detection limit of 2.5×10^{-11} M for riboflavin. The relative extent of adsorption and the resulting response were found rather sensitive to the composition of the flavin side chain e.g. with a 2 min. preconcentration, signal enhancement factors of 34, 15 and 7 were found for the analysis of riboflavin, FMN and alloxazine respectively. Relative standard deviations were found in the range 2-8%.

A unique application of AdSV relates to the molecule DNA and its reduction at mercury electrodes after adsorption and unwinding of the double helix. As a result, the adsorption accumulation has permitted investigations such as binding of antitumor antibiotics with DNA (ref. 23) or assessing the damage in native DNA by small δ-irradiation doses (ref. 24).

Wang et al. (ref. 25) have recently used AdSV to obtain detection limits of 1.6×10^{-10} M for testosterone and 2×10^{-10} M for progesterone with 15 min. preconcentration at the static mercury drop electrode.

(vi) Metal Ions . Brainina recently reviewed the use of organic reagents in stripping or inverse voltammetry (ref. 26), both processes which involved electron exchange in the adsorption step (e.g., variable valence ions such as Co (II)/Co(III) with nitroso-napthol ligands) and those that were ACE and CAE types (i.e. only adsorption (A) and chemical reactions (C) were involved in the deposition step). The latter category can produce electroanalytical methods of not only high sensitivity but also high selectivity e.g., it is possible in the presence of hematein to determine Zn(II) even with 10^5 fold excess quantities of Pb (II) (ref. 27). The reagent hematein can also be used to determine Sn(IV) (ref. 28). Sodium diethyldithiocarbamate can be used to determine Mo(VI) (ref. 29, 30), Solochrome violet RS, Al(III) (ref. 27) and V (ref. 27) and tetraphenyl porphyrin (ref. 31) for Fe(III).

REFERENCES

1   J. Koryta, Coll. Czech. Chem. Comm. 18 (1953), 206-213.
2   A.M. Bond, Modern Polarographic Methods in Analytical Chemistry, Marcel Dekker Inc., New York, 1980.
3   H. Batycka and Z. Lukaszewski, Anal. Chim. Acta. 162 (1984), 207-212.
4   J. Wang and P.A.M. Farias, J. Electroanal. Chem. 182 (1985) 211-216.
5   V. Stara and M. Kopanica, Anal. Chim. Acta. 159 (1984), 105-110.
6   J. Wang and B.A. Freiha, Anal. Chem. 56 (1984) 849-852.

7   J. Wang, D.B. Lou, A.M. Farias and J.S. Mahmoud, Anal. Chem. 57 (1985) 158-162.
8   A. Webber, M. Shah and J. Osteryoung, Anal. Chim. Acta. 154 (1983) 105-119.
9   A.P. Brown and F.C. Anson, Anal. Chem., 49 (1977), 1589-1595.
10 J. Wang, and B.A. Freiha, Anal. Chim. Acta. 148 (1983), 79-85.
11 J. Wang and B.A. Freiha, Anal. Chem. 55 (1983), 1285-1288.
12 R. Kalvoda, Anal. Chim. Acta. 162 (1984), 197-205.
13 T.B. Jarbawi and W.R. Heineman, Anal. Chim. Acta. 135 (1982), 359-362.
14 J. Wang, B.A. Freiha and B.K. Deshmukh, Bioelect. Bioenerg. 14 (1985) 457-467.
15 E.N. Chaney, Jr., R.P. Baldwin, Anal. Chem. 54 (1982), 2556-2560.
16 E.N. Chaney, Jr., R.P. Baldwin, Anal. Chim. Acta. 176 (1985), 105-112.
17 A. Webber, M. Shah and J. Osteryoung. Anal. Chim. Acta. 154 (1983) 105-119.
18 R. Kalvoda, Anal. Chim. Acta. 138 (1982) 11-18.
19 J. Wang, J.S. Mahmoud & P.A.M. Farias, Analyst. 110 (1985) 855-865.
20 H. Benadikova, and R. Kalvoda, Anal. Lett, 17 (1984), 1519-1525.
21 N.K. Lam and M. Kopanica, Anal. Chim. Acta. 161 (1984), 315-324.
22 J. Wang, D-B Luo, P.A.M. Farias, J.S. Mahmoud, Anal. Chem, 57 (1985), 158-162.
23 J.A. Plambeck and J.W. Lown, J. Electrochem. Soc. 131 (1984) 2256-2262.
24 J.M. Sequaris, P. Valenta, and H.W. Nurnberg, Int. J. Radiat. Biol. 42 (1982) 407-412.
25 J. Wang, P.A.M. Farias and J.S. Mahmoud, Anal. Chim. Acta. 171 (1985) 195-204.
26 Kh.Z. Brainina, Z. Anal. Chem. 312 (1982) 428-437.
27 H. Specker, H. Monien, B. Lendermann, Chem. Anal. (Warsaw) 17 (1971) 1003-1010.
28 H. Monien, K. Zinke, Z. Anal. Chem. 250 (1970) 178-185.
29 H. Monien, P. Jacob, B. Janisch, Z. Anal. Chem. 267 (1973) 108-114.
30 H. Monien, R. Bodenkerk, K.P. Kringe, D. Rath, Z. Anal. Chem. 300 (1980) 363-371.
31 A.P. Brown, C. Koval and F.C. Anson, J. Electroanal. Chem. 72 (1976) 379-387.

M.R. Smyth and J.G. Vos
Electrochemistry, Sensors and Analysis
© Elsevier Science Publishers B.V., Amsterdam — Printed in The Netherlands

DETERMINATION OF SOME ENVIRONMENTALLY IMPORTANT ORGANIC COMPOUNDS BY HIGH PERFORMANCE LIQUID CHROMATOGRAPHY WITH VOLTAMMETRIC DETECTION

Malcolm R. Smyth, Patrick J. Hayes and Darioush Dadgar
School of Chemical Sciences, NIHE Dublin, Glasnevin,
Dublin 9, Ireland.

ABSTRACT
In this paper we present a review of the application of HPLC with voltammetric detection for the determination of some organic compounds of environmental importance. Where possible, a comparison has been made with similar methods employing ultraviolet detection.

INTRODUCTION

The application of voltammetric techniques to the determination of molecules of biological and environmental importance has recently been surveyed (refs.1-2). As is the case with any analytical technique which is used to measure trace quantities of organic compounds in complex matrices, care must be taken not only to optimise conditions for the detection and quantification of the species of interest, i.e. the "end step", but also in the areas of sampling, initial treatment of sample, separation and derivatisation (ref.2).

With the recent advances in the development of pulse polarographic (refs.3,4) and stripping voltammetric (refs.2,4) techniques, methods employing a voltammetric "end-step" now show comparable (in some cases better) sensitivity when compared with methods employing gas chromatography (GC) or spectrofluorimetry. In terms of selectivity, voltammetric methods often show a better resolution for structurally-related compounds than say

ultraviolet spectroscopy, but this is rarely sufficient in the
field of environmental trace analysis, where there are many
naturally occurring substances and metabolites that can interfere
at the low nanogram/picogram levels commonly encountered. In
organic voltammetric trace analysis it has therefore been found
necessary to incorporate some form of chromatography to achieve
the required separation prior to voltammetric detection of the
eluate.

The concept of "chromatovoltammetry" was first introduced by
Kemula (ref.5), and since then several attempts have been made to
couple chromatographic systems e.g. gel permeation
chromatography, thin-layer chromatography and ion-exchange
chromatography with voltammetric detection (VD) with varying
degrees of success (ref.2). It was not, however, until the
upsurge in the development of high performance liquid
chromatography (HPLC), and the search for selective and sensitive
detection systems for use with this technique, that
chromatovoltammetry really "took off". This time the major
impetus was created by the work of Kissinger and co-workers, who
demonstrated the advantageous possibilities of employing a carbon
paste electrode to detect sub-nanogram quantities of organic
materials following HPLC separation. The results of their
pioneering work are summarised in several reviews (refs.6-8).

Since voltammetric methods are based on processes occurring at
electrode surfaces which are maintained in solution, the first
requirement of combining a chromatographic with a voltammetric
system is that the mobile phase used in the chromatographic
process should be a liquid which is also conducting in nature.
Hence most recent applications of chromatovoltammetry have
employed reverse-phase chromatographic systems which permit the
use of water/alcohol/acetonitrile mixtures, (which are also good
solvents for voltammetry), as eluants.

The most widely used voltammetric detection systems employed in
liquid chromatography are those based on direct current
hydrodynamic chronoamperometry where the current arising from th
oxidation or reduction of the substance of interest is measured
as a function of time at a fixed electrode whose potential is
held constant in the flowing stream of eluant.

Taking the case for the oxidation of a compound R which is
oxidised at a potential E_1 to give an initial product O:

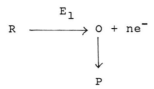

the current (i) arising from this reaction when plotted against
time (t) will give rise to the characteristic chromato
-voltammogram. In general, the initial product(s) of such
electrochemical oxidations are unstable and are rapidly converted
to more stable compounds(P) which are not reactive at the
electrode surface and thus pass out of the cell. This is
especially important for chromatovoltammetric detection systems,
because if the product(s) of the electrooxidation remained at the
electrode surface, then this would lead to "fouling" of the
active surface and thus to non-reproducibility of the response
(ref.7).

The response of a chronoamperometric detector to the oxidation
or reduction of the compound of interest is dependent on the rate
of mass transfer of that substance to the electrode surface. The
general relationship relating the limiting current, i_1, the
bulk concentration of the electroactive species, C_o, and the
volume flow rate, ν, is given by the equation:

$$i_1 = k\ n\ F\ C_o\ \nu^\alpha$$

where n is the number of electrons involved in the electrode
process, F is the Faraday and k is a constant dependent on the
kinematic viscosity of the fluid, the diffusion coefficient of
the electroactive species, the geometry and area of the electrode
surface etc. The exact value of the exponent, α, of the volume
flow rate ν is determined by the specific nature of the
hydrodynamic conditions. For a stationary electrode in a flowing
stream, α will generally be between 1/3 and 2/3. The fluctuation
in the measured current Δi_1 associated with a flow rate
change $\Delta\nu$ is given by:

$$\Delta i_1 = (\Delta v/v)i_1$$

therefore an n% increase in the flow rate will cause an (nα)% change in the measured current (ref.9).

There are of course a large number of voltammetric cell designs and systems which have been developed for use with HPLC, and the reader is referred to the reviews of Poppe (ref.10), Rucki (ref.11) and Surmann (ref.12) for further information on these. In general, most of the cells designed for use with solid carbon indicator electrode materials, e.g. "thin-layer" and "wall-jet" designs, give good sensitivity for compounds which are oxidised in the region +0.1 $\longrightarrow$ +1.2V (vs. Ag/AgCl reference electrode). At higher positive potentials, difficulties are encountered due to the blank electrolyte contribution. For reducible organic compounds, there have been numerous attempts to develop voltammetric cells for HPLC based on dropping or stationary mercury electrodes (DME, SMDE) and although much has been achieved to counteract the problems of incorporating mercury electrodes into flowing streams and for overcoming the effects of oxygen reduction, voltammetric detection for most reducible organic compounds following HPLC separation still offers little improvement in sensitivity as compared to ultraviolet detection (UVD).

In deciding whether a chromatovoltammetric procedure could provide a sensitive and selective method for the determination of a particular compound, it is necessary in the first instance to examine its voltammetric behaviour at a variety of different electrodes in a variety of supporting electrolytes (containing different salts e.g. lithium perchlorate, sodium acetate etc. and different %'s of water, alcohol, acetonitrile etc.) over a reasonable range of pH (for chromatographic purposes, column materials can usually tolerate solutions of pH in the range 3-9). It is not possible here to give the reader an insight into the variety of different techniques available for the investigation of electrochemical reactions or to the variety of compounds that are amenable to these techniques, but the interested reader need only consult standard reference texts for this purpose. For example, the books of Heyrovsky and Zuman (ref.13) and Franklin Smyth (ref.14) give relatively uncomplicated treatments for beginners to the subject.

In those cases where the chromatographic system is unable to
separate two compounds of related structure, the inherent
selectivity of the voltammetric approach may be called into
play. This has been well illustrated in the selective
determination of the growth-promoting hormones dienestrol and
diethylstilbestrol (ref.15) and the carbamate pesticides barban
and captafol (ref.16). When the oxidation or reduction
potentials of two compounds are close together (of the order of
50-100 mV), the use of direct current hydrodynamic voltammetry
may not give the required selectivity for the simultaneous
determination of the two compounds, and in this case an
improvement in selectivity may be achieved using the differential
pulse mode of operation. In some cases, the use of the
differential pulse mode may also give improved sensitivity,
especially for compounds with high potentials of oxidation or
reduction e.g. organometallic cations (ref.17).

In recent years, a major advance in chromatovoltammetry has
been the introduction of dual-electrode detection systems
(ref.18) which have been demonstrated to improve both the
selectivity and sensitivity of certain determinations as well as
helping in the identification of unknown peaks. In addition,
important advances have come about through the application of
pre- and post-column derivatisation reactions (ref.19).

DETERMINATION OF SOME ENVIRONMENTALLY IMPORTANT COMPOUNDS BY
HPLC-VD

(i) Sulphur-containing compounds

The electrochemical behaviour of sulphur-containing compounds
has been well documented in a recent review (ref.1). In general,
there has been little application of chromatovoltammetry for the
determination of compounds containing -S-S-, >C=S or -SH
moieties because their electrochemical behaviour is dependent on
the use of mercury as the working electrode. In particular,
problems have been encountered in coupling HPLC with a cathodic
stripping voltammetric mode of detection. The use of cathodic
stripping voltammetry for the determination of sulphur -
containing organic species in quiescent solution is well

documented (ref.4), but inherent problems lie in finding the
correct conditions where these substances will remain plated in a
flowing stream. Particular problems in this regard have been
illustrated in the case of thioamides using a thin-film mercury
electrode (ref.20). Application of chromatovoltammetry for the
determination of sulphur-containing compounds in environmental
samples have been cited in the determination of the potentially
carcinogenic metabolite ethylene thiourea and related substances
in rat urine (ref.21), and the pesticide dithianon in fruit
residues (ref.22).

(ii) Nitrogen-containing compounds

The application of a polarographic detector for the
determination of some nitro-containing pesticides was first
demonstrated by Koen and Huber (ref.23) who were able to
selectively determine parathion and methylparathion in lettuce
following a liquid chromatographic separation. Since then
several other papers have been published dealing with the
determination of nitro-containing thiophosphate pesticides using
liquid chromatography with polarographic detection (refs.24-27).
In the case of nitro-containing compounds, which give rise to the
most sensitive signals from an organic voltammetric point of
view, it has been demonstrated that the sensitivity of HPLC
methods employing a polarographic detector is comparable to HPLC
methods employing UVD (ref.28). This has been further
demonstrated in the case of nitroso-containing species (which are
reduced at potentials more negative than nitro-containing
compounds), where the limit of detection of the method (10^{-7}M)
is similar regardless of whether the analysis is carried out in
quiescent solution or in a flowing stream (ref.29). Improvements
in this area may come about, however with the use of thin-film
mercury electrodes or gold amalgamated mercury electrodes, as
demonstrated in a recent publication (ref.17).

In environmental analysis, the nitrogen-containing compounds
which have most been studied are those which contain an aromatic
amine function. The oxidative voltammetric behaviour of aromatic
amines at carbon electrodes is often complicated, as was
exemplified in a recent study by Hart et al. (ref.30) on 2-, 3-
and 4-chloroaniline. However, in contrast to the situation

regarding polarographic detection in HPLC, the use of carbon electrodes in the hydrodynamic chromoamperometric mode often improves the limit detection for compounds that are easily oxidised ($+0.2 \rightarrow +1.0$ V) by many orders of magnitude. This is exemplified in the case of the chloranilines where detection limits of the order of 10-20 pg have been quoted for the determination of 2-chloroaniline and 4,4'-methylenebis (2-chloroaniline) in factory atmospheres (ref.31). A method has also been reported for the determination of chlorinated anilines in urine (ref.32). In such cases, HPLC methods employing voltammetric detection offer a great improvement in sensitivity (of up to 50 times) over similar methods employing UVD. This is further exemplified in the analysis of aromatic amine carcinogens (refs.33-42). In the analysis of benzidine, for instance, a variety of different columns (both reverse-phase and ion-exchange) and detection systems (employing carbon paste, carbon black/polythylene or reticulated vitreous carbon electrodes) have been used for its determination in waste water in the presence of 3,3' -dichlorobenzidine (ref.34), in the presence of its acetylated metabolites in urine (ref.35), in various effluents (ref.35), in soil (ref.40) and in a 1000-fold excess of aniline (ref.41). In all cases, the methods described were able to determine benzidine and related aromatic amines in the low picogram range with good precision. Other aromatic amines that have been determined by HPLC-VD include aniline (ref.36), 1- and 2-naphthylamine (refs.32,36), hydrazine and its monomethyl-, 1,1-dimethyl- and 1,2-dimethyl analogues (ref.37), anthracene, 2-aminoanthracene and acridine (ref.42).

(iii) Oxygen-containing compounds

The oxygen-containing compounds that have most been investigated using chromatovoltammetry are those containing a phenolic moiety. Although compounds containing this group are less easily oxidised at carbon electrodes than aromatic amines, this oxidation process still occurs at a potential ($+1.2 \rightarrow +1.3$V) which can give rise to a sensitive method of analysis. This is exemplified in the analysis of nine estrogenic and pseudo-estrogenic hormones in animal meat (ref.43) where the use of voltammetric detection permitted the simultaneous determination of eight of these compounds down to the picogram

g^{-1} range. In addition, the voltammetric approach was shown to exhibit much less interference from natural constituents in the meat and could selectively determine diethylstilbestrol in the presence of either dienestrol of hexestrol (refs.15,43). An alternative procedure has also been described for the determination of diethylstilbestrol in animal tissue using a carbon paste instead of a glassy carbon working electrode (ref.44). In addition, a method has been developed for the determination of the cis- and trans-forms of the phenolic-containing mycotoxin zearalenone in cereal products (ref.45).

The determination of individual phenols in water is of course of great environmental importance, and a recent EEC publication highlights many of the methods currently being investigated for this purpose. The application of HPLC with voltammetric detection for the determination of phenols in ground and waste waters has been reported by several authors (refs.47-49) and the limits of detection for compounds such as resorcinol, phenols, xylenols and chlorinated plenols quoted to be in the low picogram range. The sensitivity is thus much better than corresponding methods employing UVD (phenols generally have low extinction coefficients), and should provide a complimentary means to capillary GC, which is recognised as the forerunner in this area of analysis (ref.50). Chromatovoltammetric methods have also been described for the determination of chlorophenols in urine (ref.51) and 2-phenylphenol in orange rind (ref.52).

Because of the high sensitivity that voltammetric detection affords for phenolic compounds, several methods have been developed for the determination of carbamate pesticides following conversion to the corresponding phenolic derivative. This conversion is best carried out in alkaline media (pH 11-12) and is an improvement over existing gas chromatographic procedures which require further derivatisation of the liberated phenol prior to analysis (refs.39,53). Several carbamate pesticides can however, be determined directly using chromatovoltammetry, but these compounds rely on the presence of secondary or tertiary amino moieties in their molecular structure for their electroactivity. In a theoretical study on 13 carbamate pesticides, Batley and Afghan (ref.54) reported that only two compounds, aminocarb and mexacarbate, were amenable to voltammetric detection following HPLC, and several workers have been able to develop working methods for the determination of aminocarb based on its oxidation behaviour (refs.55,56). In another study of carbamate pesticides, Mayer and Greenberg (ref.16) showed that voltammetric detection was more selective than UVD, but stated that UVD had in general the better sensitivity for this particular group of compounds. Voltammetric detection has also been found useful for the determination of uric acid in cereal products as a means of monitoring insect infestation (ref.57).

(iv) Organometallic compounds

The electrochemical behaviour of organometallic species is well documented in a recent review (ref.58), and can be determined by a variety of voltammetric procedures in quiescent solution. For their determination in a flowing stream, MacCrehan and Durst (refs.17,59) have employed a gold amalgamated mercury electrode (GAME) which has the great advantage over the DME or SMDE for chromatographic purposes in that the GAME can be adapted successfully for use with oxidative voltammetric detector designs. It also gives reproducible results over a full day's operation. Using HPLC in conjunction with a GAME, it has been found possible to selectively determine compounds such as methyl-, ethyl- and phenylmercury in fish material down to the 40 picogram level (ref.59). In addition, when the GAME is operated in the DPP mode, it has been found to offer a greater selectivity for the determination of both organomercury and organotin compounds (ref.17) by improving the signal to noise ratio of the resulting chromatovoltammograms. Another method has been described for the determination of methylmercury in fish, but the polarographic detection was not found to be as sensitive as cold-vapour atomic absorption spectroscopic detection (ref.60).

(v) Miscellaneous compounds

A method has recently been described for the determination of aromatic and aliphatic isocyanates based on a derivatisation reaction with the electrogenic reagent 1-(2-methoxyphenyl) piperazine (ref.61). The resulting derivatives were all found to give an oxidation wave at +0.75 V, and the chromatovoltammetric procedure thus developed was found to be 20 times more sensitive than the corresponding method employing UVD. HPLC-VD methods have also been described for the determination of phenoxy-acid herbicides (ref.62), 2,4, di-isocyanotoluene in air (ref.63), pesticides in air (ref.64) and nitro-substituted polynuclear hydrocarbons in diesel soot (refs.65,66).

CONCLUSION

The combination of high performance liquid chromatography with voltammetric detection has proved to be a powerful technique for the analysis of trace amounts of phenolic - and amino - containing compounds of environmental significance, where the sensitivity, (and in some cases the selectivity), of resulting methods has been shown to be better than corresponding methods employing ultraviolet detection. A problem remains, however, in making this technique more universal with respect to other compounds of environmental significance e.g. sulphur-, nitro-, nitroso- containing compounds which give rise to reduction processes at mercury electrodes, but where the combination of HPLC with a voltammetric detector, operated in the reductive mode, does not often offer any significant advantages over a conventional UV detector.

REFERENCES

1.   M.R.Smyth and W.Franklin Smyth, Analyst, 103 (1978) 529.
2.   W.Franklin Smyth and M.R. Smyth, in W.Franklin Smyth (Ed.), Polarography of Molecules of Biological Significance, Academic Press, N.York and London, 1979, pp3-36.
3.   J.G.Osteryoung and K.Hasebe, Rev.Polar. (Japan), 22 (1976)1.
4.   W.Franklin Smyth and I.E. Davidson, in W.Franklin Smyth (Ed.) Electroanalysis in Hygiene, Environmental, Clinical and Pharmaceutical Chemistry, Elsevier, Amsterdam and N.York,1980, pp271-286.
5.   W.Kemula, Rocz.Chem.,26 (1952)281.
6.   P.T.Kissinger, Anal.Chem.,49 (1977)447A.
7.   P.T.Kissinger, Methodol.Surv.Biochem.,no 7,pp 213-226.
8.   P.T.Kissinger, L.J.Felice, D.J.Miner, C.R.Preddy and R.E. Shoup, "Advances in Analytical and Clinical Chemistry", Plenum Press,(1978), pp 55-175.
9.   D.G.Swartzfager, Anal.Chem.,48(1976) 2189.
10.   H.Poppe, J.Chromat.Libr.,no 13, (1978), pp 131-149.
11   R.J.Rucki, Talanta, 27 (1980) 147.

12. P.Surmann, Fres. Z. Anal. Chem., 316 (1983) 373.
13. J.Heyrovsky and P.Zuman, Practical Polarography, Interscience, N.York and London, 1968.
14. W.Franklin Smyth (Ed.) Polarography of Molecules of Biological Significance, Academic Press, N.York and London, 1979.
15. M.R.Smyth and C.G.B.Frischkorn, Fres.Z.Anal.Chem.,301 (1980)220.
16. W.J.Mayer and M.S.Greenberg, J.Chromat.,208 (1981) 295.
17. W.A.MacCrehan, Anal.Chem.,53 (1981)74.
18. D.A.Roston, R.E.Shoup and P.T.Kissinger, Anal.Chem., 54 (1982) 1417A.
19. P.T.Kissinger, K.Bratin, G.C.Davis, and L.Pachla, J.Chromat. Sci., 17 (1979)137.
20. W.Franklin Smyth, A.Ivaska, J.S.Burmicz, I.E.Davidson and Y.Vaneesorn, Bioelectrochem.Bioenerg., 8 (1981)459.
21. J.F.Lawrence, F.Iverson, H.B.Hanekamp, P.Bos and R.W.Frei, J.Chromat., 212 (1981)245.
22. W.Buchberger and K.Winsauer, Mikrochim.Acta, (1980)257.
23. J.G.Koen and J.F.K.Huber, Anal.Chim. Acta, 51 (1970)303.
24. J.G.Koen, J.F.K.Huber, H.Poppe and G.den Boef, J.Chromat. Sci.,8 (1970) 192.
25. R.Stillman and T.S.Ma, Mikrochim.Acta, (1973)491.
26. H.A.Moye, J.Chromat.Sci., 13(1975)268.
27. R.C.Buchta and L.J.Papa, J.Chromat, Sci., 14 (1976)213.
28. H.B.Hanekamp, P.Bos, V.A.Th.Brinkman and R.W.Frei, Fres.Z.Anal.Chem., 297(1979)404.
29. R.Samuelsson and J.G.Osteryoung, Anal. Chim. Acta, 123(1981)97.
30. J.P.Hart, M.R.Smyth and W.Franklin Smyth, Analyst, 106 (1981)146.
31. C.J.Purnell and C.J.Warwick, Analyst, 105 (1980)
32. E.M.Lores, F.C.Meekins and R.F.Moseman, J.Chromat., 188 (1980)412.
33. I.Mefford, R.W.Keller, R.N.Adams, L.A.Sternson and M.A. Yllo, Anal. Chem., 49 (1977)683.
34. D.N.Armentrout, and S.A. Cutie, J.Chromat. Sci., 18 (1980)370.
35. J.R.Rice and P.T.Kissinger, J.Anal.Toxicol., 3 (1979)64.
36. R.M.Riggin and C.C.Howard, Anal.Chem., 51 (1979)210.
37. H.R.Hunziker and A.Miserez, Mitt.Geb. Lebensmittelunters. Hyg., 72 (1981)216.
38. E.S.Fiala and C.J.Kulakis, J.Chromat., 214(1981)229.
39. P.T.Kissinger, K.Bratin, W.P.King and J.R.Rice in Pesticide Analytical Methodology, American Chemical Society, 1980, pp 57-88.
40. J.R.Rice and P.T.Kissinger, Environ. Sci. Technol., 16(1982)263.
41. V.Concialini, G.Chiavari and P.Vitali, J.Chromat, 258(1983) 244.
42. W.L. Caudill, M.V. Novotny and R.M. Wightman, J.Chromat, 261 (1983) 415.
43. C.G.B.Frischkorn, M.R.Smyth, H.E.Frischkorn and J.Golimowski, Fres.Z. Anal.Chem., 300 (1980)407.
44. T.H.Kenyhercz and P.T.Kissinger, J.Anal.Toxicol., 2 (1978)1.
45. M.R.Smyth and C.G.B.Frischkorn, Anal.Chim.Acta, 115 (1980)293.
46. A.Bjorseth and G.Angeletti (Ed's), Analysis of Organic Micropollutants in Water, D.Reidel, Dordrecht, Boston and London, 1982.

47. D.E.Weisshaar, D.E.Tallman and J.L.Anderson, Anal.Chem. 53(1981)1809.
48. D.N.Armentrout, J.D.McLean and M.W.Long, Anal. Chem., 51(1979)1039.
49. R.E.Shoup and G.S.Mayer, Anal.Chem. 54(1982)1164.
50. L.Renberg, in reference 42, pp. 286-297.
51. E.M.Lores, T.R.Edgerton and R.F.Moseman, J.Chromat. Sci. 19(1981)466.
52. D.E.Ott, J.A.O.A.C., 61 (1978)1465.
53. W.P.King, Current Separations, 2(1980)6.
54. G.E.Batley and B.K.Afghan, J. Electroanal. Chem., 125 (1981)437.
55. J.L.Anderson and D.J.Chesney, Anal. Chem., 52(1980)2156.
56. M.Lanouette and R.K.Pike, J.Chromat., 190(1980)208.
57. L.A.Pachla and P.T.Kissinger, Anal. Chim. Acta, (1977) 88 385.
58. N.B.Fouzder and B.Fleet, in reference 14, pp 261-293.
59. W.A. MacCrehan and R.A. Durst, Anal. Chem., 50(1978)2108.
60. W. Holak, Analyst, 107 (1982) 1457.
61. C.J.Warwick, D.A.Bagon and Purnell, C.J., Analyst, 106 (1981)676.
62. M.Akerblom and B.Lindgren, J.Chromat. 258 (1983) 302.
63. S.D.Meyer and D.E.Tallman, Anal. Chim. Acta. 146 (1983) 227.
64. D.A.Bagon and C.J.Warwick, Chromatographia, 16(1982)290.
65. Z.Jin and S.M.Rappaport, Anal.Chem. 55 (1983)1778.
66. W.A.MacCrehan and S.D.Yang, in Abstracts of "Electroanalysis na h'Eireann", Dublin, June 10-12, 1986.

M.R. Smyth and J.G. Vos
Electrochemistry, Sensors and Analysis
© Elsevier Science Publishers B.V., Amsterdam — Printed in The Netherlands

IMPROVEMENT OF PEAK RESOLUTION IN AC POLAROGRAPHY BY DECONVOLUTION USING FOURIER TRANSFORMATION

S.O. ENGBLOM and A.U. IVASKA

Laboratory of Analytical Chemistry, Åbo Akademi, SF-20500 Turku (Åbo) Finland

SUMMARY

Resolution enhancement using deconvolution is discussed. The enhancement is accomplished by division of the Fourier transform of the experimentally obtained polarogram by the transform of a suitable deconvolution function. The method's major disadvantage, division by zero, can be avoided by using digital filters. Two filters are compared, the rectangular filter and a method known as Tikhonov regularization. Simulated polarograms are used to demonstrate the usefulness of the method.

INTRODUCTION

The ac polarographic analysis of a multi-component system is often complicated by the presence of overlapping peaks. Overlapping hinders the precise determination of, e.g., peak position, width at half height and height of the individual peaks. There exists a considerable need for methods to overcome this problem. The use of complexing or masking agents is common, but it is possible only when the electroactive species in the solution are known and depends also on whether such agents exist or not. Applications of computers and mathematical methods have recently gained an increasing interest. A range of different methods are now available. The simplest one is perhaps the one described for a two-component system by Bond and Grabarić [1]. Curves for different concentrations of one component are recorded and subtracted from the original curve until a visual inspection of the result confirms the correct concentration of that component. This method requires of course that the components are known or can be identified and it may be time consuming if the number of necessary recordings becomes large. Another, more advanced, method is the one given by Perone *et al.* [2,3] which is based on a curve-fitting algorithm that uses least-square regression analysis. This method, however, requires knowledge of the number of components and a library of standard polarograms/voltammograms. This is also the case for the Kalman filter that has been used by Brown *et al.* [4,5] for the same purpose.

The method described here is based on the theoretical assumption that a polarographic peak is the convolution of a single pulse and a "broadening" function. By performing the inverse operation, i.e., a deconvolution, one can theoretically obtain a peak with zero width. One can thus obtain a "complete" resolution enhancement, i.e., all peaks in the polarogram will be visible unless two or more peaks have the same half-wave potential. Complete enhancement can not be achieved in practice, but considerable improvement is possible as will be shown in this work. The deconvolution is performed by using Fourier transformation. A deconvolution in the time domain is equivalent with a division in the frequency domain.

The deconvolution can also be performed in the time domain as shown by den Harder and de Galan [6], but the possibilities of performing, e.g., interpolation and filtering in the frequency domain [7,8] make the use of Fourier transformation more attractive. This method, introduced by Smith *et al.* [9], suffers from one disadvantage. The division in the frequency domain is unstable because the Fourier transform of the deconvolution function is rapidly approaching zero when going from lower to higher frequencies. The decrease is mostly faster for the deconvolution function than for the experimental curve. This results in repeated divisions by numbers close to or equal to zero. This disadvantage can to a certain degree be overcome by using digital filters as will be shown below.

THEORY

The ac polarographic response to a reversible system is an ac current with amplitude

$$I = \frac{n^2 F^2 A C_o^* \sqrt{\omega D_o} \Delta E}{4RT \cosh^2(j/2)} \tag{1}$$

where

$$j = \frac{nF}{RT}(E_{dc} - E_{1/2}) \tag{2}$$

All symbols have their usual meanings, see, e.g., chapter 9 in [10]. When equation (1) is plotted as a function of E_{dc} one obtains a bell-shaped curve, where the maximum occurs at $E_{dc} = E_{1/2}$. At this point $\cosh(j/2) = 1$ and the peak current is thus

$$I_p = \frac{n^2 F^2 A C_o^* \sqrt{\omega D_o} \Delta E}{4RT} \tag{3}$$

Since a computer uses discrete mathematics one can rewrite equations (1) and (2) as follows

$$I(m) = \frac{n^2 F^2 A C_o^* \sqrt{\omega D_o} \Delta E}{4RT \cosh^2(\Delta j/2)} \tag{4}$$

where

$$\Delta j = \frac{nF}{RT}(E_{dc}^{in} + m\Delta E_{dc} - E_{1/2}), \qquad m = 0, 1, \ldots, N-1 \tag{5}$$

ΔE_{dc} is the small potential increment in the dc scale between sampling and E_{dc}^{in} is the initial potential of the sweep. N is the number of measurements. To simplify the expressions below the potential sweep is considered to start at 0 V and to be directed towards more negative potentials, i.e., $E_{dc}^{in} = 0$ V and $\Delta E_{dc} < 0$. The desired pulse function, mentioned above, can now be written

$$P(m) = \begin{cases} I_p = \dfrac{n^2 F^2 A C_o^* \sqrt{\omega D_o} \Delta E}{4RT}, & \text{if } m\Delta E_{dc} = E_{1/2} \\ 0, & \text{otherwise} \end{cases} \tag{6}$$

The ac polarographic curve, equation (4), can now be considered as the convolution of the single peak, $P(m)$, and a "broadening" function, $D(m)$, given by

$$D(m) = \frac{N}{\cosh^2(\Delta j'/2)} \tag{7}$$

where

$$\Delta j' = \frac{n_d F}{RT} m \Delta E_{dc} \qquad m = \left[-\frac{N}{2} \right], \ldots, 0, \ldots, \left[\frac{N}{2} - 1 \right] \tag{8}$$

and letting $n_d = n$. Here $\lceil k \rceil$ means the smallest integer $\geq$ k. The function $D(m)$ has the same shape as $I(m)$ but is always centered over the origin (if $E_{dc}^{in} \neq 0$ V then the maximum will occur at this potential). This mathematical restriction was not clearly pointed out in an earlier work [9]. Before the convolution can be calculated, the following assumptions must be made, the potential range used is denoted by E_{ra},

$$I(m) = \begin{cases} I(m), & \text{if } m\Delta E_{dc} \in [0, E_{ra}] \\ 0, & \text{otherwise} \end{cases} \tag{9}$$

$$P(m) = \begin{cases} P(m), & \text{if } m\Delta E_{dc} \in [0, E_{ra}] \\ 0, & \text{otherwise} \end{cases} \tag{10}$$

$$D(m) = \begin{cases} D(m), & \text{if } m\Delta E_{dc} \in \left[\left[-\frac{N}{2} \right] \Delta E_{dc}, \left[\frac{N}{2} - 1 \right] \Delta E_{dc} \right] \\ 0, & \text{otherwise} \end{cases} \tag{11}$$

One must also assume that there exists an ℓ, $\ell = 0, 1, \ldots, N - 1$, such that $\ell \Delta E_{dc} = E_{1/2}$. That $I(m)$ is the convolution of $P(m)$ and $D(m)$ can now be confirmed through a straightforward calculation:

$$\begin{aligned}
\frac{1}{N} \sum_{k=0}^{N-1} P(k) D(m-k) &= \frac{1}{N} \sum_{k=0}^{N-1} P(k) \frac{N}{\cosh^2 \left(\frac{nF}{2RT} (m-k) \Delta E_{dc} \right)} \\
&= \frac{n^2 F^2 A C_o^* \sqrt{\omega D_o} \Delta E}{4RT \cosh^2 \left(\frac{nF}{2RT} (m-\ell) \Delta E_{dc} \right)} \\
&= \frac{n^2 F^2 A C_o^* \sqrt{\omega D_o} \Delta E}{4RT \cosh^2 \left(\frac{nF}{2RT} \left(m\Delta E_{dc} - E_{1/2} \right) \right)} \\
&= I(m) \qquad m = 0, 1, \ldots, N - 1
\end{aligned} \tag{12}$$

The function $D(m)$, with $n_d = n$, is thus the wanted deconvolution function. Since convolution in the time domain is equivalent with a multiplication in the frequency domain [11], one can now obtain $P(m)$ by calculating the inverse Fourier transform of the quotient of the transforms of $I(m)$ and $D(m)$, i.e.,

$$P(m) = \mathcal{F}^{-1} \left[\frac{\mathcal{F}[I(m)]}{\mathcal{F}[D(m)]} \right] \qquad m = 0, 1, \ldots, N - 1 \tag{13}$$

where $\mathcal{F}$ stands for the Fourier transform and $\mathcal{F}^{-1}$ for the inverse transform. A simulated polarogram and the theoretical result of a deconvolution is shown in fig. 1.

What happens then if we choose another n_d? This is often the case when deconvolution is done on experimental data. If $n_d > n$ the resulting peak will be thinner but the peak height will also be affected, see fig. 2. This is probably the reason why Smith *et al.* [9] defined their deconvolution function essentially as equation (7) but instead of N in the numerator they

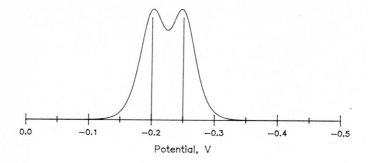

Fig. 1. Simulated polarogram consisting of two peaks with $\Delta E_{1/2} = 50$ mV, $n_1 = n_2 = 2$, $C_1^* = C_2^* = 10^{-4}$ M and $A\sqrt{\omega D_o}\Delta E = 1$. F, R and T have their usual values (25 °C). The corresponding pulse functions are shown in the figure as well.

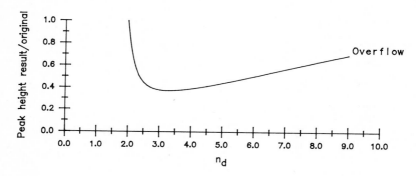

Fig. 2. Peak height deconvoluted/original as a function of n_d when $n = 2$ and $N = 64$. Overflow is due to the limited computing capacity of the computer used (Sirius, Intel 8088).

used an "arbitrary constant" to rescale the results. If n_d becomes very large the function $D(m)$ can be approximated with

$$D(m) \approx \delta(m) = \begin{cases} N, & \text{if } m = 0 \\ 0, & \text{otherwise} \end{cases} \qquad (14)$$

The Fourier transform of this function is

$$\mathcal{F}[\delta(m)] = \frac{1}{N} \sum_{m=0}^{N-1} \delta(m) e^{-2\pi ikm/N}$$

$$= 1 \qquad k = 0, 1, \dots, N - 1 \qquad (15)$$

Deconvolution in this case does not affect the polarogram at all. The case $n_d < n$ is theoretically not meaningful, it's the same as trying to obtain a peak with a width that is less than zero. One can thus affect the peak width by letting n_d vary, $n \leq n_d < \infty$. The peak width will then vary between zero and the width of the original peak.

The Fourier transform of the deconvolution function is composed mainly of low frequencies. When the transform of a polarogram is divided by the transform of the deconvolution function, several cases of division by zero will probably occur. This means that the deconvolution of experimentally obtained polarograms in many cases will be unsuccessful, the result is either an error message from the computer or a very "noisy" polarogram, see fig. 3. There are, however, methods available to stabilize the deconvolution. Two methods will be considered here, the simplest one, which can be seen as a rectangular filter, and the Tikhonov regularization method [12].

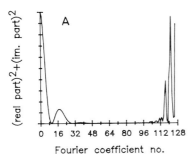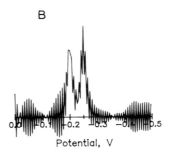

Fig. 3. (a) Result after division of the transform of the polarogram in fig. 1 and the transform of the deconvolution function with $n_d = 2.8$. (b) The result after inverse transformation.

The most obvious way in dealing with the problem is of course to stop the division when the transform of the deconvolution function comes close to zero. The rectangular filter can be used for this purpose. The division proceeds normally up to a cut-off frequency chosen by the user or the computer. The higher frequency components are then set to zero. This method is in many cases sufficient, but it fails to give acceptable results in the case where the cut-off frequency is very low. The use of the rectangular filter in the frequency domain is equivalent with the convolution of the polarogram and a function of $\sin(x)/x$-type in the time domain [11]. This means that side lobes can hardly ever be avoided and the peak height and width will both be affected.

In the Tikhonov regularization method the division is stabilized in the following way:

$$P_{p,\lambda}(m) = \mathcal{F}^{-1}\left[\left(\frac{|\mathcal{F}[D(m)]|^2}{|\mathcal{F}[D(m)]|^2 + N^2\lambda\omega^{2p}}\right)\frac{\mathcal{F}[I(m)]}{\mathcal{F}[D(m)]}\right]$$

$$= \mathcal{F}^{-1}\left[\frac{\mathcal{F}[I(m)]\,\mathcal{F}[D(m)]^*}{|\mathcal{F}[D(m)]|^2 + N^2\lambda\omega^{2p}}\right] \tag{16}$$

where $*$ denotes the complex conjugate. This method can be seen as a filter where λ determines the filter's cut-off frequency and p how fast the filter will decrease towards zero. By varying p and λ it is possible to a certain degree to minimize the unwanted effects of using a filter. It is however impossible to eliminate these effects totally. There exist some methods for chosing a statistically optimal p and λ [13], but these were not used here. A few trials will soon give the user a feeling for how different values will affect the resulting peak. For example very small values of λ may result in instability, which is understandable when one notices that the case $\lambda = 0$ is equivalent with equation (13). The effects of the filters is demonstrated in fig. 4. As can be seen, the background noise is more effectively removed when the Tikhonov method is used.

Since the total current measured in a multicomponent system is assumed to be the sum of the individual currents of the species, and furthermore since the Fourier transformation is a linear operation, every single peak in the polarogram is treated as if it were the only peak present. This fact should be considered when choosing n_d. If the number of electrons involved in the different reactions is denoted by $n_1, n_2, \ldots$, the following constrain on n_d

$$n_d \geq \max\{n_1, n_2, \ldots\} \tag{17}$$

should be fulfilled. In the practical applications n_d values smaller than the individual n values can be used due to the damping effects of the filter. This, however, results in increased side lobes, making the interpretation of the obtained curve more difficult.

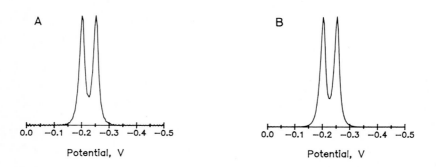

Fig. 4. Result after deconvolution ($n_d = 2.8$) of the polarogram in fig. 1, when (a) the rectangular filter and (b) the Tikhonov method is used.

REFERENCES

1 A. M. Bond and B. S. Grabarić, Anal. Chem., **48** (1976) 1624-1628.
2 W. F. Gutknecht and S. P. Perone, Anal. Chem., **42** (1970) 906-917.
3 P. A. Boudreau and S. P. Perone, Anal. Chem., **51** (1979) 811-817.
4 T. F. Brown and S. D. Brown, Anal. Chem., **53** (1981) 1410-1417.
5 C. A. Scolari and S. D. Brown, Anal. Chim. Acta, **178** (1985) 239-246.
6 A. den Harder and L. de Galan, Anal. Chem., **46** (1974) 1464-1470.
7 R. J. O'Halloran and D. E. Smith, Anal. Chem., **50** (1978) 1391-1394.
8 J. W. Hayes, D. E. Glover, D. E. Smith, M. W. Overton, Anal. Chem., **45** (1973) 277-284.
9 B. S. Grabarić and R. J. O'Halloran, D. E. Smith, Anal. Chim. Acta, **133** (1981) 349-358.
10 A. J. Bard and L. R. Faulkner, Electrochemical Methods. Fundamentals and Applications, John Wiley & Sons 1980.
11 D. F. Elliot and K. R. Rao, Fast Transforms. Algorithms, Analyses and Applications, Academic Press 1982.
12 A. N. Tikhonov and V. Y. Arsenin, Solutions of ill-posed problems, Winston & Sons 1977.
13 A. R. Davies, M. Iqbal, K. Maleknejad and T. C. Redshaw, in P. Deuflhard and E. Hairer (eds), Numerical Treatment of Inverse Problems in Differential and Integral Equations, Birkhäuser Verlag 1983, pp. 320-334.

M.R. Smyth and J.G. Vos
Electrochemistry, Sensors and Analysis
© Elsevier Science Publishers B.V., Amsterdam — Printed in The Netherlands

INTELLIGENT INSTRUMENTATION FOR ELECTROANALYSIS

P.R.FIELDEN, R.N.CARR and C.F.ODUOZA

Department of Instrumentation and Analytical Science, University of Manchester
Institute of Science and Technology, P.O.Box 88, Manchester M60 1QD (United Kingdom).

ABSTRACT
The essential elements of intelligent instrumentation for electroanalysis and how
they may be implemented, are described. Steps in the construction of an intelligent
polarograph are used to illustrate how data may be obtained, and how parameters may be
estimated from experimental data. Conversion of data into useful information through
processes such as interpretation and inference relative to a knowledge base is discussed
with the aim of employing an expert system. Intelligence may be imparted to the
polarograph through using the expert system to mimic the equivalent procedures that
might be adopted by a skilled operator.

INTRODUCTION

In recent years, much progress has been made in the design of instrumentation for
electroanalysis. Enhanced precision and reproducibility of measurements and simple
control of complicated timing sequences have contributed towards a renewed interest in
electrochemical methods of analysis. Both potentiometric and voltammetric techniques
have benefitted from the computer control of the measurement instrumentation.
Commercially available systems can offer a combination of features such as: automatic
sampling, and reagent delivery (in the case of titrimetric analysis); temperature
measurement and correction; in-built calibration routines; data storage for background
and blank correction with the provision for detailed data processing; and full software
control of instrument parameters.

In spite of these advances, most instruments, whilst offering a degree of automation,
rely on careful setting up by a skilled operator and lengthy method development. The
removal of this pre-requisite for highly trained operators would, therefore, represent a
significant advance in electroanalysis such that intelligent instrumentation could be used
by individuals with less experience.

Conventional instruments usually consist of an electrochemical analyser which
outputs measured data from a sample (which may be collected manually or automatically). These data are subsequently interpreted to give the desired information. The
procedures in each of these steps are specified by the operator of the instrument. For
intelligent instrumentation, as shown in Figure 1, additional units must be present.
Decision analysis and overall management, usually performed by the operator, must be
implemented in some way. A knowledge base of technical information is necessary in

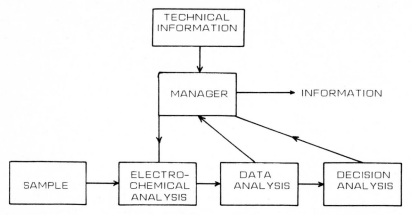

Fig.1. Block diagram showing the essential elements of an intelligent instrument for electroanalysis.

order that meaningful information may be obtained through the selection of a single, or a series, of appropriate measurement procedures.

EXPERIMENTAL

The design of intelligent instrumentation for electroanalysis is illustrated through the authors' approach to the development of intelligent polarography. The requirements of the proposed instrument may be summarised as follows:

- flexible software generation of method (waveforms, sampling, stirring, purging).
- automatic correction routines (background, blank, "bad" drop removal, scaling).
- data processing (filtering, deconvolution, parameter estimation).
- self-optimisation.
- interpretation (data transformed into information).

Software control has been reported previously (refs. 1-2) and automatic correction routines are available on some commercially available instruments (ref.3). Data processing methods have been well explored and include data manipulation in the frequency domain through the use of the digital fast Fourier transform (ref.4). The problem of parameter estimation, however, is in need of significant research input if the intelligent analysis of complex mixtures of species is to be realised.

The instrument used by the authors has been described by Miller and Thomas (ref.5) and consists of a static mercury drop electrode (SMDE – P.A.R.303) linked to a house-built computer controlled potentiostat. Waveform generation and data collection is performed through software routines run on a microcomputer.

Parameter Estimation

Parameter estimation is usually performed by an operator, who through previous experience, knows what the response to a given electroactive species should look like. The human process of comparing actual data with the expected response is often taken for granted. For intelligent polarography, it is necessary to reproduce this process. One approach is to simulate the expected response and compare the model data with the experimental data. In this way, parameters may be taken from the fitted model as reasonable estimates of the experimental parameters.

In the case of sampled DC polarography, the expected response of the polarographic wave for a reversible system is described by

$$E = E^* + RT/nF \ \ln(i_\ell - i/i) \tag{1}$$

where E and i are the applied potential and measured current respectively. i_ℓ is the limiting current and E^* the half wave potential. It is necessary to obtain values for E^*, i_ℓ and also the background current i_B, in order to characterise the polarographic response to an analyte. A 200 point model was generated and compared to a 200 point experimental polarogram (10ppm Cd; sampled DC polarography). The model was adjusted using a modified SIMPLEX with factors i_ℓ, i_B and E^*. The degree of fitting was estimated from a mean square error value (MSE) which was minimised for an optimum fit. This approach is fairly rapid; Table 1 shows the progress of the SIMPLEX which fitted the model to the experimental data in five moves. The resultant fit is shown in Figure 2.

TABLE 1

Move	MSE	E^* mV	i_ℓ μA	i_B μA
1	2.3×10^{-3}	−660	2.0	1.66
2	1.2×10^{-3}	−680	1.5	2.0
3	2.8×10^{-3}	−700	1.4	2.5
4	4.7×10^{-4}	−710	1.35	1.99
5	3.9×10^{-4}	−725	1.30	1.98

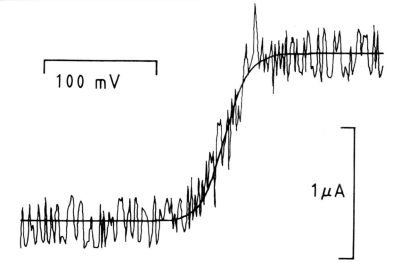

100 mV

1 μA

Fig.2. Diagram showing the optimised fit of a model polarographic wave to experimental data. This corresponds to the fifth move of the SIMPLEX (see Table 1 for parameter estimation).

DISCUSSION

In order to fulfill the requirements of intelligent polarography, it is necessary to incorporate a means of data interpretation and a regime for optimisation of the analytical measurement. If it is assumed that the parameter estimation is successful, boundaries must be set before the process of interpretation can be initiated. Constraints such as the working potential range of the electrode and electrolytes are important so that anomalous results, that could be generated by searching outside the set boundaries, are rejected.

Most importantly, a knowledge input is required and an inference made to link the estimated parameters with knowledge of the electrochemical behaviour of the analyte solution. Similarly, optimisation requires definition of the measurement being undertaken. Depending on the measurement circumstances, it may be that to maximise resolution in a multicomponent mixture is of greater value than the optimisation of sensitivity. Both interpretation and optimisation processes assume a knowledge input. For conventional instrumentation, this knowledge input has usually come from the operator.

The authors' are currently assessing the application of an expert system approach. The expert system chosen (Xi-Expertech) enables the experimenter to build a knowledge base which contains information about the types of samples that may be encountered in future analyses. It is carefully written to include default mechanisms to avoid erroneous interpretation. That is, if the information required or the data forwarded from the experiment is outside the scope of the knowledge base, further knowledge will be requested before the interpretation process is allowed to proceed. The Xi package is interactive and operates on a tutor/tutee basis so that an operator is permitted to interrogate the expert system and ask why certain decisions have been taken or particular interpretation routes followed.

Whilst the expert system approach will not take over complete control from the operator, it will enable more complicated and intelligent experimental control and data interpretation to be achieved by operators with a reduced expertise in electroanalysis.

CONCLUSIONS

The application of an expert system to assist the intelligent control of electro-chemical experiments may be a useful addition to present instrument design. Careful interfacing of computer based instrumentation is a prerequisite if useful results are to be achieved. In particular, reliable methods of parameter estimation must be developed for complex, multicomponent samples in order to achieve useful interpretation. When these difficulties are overcome, electroanalysis may well be usable by unskilled or semi-skilled operators.

ACKNOWLEDGEMENTS

The authors' wish to acknowledge the work of Dr.R.M.Miller and Miss K.Thomas in initiating this research programme, the guidance of Dr.M.Turpin in the development of expert system based instrumentation and the financial support of Unilever Research/SERC and British Council.

REFERENCES

1.   M.Bos, The development of a fully computerised system for sampled D.C.
     polarography with standard interfacing, Anal.Chim.Acta, 81 (1976) 21-30.
2.   P.He, J.P.Avery and L.R.Faulkner, Cybernetic control of an electrochemical
     repertoire, Anal.Chem., 54 (12) (1982) 1313A-1326A.
3.   A.M.Bond, Modern polarographic methods in analytical chemistry, Marcel Dekker,
     Inc., New York, 1980, pp.473-492.
4.   D.E.Smith, Data processing in electrochemistry, Anal.Chem., 48 (2) (1976) 221A-240A.
5.   R.M.Miller and K.E.Thomas, Cybernetic Voltammetry, Poster 1P.A6, 30th IUPAC
     Congress, Manchester, UK, September 1985.

M.R. Smyth and J.G. Vos

Electrochemistry, Sensors and Analysis

© Elsevier Science Publishers B.V., Amsterdam — Printed in The Netherlands

THE APPLICATION OF CONVOLUTION TRANSFORM TECHNIQUES TO THE STUDY OF

ORGANOMETALLIC ELECTRODE REACTION MECHANISMS

I.D. Dobson[1], N. Taylor*[1] and L. R. H. Tipping[2]

[1] Department of Physical Chemistry, University of Leeds, Leeds, LS2 9JT.

[2] EG&G Instruments Ltd, Doncastle House, Doncastle Road, Bracknell, Berkshire, RG12 4PG.

* Author to whom enquiries should be addressed.

SUMMARY

The use of convolution transforms of current, as an approach to extracting information from cyclic voltammetric data, is discussed. Experimental examples are presented using the complexes cis - [Cr $(CO)_2$ (dppma PP')$_2$] which exhibits an EC$_{(irrev)}$ electrode process and mer - [Cr $(CO)_3$ (dppm PP') (dppm P)] which exhibits complicated redox behaviour. (1)

INTRODUCTION

With the present availability of low cost laboratory

microcomputers, there is a trend towards performing sophisticated data

treatments to allow the easier and/or more accurate assessment of

results. The recent upsurge of interest in FTir techniques is an

obvious example. Electrochemistry has perhaps tended to lag behind

other fields in this respect, with computerisation having extended

little further than data capture and background subtraction

facilities. Now, however, a new software package (designed for use

with IBM PC microcomputers) has been introduced by EG&G Instruments

Ltd which allows not only the capture and storage of experimental data

with optional background subtraction but also the

(1) dppma = bis (diphenylphosphinomethylamine)
    dppm = bis (diphenylphosphinomethane)

comprehensive analysis of such data using a range of convolution and deconvolution transform functions.

It is the aim of this paper to give some indication of how such a package can aid the analysis of cyclic voltammetric data(2) both in terms of parameter extraction and mechanism determination.

TRANSFORMS OF THE CURRENT DATA

The analysis of cyclic voltammograms has traditionally relied on following the magnitude and position of particular features, most often the current peaks, whilst varying the potential scan rate in a series of experiments [1]. There are, however, severe limitations with this approach to testing the experimental data for agreement with a particular mechanistic/kinetic model. Firstly, very little of the total data obtained is actually used. This limitation is often compounded by the difficulty encountered in deciding the true position of the reverse peak if indeed such a peak is even evident. Secondly, before any current, potential analysis can be performed it is necessary to make an a priori assumption of the electron transfer kinetics. (Often for example, the assumption of Nernstian behaviour is made). A third limitation lies in the extremely complex nature (in mathematical terms) of the cyclic voltammetric curve.

All of these difficulties can be alleviated by using, instead of the original current function, an appropriate transform of the current.

Consider the electron transfer reaction

$$A \rightleftharpoons A^+ + e^- \tag{1}$$

(2) The EG&G software also allows analysis of potential step experiments to be performed. This aspect is outside the scope of this discussion.

In such a case it has been shown [2] that the solution of the appropriate diffusion equations via Laplace transformation, manipulation in Laplace space and back transformation aided by the convolution theorem, gives expressions for the concentrations of each species at the electrode surface i.e.

$$[A]_{(o,t)} = [A]_{bulk} - \frac{1}{(\pi D_A)^{\frac{1}{2}}} \int_0^t \left[\frac{i(u)}{nFa} \cdot \frac{1}{(t-u)^{\frac{1}{2}}} \right] du \qquad (2)$$

and

$$[A^+]_{(o,t)} = \frac{1}{(\pi D_{A^+})^{\frac{1}{2}}} \int_0^t \left[\frac{i(u)}{nFa} \cdot \frac{1}{(t-u)^{\frac{1}{2}}} \right] du \qquad (3)$$

These expressions are arrived at without recourse to electrode boundary conditions and in consequence no restrictions whatsoever are imposed with regard to either electron transfer rate or the particular potential waveform imposed. For convenience the convolution integral in expressions (2) and (3) can be expressed as a transformed current

$$I_1 = \frac{1}{\pi^{\frac{1}{2}}} \int_0^t i(u) \cdot \frac{1}{(t-u)^{\frac{1}{2}}} \cdot du \qquad (4)$$

allowing us to rewrite expressions 2 and 3 as

$$[A]_{(o,t)} = [A]_{bulk} - \frac{I_1}{nFaD_A^{\frac{1}{2}}}$$

64

and

$$[A^+]_{(o,t)} \quad = \quad \frac{I_1}{nFaD_A^{\frac{1}{2}}} \qquad\qquad (6)$$

Note that with expression (5) by adopting a potential at which the electrode surface concentration is driven to zero then I_1 adopts a maximum value, which we shall term I_{lim}. Rearranging (5) gives

$$I_{lim} = nFaD_A^{\frac{1}{2}} [A]_{bulk} \qquad\qquad (8)$$

In this form, two important points are immediately apparent. The limiting (plateau) value of the convolution transform offers easy access to the diffusion coefficient of the starting species. Secondly, knowing the diffusion coefficient, the proportionality of I_{lim} and the bulk concentration of A can be exploited for analytical purposes as suggested by Oldham [3].

In the area of mechanistic studies, the benefit of using this convolution transform arises from the fact that we are now dealing with a variable related to concentration rather than to reaction rate (as is the case with current). For example, with an electrode reaction uncomplicated by homogeneous chemical reaction and exhibiting fast electron transfer, then I_1 follows the simple mathematical form predicted by the Nernst equation. This means that the forward and reverse sweeps are superimposable (in contrast to current) and additionally the response is independant of the potential sweep rate (again unlike the current function). The behaviour of the I_1 convolution transform, in the case of the more common electrode mechanisms, has been extensively discussed by Saveant and co-workers [4 - 8].

Whilst the I_1 transform gives improved data analysis in the vast majority of electrode processes, its utility can be greatly enhanced by using other convolution transforms in certain specific cases. The so called EC$_{(irrev)}$ electrode process (i.e. electron transfer followed by an irreversible homogeneous chemical reaction) is very commonly found. Naturally it is often of academic or commercial interest to obtain accurate measurements of the chemical reaction rate.

We are now considering the scheme

$$A \rightleftharpoons A^+ \xrightarrow{\quad k_c \quad} B \qquad (9)$$

In this case, expression (5) remains a valid expression for the surface concentration of the starting species, A. Expression (6) in contrast, now becomes

$$[A^+]_{(o,t)} + [B]_{(o,t)} = \frac{I_1}{nFaD_A^{\frac{1}{2}}} \qquad (10)$$

To evaluate the rate constant k_c in expression (9), we must have access to either the surface concentration of A^+ alone or of B alone. This is possible with the use of a second convolution integral I_2 where

$$I_2 = \frac{1}{\pi^{\frac{1}{2}}} \int_o^t i(u) \cdot \frac{e^{-k_c(t-u)}}{(t-u)^{\frac{1}{2}}} \cdot du \qquad (11)$$

Now we have

$$[A^+]_{(o,t)} = \frac{I_2}{nFaD_{A^+}^{\frac{1}{2}}}$$

Note, however, that I_2 is calculated using the as yet unknown parameter, k_c. This does not present any difficulty, since we know an experiment commenced and terminated at a potential before the start of a cyclic voltammetric wave (typically 100mV before $E_{\frac{1}{2}}$ is sufficient) both starts and finishes in the absence of the species A^+ so long as we arrange the bulk concentration of A^+ to be zero. It follows that I_2 when correctly calculated, will start and finish at zero. Trial values of k_c must be inserted into the calculation and a simple simplex procedure allows k_c to be accurately determined. Figure 1 shows the appearance of the I_2 function calculated with the correct value of k_c and also with higher and lower values.

A third type of transform offered by the EG&G software, is a deconvolution of current with the function $(\pi t)^{-\frac{1}{2}}$. This gives a significant improvement in the visual presentation of cyclic voltammetric data. We have found this transform to be particularly useful where the "reverse wave" is poorly defined in the original current function. This point and other advantages will be apparent in the following discussion.

RESULTS AND DISCUSSION

Electrode behaviour of the complex cis - [Cr (CO)$_2$ (dppmaPP')$_2$]

Bond and co-workers have discussed the electrode reactions of complexes of this general type [9, 10, 11]. It has been established

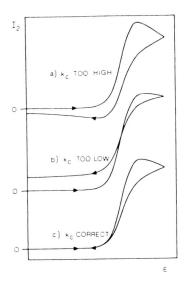

Fig. 1.   Establishment of a correct value for the homogeneous
          rate constant.

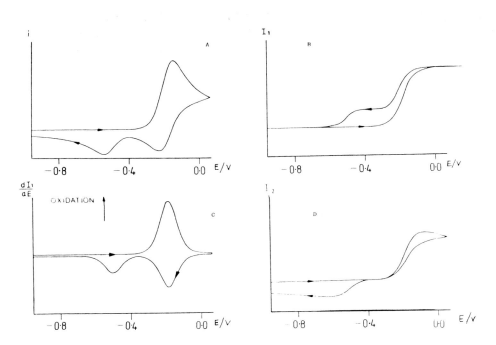

Fig. 2. Electrochemical data for the isomerisation of cis - [Cr(CO)$_2$ (dppmPP')$_2$]:
        (a)   cyclic voltammogram,   (b) I$_1$ convolution, (c) deconvolution of I$_1$
        and (d) I$_2$ convolution.

68

that the redox behaviour can be summarised by the scheme.

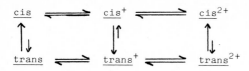

Here we shall be concerned only with the electrochemistry of the first electron transfer couples. This is, with the 18 and 17 electron complexes. Since the standard electrode potential for the trans isomer of this particular complex lies less positive than that for the cis isomer [12] then on a forward sweep only the $EC_{(irrev)}$ reaction (13)

$$cis \rightleftharpoons cis^+ \longrightarrow trans^+$$

is seen assuming the trans species to be absent initially. Knowing the starting bulk concentration of the cis isomer we observe that the concentration of the 18 electron species is followed (at the electrode surface) by the I_1 convolution transform. This is equally true for the reverse sweep.

Figure 2b shows this transform as derived from the cyclic voltammogram (figure 2a). Points to note include the simple sigmoidal shape of the forward sweep, varying between $I_1=0$ (where $cis_{(0,t)} = cis_{(bulk)}$ and $I_1 = I_{lim}$ (where $cis_{(0,t)}$ has been driven to zero by the forward electron transfer reaction).

Time spent in regions of potential where I_1 is non zero, implies the formation of the cis^+ species and consequently of the isomeric $trans^+$ form. On the return sweep, we see evidence of this. Note how as the standard oxidation potential of the cis isomer is reached, the

I_1 convolution falls towards zero (as the 18 electron cis species is regenerated) but because of the homogeneous production of the trans[+] isomer, the step is smaller than on the forward sweep. This simply reflects the failure to return to the bulk 18 electron species concentration in the course of such an experiment until the standard electrode potential of the _trans_ isomer is traversed. In figure 2b, we see a second step which corresponds to regeneration of 18 electron _trans_ complex from its 17 electron counterpart. It is an important diagnostic point that the I_1 transform has at this stage returned to zero in that this indicates complete conservation of material within the proposed scheme. Side reactions to non electroactive species would result in a failure of I_1 to return to zero at the end of the experiment.

Turning attention now to figure 2c., this shows the deconvolution of the current, potential data with $1/(\pi t)^{\frac{1}{2}}$. This is presented as $\frac{dI_1}{dE}$ rather than as $\frac{dI_1}{dt}$ to allow the easy correlation of features between the forward and return sweeps. A useful property of this transform is that in cases of fast electron transfer (3) the forward and reverse peaks for an electron transfer couple are located at $E_{\frac{1}{2}}$.

(3) Measurement of electron transfer rates is possible using the potential step convolution transform treatment offered by the EG&G software. A separate calculation is performed to extract $E_{\frac{1}{2}}$ in cases where electron transfer is not classed as fast.

This is exactly so for a chemically uncomplicated reaction and very nearly exact where chemical complication intrudes [13]. Clearly this gives a rapid assessment of either the electron transfer regime or the value of $E_{\frac{1}{2}}$.

The final data presentation in figure 2(d) is of the I_2 convolution transform. This has been calculated using the correct value for the chemical rate constant as discussed above. Notice that in the potential region positive of the trans/trans$^+$ activity in the I_1 transform, we observe the I_2 transform to have the shape seen in figure 1 and indeed this is the region for which I_2 is an appropriate function, the electrochemistry following an $EC_{(irrev)}$ scheme. On the reverse sweep, at potentials where trans$^+$ complex is reduced then we have a deviation of I_2 below zero. This is indicating that we have now deviated from the EC scheme and must invoke the square scheme previously described to fully encompass the electrode behaviour. The test for a correct assessment of the rate constant is the overlay of I_2 in the region where trans$^+$ and cis^0 are the predominantly stable forms. Using the above techniques the parameters listed in table 1 were extracted.

TABLE 1

Parameters for the electrode processes of cis - [Cr (CO)$_2$ (dppma PP')$_2$]

$E_{\frac{1}{2}}$ cis/cis$^+$	− 0.194	V
$E_{\frac{1}{2}}$ trans/trans$^+$	− 0.511	V
$E_{\frac{1}{2}}$ cis$^+$/cis^{++}	+ 0.615	V
$E_{\frac{1}{2}}$ trans$^+$/trans^{++}	+ 0.810	V
k_c measured as a function of temperature giving		
$\Delta H^{\ddagger}$	30.0	KJ mol
$\Delta S^{\ddagger}$	−134	JK mol
$\Delta G^{\ddagger}_{298}$	69.9	KJ mol

Electrode behaviour of the complex mer - [Cr (CO)$_3$(dppm PP') (dppm P)]

It is interesting to compare the cyclic voltammogram obtained from
this tricarbonyl complex (0.1 Vs^{-1} sweep rate) at -65°C with the
deconvolution transform obtained using a high scan rate (10Vs^{-1}) at
20°C. These are shown in figure 3a and 3b respectively. Visual
assessment is aided greatly by the application of the transform.
Particular benefit is noted in the case of the most positive wave, in
which evidence of the reverse reaction (mer^{2+} + e $\rightarrow$ mer^{+}) is only
observable using the transformed presentation.

Superficial interpretation of the data shown in figure 3,

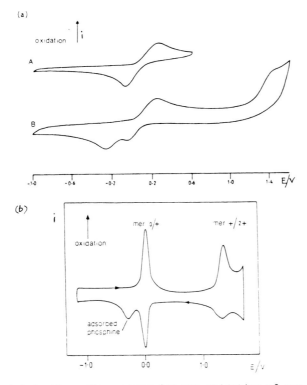

Fig. 3. (a) Cyclic voltammograms for the oxidation of mer - [Cr(CO)$_3$
(dppmPP')(dppmP)] in dichloromethane at -65°C (scan rate 0.1Vs^{-1}):
A showing the first oxidation and B with the second oxidation included.
(b) Deconvolution of cyclic voltammogram B.

indicates the following minimum scheme to describe the electrode behaviour:

$$\underline{mer}^0 \rightleftharpoons \underline{mer}^+ \rightleftharpoons \underline{mer}^{2+} \longrightarrow products$$

Clearly decomposition of the $\underline{mer}^{2+}$ species if rapid, in view of the reduced size of the return peak in the deconvolution response. The wave at -0.25V, characteristic of the uncoordinated phosphine ligand, appears to be associated with decomposition of the $\underline{mer}^{2+}$ species. As previously mentioned, we can immediately say electron transfer is fast for both oxidation steps, since in each case the corresponding forward and reverse deconvoluted peaks have an equivalent position on the potential axis. This further implies that these peaks are centered on the appropriate $E_{\frac{1}{2}}$ values of potential.

Focussing attention on the 18/17 electron redox couple, the cyclic voltammetric behaviour at 20°C is shown in figure 4. An assessment of this data by traditional methods would be likely to classify the process as a chemically uncomplicated electron transfer. Use of the I_1 convolution transform allows a more quantitative assessment to be made. In figure 5 we present this transform for a potential sweep rate of $1Vs^{-1}$ and also of 0.05 Vs^{-1}; a difference in behaviour is immediately apparent. It transpires that correlations exist between the timescale of the experiment and the values of both the limiting (plateau) value of I_1 and its final value at the end of the cycle.

These are shown in figure 6, where the behaviour at -65°C is also shown (4). The correlations are seen to disappear as temperature is lowered.

(4) The lower magnitude of I_{1im} at -65°C is a reflection of the lower diffusion coefficient at this temperature. Solvent viscoscity plays a major role here.

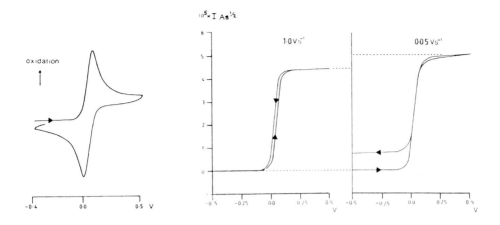

Fig. 4. Cyclic voltammogram of the mer/mer⁺ electrode
process. Potential scan rate 0.1Vs⁻¹.

Fig. 5. I_1 convolution transform of the mer/mer⁺ redox couple at
two potential scan rates.

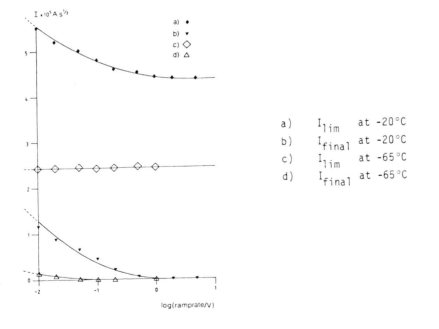

a) I_{lim} at -20°C
b) I_{final} at -20°C
c) I_{lim} at -65°C
d) I_{final} at -65°C

Fig. 6. Plots of the limiting [I_{lim}] and final [I_{final}] values of the
convoluted current (I_1) as a function of scan rate for a series of oxidative
cyclic voltammograms of mer - [Cr(CO)₃(dppmPP') (dppmP)] at a platinum electrode

Explanation of these results is provided by the addition to the reaction scheme, of a disproportionation reaction

$$\underline{mer}^+ \quad + \quad \underline{mer}^+ \longrightarrow \underline{mer}^O \quad + \quad \underline{mer}^{2+}$$

During the forward sweep, regeneration of the $\underline{mer}^O$ species enhances the value of I_1 the degree of enhancement being time dependant. On the reverse part of the scan we are observing the reduction of $\underline{mer}^+$ to $\underline{mer}^O$ and here the loss due to disproportionation results in a lower amplitude for the convoluted response which hence does not return to zero. This explanation predicts that the adsorbed phosphine wave associated with the $\underline{mer}^{2+}$ decomposition, should appear even if the $\underline{mer}^+/\underline{mer}^{2+}$ wave is not encompassed in the sweep. On slow scans, this is indeed found to be the case.

One other process was identified in the electrochemical behaviour of this complex. Controlled potential electrolysis at -70^OC produced a stable solution of the 17 electron $\underline{mer}^+$ complex. Warming this solution to room temperature, produced decomposition products via the disproportionation route, but additionally an intramolecular elimination of carbon monoxide was observed, giving dicarbonyl species as product. This was a slow reaction, and was monitored over a period of 2 hours using e.s.r. spectroscopy [12].

EXPERIMENTAL

Preparation of the complexes is described in reference [12].
Electrochemical measurements were made at a platinum working electrode in dichloromethane solution using 0.1M tetrabutylammonium perchlorate

as the background electrolyte. Potential control was achieved using

an EG&G PAR 363 potentiostat and model 175 universal waveform

programmer. Data collection was accomplished using an EG&G Analogue

to Digital Interface and an IBM PC compatible microcomputer. Data

treatment was performed using the previously mentioned EG&G PAR

Condecon 300 software package.

CONCLUSIONS

In this paper we have described some of the possibilities which

exist for extracting information from cyclic voltammetric data using

EG&G's new software package. Clearly advantages over the traditional

methods of data analysis exist. Note that with the transformed

functions each point incorporates information from all earlier points,

as captured and this helps in increasing the accuracy of any

interpretation. Many features of the convolution approach have not

been described here, due to lack of space, but it is important to

mention such aspects as software uncompensated resistance compensation

and logarithmic functions which linearise the data for particular

mechanisms, and allow the assessment of the number of electrons

transferred in the particular redox reaction. The chronoamperometric

section of the package is particularly useful for measuring electron

transfer rates.

76

REFERENCES

[1] R.S. Nicholson and I. Shain Anal. Chem. 36, 706, 1964

[2] D.D. MacDonald, Transient Techniques in Electrochemistry,
     Plenum Press, New York, 1977

[3] M. Grenness and K.B. Oldham Anal. Chem. 44 (1972) 1121

[4] J.C. Imbeaux and J.M. Saveant J. electroanal and Interfac.
     Electrochem. 44 (1973) 169

[5] F. Ammar and J.M. Sàveant J. electroanal chemistry 47 (1973)
     215

[6] L. Nadjo, J.M. Saveant and D. Tessier J. electroanal chemistry
     52 (1974) 403

[7] J.M. Saveant and D. Tessier   J. electroanal chemistry 65
     (1975) 57

[8] J.M. Saveant and D. Tessier   J. electroanal chemistry 61 (1975)
     251

[9] A.M. Bond, R. Colton and J.J. Jackowski Inorg. Chem. 14 (1975)
     2526

[10] F.L. Wimmer, M.R. Snow and A.M. Bond Inorg.Chem 13 (1974) 1617

[11] A.M. Bond, R. Colton and J.J. Jackowski Inorg.Chem 14 (1975)
     274

[12] A. Blagg, S.W. Carr, G.R. Cooper, I.D. Dobson, J.B. Gill,
     D.C. Goodall, B.L. Shaw, N. Taylor and T. Boddington.
     J.Chem.Soc. Dalton (1985) 1213

[13] I.D. Dodson and N.K. Blackwell unpublished results, 1984

M.R. Smyth and J.G. Vos
Electrochemistry, Sensors and Analysis
© Elsevier Science Publishers B.V., Amsterdam — Printed in The Netherlands

FLOW INJECTION ANALYSIS FOR ELECTROCHEMICAL STRIPPING
- COMPARISON OF DIFFERENT TECHNIQUES

W. FRENZEL, G. WELTER and P. BRÄTTER

Hahn-Meitner-Institut für Kernforschung GmbH,
Glienickerstr. 100, 1000 Berlin 39, FRG.

SUMMARY
    A comparison of the performance characteristics of various electrochemical
stripping techniques in Flow Injection Analysis is presented with special emphasis to the properties of the glassy-carbon mercury film electrode. Under
favourable experimental conditions fast linear scan Anodic Stripping Voltammetry
and Potentiometric Stripping Analysis turned out to offer exellent sensitivity
and signal-to-background ratios which could not be obtained by differential-
pulse and hardly improved by square-wave Anodic Stripping Voltammetry.

INTRODUCTION

    Flow injection systems for electrochemical stripping analysis (i.e. Anodic
Stripping Voltammetry (ASV), Potentiometric Stripping Analysis (PSA)) have
gained increasing interest since they offer distinct advantages when compared to
batch or continous flow procedures [1-10]. Various flow-through detector cells
with mercury film working electrodes have been designed including wall-jet
[1,4,6] and thin film [3] geometries. The voltammetric stripping step is usually
performed with linear [3,7], differential-pulse [1,2,4] or square-wave [4,8]
potential ramps. However, PSA has also been reported in combination with Flow
Injection Analysis (FIA) [5,6,9,10].

    In each of these papers the respective performance characteristics (e.g. sen-
sitivity, selectivity, sample throughput, system's stability) are given but the
experimental conditions used differ to such an extent that a realistic intercomparison of the particular techniques is impossible. Therefore it seemed
appropriate to establish the analytical properties of the various electrochemical stripping techniques mentioned above under comparable conditions.

EXPERIMENTAL

Instrumentation

    A PAR 174 A polarographic analyzer and a PAR 173/276 interfaced to an APPLE
II e computer and appropriate software were used for voltammetric measurements.

Thus, fast LSASV (up to 10 V·s^{-1} scan-rate) and SWASV [11] with flexible choice of frequency (1-1000 Hz), pulse height (1-100 mV) and scan increment (2-10 mV) were accessible. DPASV parameters (sweep rate 5 mV·s^{-1}, pulse repetition time 0.5 s, pulse-height 50 mV) were used without further optimization. The computerized PSA instrument (GÖTALAB, Sweden) consists of an ABC 80 8-bit microcomputer (Luxor, Sweden) interfaced to a potentiostat via an analog to digital converter. The software enables the subsequent recording of the original stripping and background curves with a conversion rate of 9.3 kHz and 5 mV potential resolution. The stripping times of the respective elements are determined by integration of the peak signal after background substraction [12,13].

The basic flow system has been described previously [5]. However, modifications in the construction of the wall-jet cell were made (see ref. 6) with the aim to reduce ohmic drop problems and an additional switching valve upstream of the injection valve was inserted to allow for a convenient change between different carrier solutions. Special precautions were exercised to prevent reaeration of the carrier solution on its way to the detector cell. This was accomplished by inserting the teflon connecting tubes into thick walled tygon tubes which were continuously flushed with argon. The 2 mm diameter Tokai glassy-carbon working electrode was inserted into a perspex body using epoxy glue resin and thoroughly polished with diamond paste (down to 0.025 μ). The condition of the actual electrode surface was examined by electron scanning microscopy. All potentials reported are vs. saturated Ag/AgCl.

Chemicals and Solutions

Metal ion stock solutions containing 1 g·l^{-1} Cd^{2+}, Pb^{2+} and Hg^{2+}, respectively, were prepared from Merck Titrisol. All other reagents were of highest available purity. Appropriate standards and carrier solutions were freshly prepared by dilution with demineralised and twice distilled water. Supporting electrolyte for all samples and standards was 0.1 mol·l^{-1} KNO$_3$ which was acidified to pH 4-4.5 with nitric acid. This solution also served as the carrier. Mercury ions were added to the carrier to give a final concentration of 20 mg·l^{-1} Hg^{2+}.

Procedure

The thoroughly deaerated carrier solution was propelled continuously through the system at a flow-rate of 0.6 ml·min^{-1}. Before the analytical runs a mercury film was formed on the electrode surface in-situ by setting a potential of -1.0 V for at least 3 min. To remove metallic contaminants the potential was then scanned to -0.1 V and held for 10 s. In all subsequent runs, a standard solution containing 500 μg·l^{-1} Cd^{2+} and Pb^{2+}, respectively, are injected immediately after the reduction potential was applied. The plating time was selected

according to the residence time of the sample which was determined amperometri-
cally. For the flow parameters used (125 µl injection volume, 0.6 ml·min^{-1}
flow-rate) the residence time is about 25 s. To ensure complete washout of the
sample and to make use of the matrix exchange technique, a cleaning step of additional 15 s was introduced. The stripping step was initiated after sample
passage in flowing solution.

RESULTS AND DISCUSSION
Electrode properties

Mercury films on glassy-carbon supports are widely used in electrochemical
stripping analysis [14]. However, care must be taken to avoid oxidation and
calomel formation at the mercury surface. This occurs whenever the potential
control of the electrode is lost (open circuit) or potentials more positive than
+0.1 V are applied [15, 16]. An important property of voltammetric electrodes is
the background current which originates from electrode capacity, residual
currents of electroactive contaminants (e.g. remaining oxygen, Hg^{2+}-ions in the
case of in-situ deposition) and electrochemical charge transfer reactions of
glassy-carbon surface groups. The latter was confirmed investigating glassy-
carbon materials from diverse manufacturers with respect to different ways of
production, various diameters of the rods and polishing procedures, respectively
[16]. To judge the benefits of the particular stripping techniques and make a
realistic comparison of the performance it is neccessary to consider the base
current slope of the MFE used. The properly maintained electrodes used in the
present work exhibit low slopes when compared to literature data [16].

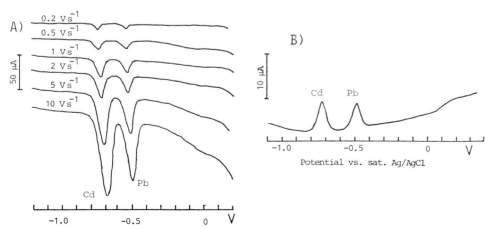

Fig.1: Comparison of LSASV (A) and DPASV (B) in FIA
Experimental details are given in the text.

Voltammetric Stripping

Linear scan voltammograms obtained at increasing scan rates are shown in Fig. 1a. The peak-height increases considerably with increasing scan rate, however above 5 V·s⁻¹ poor resolution of the Cd and Pb signals occurs. The background slope is surprisingly low even at the highest scan rate employed. In Table 1 the peak currents, the peak-width at half height, base current slopes and the ratio between peak currents and the background current slopes measured at the peak potential for Cd and Pb, respectively, are given and compared to DPASV and SWASV. The DPAS-voltammogram obtained for the same sample and identical flow conditions is shown in Fig. 1b. The absolute peak-current and signal-to-background ratio is lower than that for fast LSASV as was also found by Batley and Florence in batch procedures [15,17].

Prior to SWAS-measurements in FIA a comprehensive parameter examination was made in the batch mode at the MFE [18]. It is, however, difficult to give optimal values of frequency, scan increment and pulse-height since they are mutually influencing and in fact strongly depend on the film properties. Typical SW-voltammograms recorded at increasing frequency are shown in Fig. 2. The peak current measured at 100 Hz is substantially higher than that for DPASV but nearly agree with linear-scan peak currents obtained at 2 V·s⁻¹ scan rate (see also Table 1). Higher frequencies, although resulting in increased peak currents, are inappropriate, since background compensation is insufficient (Fig. 2b).

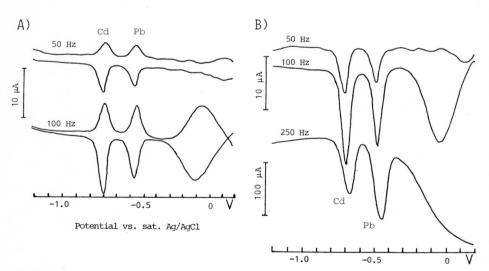

Fig.2: SWASV-electrode response in FIA

A) Forward and reverse scan obtained with 50 and 100 Hz, respectively. Scan increment 2 mV, pulse heigth 25 mV.

B) Current difference at increasing frequency.

Table 1: Comparison of the reponse characteristics of LSASV,
DPASV and SWASV under identical experimental conditions.

	LSASV (scan rate,$V \cdot s^{-1}$)				DPASV	SWASV (frequency,Hz)		
	0.2	1	5	10		50	100	250
Peak-current μA	5	18	48	80	5	6	14	130
Slope μA/V	2.8	9	20	30	7	1	5	120
S/B-Ratio	1.8	2	2.4	2.7	0.7	6	2.8	1.1
Peak-width mV	30	50	65	70	40	50	55	80

Potentiometric Stripping

The computerized PSA signals obtained on injection of the standard solution
are shown in Fig. 4. The stripping times calculated after background substraction are 260 ms and 204 ms for Cd and Pb, respectively. The signal-to-background
ratio in PSA is strongly dependent on the quality of the background substraction, that means how good the background curve fits the sample background [19].
Furthermore, the signal-to-background ratio can be improved by higher conversion rates [13]. The CPSA-system used in the present work allows for the
measurement of down to 10 ms stripping time with appropriate precision.

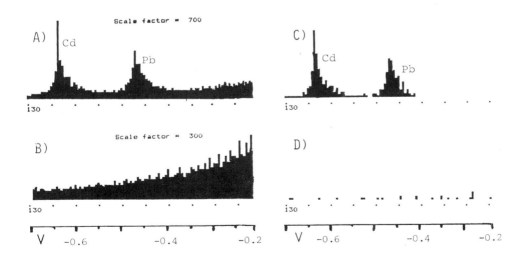

Fig.3: Potentiometric Stripping Analysis in FIA

        Signals obtained at 9.3 KHz conversion rate
        A) Sample reading, B) Background reading,
        C) Background subtracted signal, D) Remaining background

CONCLUSIONS

The benefits of the various voltammetric techniques cannot fairly be judged without considering the background slopes of the particular electrodes used. Well maintained glassy-carbon electrodes exhibit low slopes which makes the common LSASV more attractive with respect to sensitivity, selectivity and instrumental performance. Because of their basic differences PSA and ASV cannot be directly compared. However, PSA offers high sensitivity and after background substraction the signal-to-background ratio is exellent. The simpler instrumentation required and the low susceptibility to interferences [20] are further advantages of PSA.

RERERENCES

1 J. Wang, H.D. Dewald, B. Greene, Anal. Chim. Acta, 146 (1983) 45.
2 J. Wang, H.D. Dewald, Anal. Chim. Acta, 162 (1984) 189.
3 J. Wise, W.R. Heineman, P.T. Kissinger, Anal. Chim. Acta, 172 (1985) 1.
4 C. Wechter, N. Sleszynski, J.J. O'Dea, J. Osteryoung, Anal. Chim. Acta, 175 (1985) 45.
5 G. Schulze, M. Husch, W. Frenzel, Microchim. Acta, I (1984) 191.
6 W. Frenzel, P. Brätter, Anal. Chim. Acta, 179 (1986) 389.
7 J. Wang, H.D. Dewald, Anal. Chem., 56 (1984) 156.
8 C.A. Scolari, S.D. Brown, Anal. Chim. Acta, 178 (1985) 239.
9 A. Hu, R.E. Dessy, A. Granéli, Anal. Chem., 55 (1983) 320.
10 W. Frenzel, Thesis, Technical University Berlin, FRG, 1984.
11 J.G. Osteryoung, R.A. Osteryoung, Anal. Chem., 57 (1985) 101A.
12 M. Josefson, Thesis, University of Gothenburg, Sweden, 1983.
13 A. Granéli, D. Jagner, M. Josefson, Anal. Chem, 52 (1980) 2220.
14 G. Subramanian, G.P. Rao, Rev. Anal. Chem., 4 (1979) 95.
15 T.M. Florence, Anal. Chim. Acta, 119 (1980) 217.
16 W. Frenzel, in preparation.
17 G.E. Batley, T.M. Florence, J. Electroanal. Chem., 55 (1974) 23.
18 W. Frenzel, G. Welter, to be published.
19 W. Frenzel, P. Brätter, poster presentation, SAC 86, Bristol.
20 E. Han, Teachers Diploma Thesis, Technical University Berlin, 1985.

M.R. Smyth and J.G. Vos
Electrochemistry, Sensors and Analysis
© Elsevier Science Publishers B.V., Amsterdam — Printed in The Netherlands

THE APPLICATION OF AUTOMATION TO THE ELECTROANALYSIS OF INDUSTRIAL
AND ENVIRONMENTAL PROBLEMS

J.S. BURMICZ, U. HUTTER, V. PFUND

Metrohm AG, CH-9100 Herisau, Switzerland

SUMMARY
Automation and Electroanalysis are today two subjects which have
found in each other a certain "marriage of convenience". To gain
overall acceptance, electroanalytical techniques have had to provide not only the possibility for discrete measurements in the
laboratory, but have had to offer, perhaps even more so than other
analytical methods, the opportunity to perform automated, sequential sample analysis (i.e. with Sample-Changers) or even automated,
repetitive analysis on samples with more or less fixed composition (On-Line). The scope of this presentation will be conveniently subdivided under the headings of Instrumentation, Fields of
Application and Automation.

INTRODUCTION
    On considering the possibilities with respect to the Automation
of Voltammetric (and Polarographic) analyses, certain basic principles must always be observed from a realistic point of view. These
basic principles are related to the process mechanics of Polarographic (or Voltammetric) Analysis, as well as those related
to the process of generating meaningful results from a sample
remotely displaced from the instrument itself. Unfortunately,
for a lot of the dedicated advocates of Polarography, automation
will only encompass the mechanics of the sequential steps involved
in Stripping Analysis. This type of automation is, of course,
easily attained today thanks to basic electronic advances. If
these techniques however are to attain a wider acceptance in
analytical circles, then the manufacturers must address themselves
to the reality of modern Industrial processes and to a much
more realistic definition of "Automation". In this respect the
Polarograph must become a tool usable by everyone. The Modern
Polarographic systems however offer so much and are able to
be automated with reasonable effort, to provide answers often
either at levels or in environments which would present many
problems for alternative analytical "finishes".

INSTRUMENTATION

The limiting factor with regard to the application of Polarographic (and associated) techniques has always been connected
with the routine use of Mercury and Mercury electrodes. Mercury
however is only as toxic as the user permits. Routine Good Laboratory (or Analytical) Practice will eliminate all cause for concern
under the normal conditions for instrumental operation.

In order to make a truly automated system capable of sustained
operation under the normal industrial conditions,one of the most
important pieces that required significant redesigning was the
mercury electrode itself. Although attempts have been made in the
recent past to produce a modern design for mercury electrodes,
sources of concern had been raised,and thus a design which was
felt to be more industrially orientated has been developed. A
positive improvement in this latest design is that of the safety
features. The design is refered to as the MME or the Multi-Mode
Electrode.

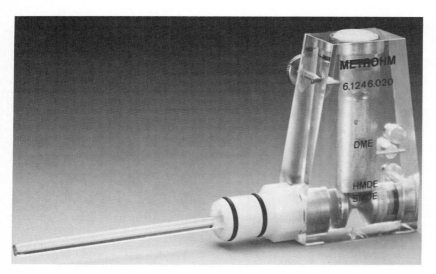

Figure 1 shows the electrode system. The most significant
feature of this system is the assembly: the Mercury cannot
flow without the Nitrogen pressure required for the overall system functioning.

In this respect another significant advantage is that the Mercury
flow can be interrupted (even for DME or Dropping Mercury Electrode operation), between each analytical scan. This of course,
allied with a design of low flow-rate capillary, greatly reduces
the amount of Mercury required for system operation. The reservoir is consequently only of 6 ml capacity. Unlike other systems
were the Mercury is exposed to the environment (even though it
may appear to be under cover), the environment within the MME is
Nitrogen which also limits the formation of oxides etc., which
under circumstances of normal exposure would be a significant contribution to system failure. This is due to oxide coating of
electromechanical moving parts. To this end, the MME contains no
electromechanical mechanism for drop generation within the MME.
The microvalve assembly is made of stainless steel and thus has
a further advantage that the contact between mercury reservoir
and instrument is always reliably maintained. Drop generation itself as well as the drop dislodge funtion are all pneumatically
controlled. The capillary mounting is also flexible and not
rigid. The capillaries themselves are also disposable and are of
conventional low-cost design. The system functions as HMDE,
SMDE and DME.

The most significant detail of course is the reproducibility.
Data outlining this parameter based on a medium level Cadmium
and Lead analysis are shown in Table 1.

Table 1: The reproducibility of the MME in the HMDE mode is
shown with a DPASV measurement from Cd and Pb in a thermostatted vessel:
Deposition Potential: -800 mV vs. Ag/AgCl/KCl(3M)
Deposition Time: 120 s

	Cadmium	Lead
Content	$p(Cd)=30 \mu g/l$	$p(Pb)=5 \mu g/l$
Automated evaluation values (nA)	122.9 121.4 124.4 122.1 122.3 123.2 123.3 124.1 123.5 122.8	27.42 26.96 27.76 27.42 27.18 27.16 27.29 27.42 27.13 26.91
Peak height (mean)	123.0 ± 2.1 nA $(\pm 0.9; 95; 10)$ $s_r = 0.7\%$	27.27 ± 0.6 nA $(\pm 0.3; 95; 10)$ $s_r = 0.9\%$

One of the original criticisms of the MME assembly was the relative
small size of the drops obtainable. However it has proved an
advantage to have drops of a smaller size. One of the most significant features is an obvious one of drop stability. The proposi-
tion was of course that the overall system sensitivity would
suffer as a result of these small drops. The key to system response lies in two main areas. These are both related to the
S/N ratio. The first source of noise is of course that due to
the drop generation system. Under normal conditions, a miscel-
lany of system solenoids produce much noise by virtue of their
function, all of which may be minimised, but never eliminated.
The MME contains no such possibilities for electromechanical
interference as the pneumatic switching is remote from the actual
capillary and mercury drop. The second source of noise generation
is that associated with the transfer of the signals from the
electrode assembly to the first stages of the polarograph itself.
In most cases, this requires the transfer of low level current
signals along lengths of wire which are in turn prone to all
types of interference.

Fig. 2

In the system illustrated in Figure 2, the "conventional" components of the Polarograph are enclosed in the stand 647. Once
the signal has undergone its initial registration and conversion
to a voltage, it is then transferred to the central processing

unit, the 646. It is sufficient to say also that several refine-
ments have been made to some of the "conventional" parts of
the Polarographic system.

This system offers the opportunity for operation at much
higher current sensitivities than in more conventional instrumentation. It is therefore not correct to extract any single element
of the instrument configuration but to look at the system performance as a whole.

Analytical Applications and Automation

The easy automatic switching between electrode systems also
gives another very significant advantage. This lies with the
possibilities now afforded for multicomponent analysis where
the levels of the components may be significantly different,
e.g. 800 ppb Zinc, 2 ppb Cadmium, 20 ppb Lead and 50 ppb Copper.
The new approach offered with the 646/647 combination will permit
in the space of one programme what is refered to as "Method
Mix" and "Electrode Mix". This means that in the above quoted
example, the Zinc may be analysed using a DME with Differential
Pulse Polarography followed by an automatic change to HMDE and
subsequent analysis of the remaining components (using Differential Pulse or Square Wave Anodic Stripping Voltammetry) either
in a single scan with one sensitivity (governed by the height
of the largest signal), or, for instance, where overlapping potential ranges are required, as a function of a potential "window",
where each element to be determined has its own potential range
and associated sensitivity . This is refered to as "segmented
analysis". This in essence provides each element with optimised
analytical conditions, and takes no longer timewise than if
the whole analysis were run at one sensitivity. This, coupled
with the possibility of automatic standard addition via dosimats
(or Calibration Curve), means that the entire analysis can be
executed from a single user designed program. In addition to
this, the segmented analysis may be further extended, again with
the use of the Dosimats (or autoburettes) to change the conditions
of the supporting electrolyte. This may be to eliminate or highlight certain other elements in the sample. For example, following
digestion of a solid sample, the pH may initially be adjusted
to pH 7. A collection of samples may then be transfered to a
sample changer. The first analysis would be for Zinc, Lead,

Cadmium and Copper. An auto-pH adjustment and auto-additon of DMG (Dimethylglyoxime), would then facilitate the determination of Nickel and Cobalt. The calculation of results would be achieved either with Standard Addition or Standard Curve. The only command for a batch of such analyses being an initial "Start".

As indicated above, the concept of Automation for many existing users of Polarographic (or Voltammetric) techniques is limited to the sequential control of mechanical subroutines in a determination using Anodic (or Cathodic) Stripping Voltammetry. These steps include the degassing of the solution, dispensing of a Mercury drop, activation of a stirring mechanism, deposition (in situ preconcentration), automatic cessation of the stirring mechanism, initiation of the scan followed by various posssibilities such as electrode conditioning, etc. Each one of these steps is accompanied by strict time control, especially that with regard to the preconcentration step, which has a direct effect on the signal height. These steps are relatively easy to control, with simple logic circuitry. Automation for the majority of people goes a lot further than this basic concept. The system encompassing the 646/647 combination has therefore striven to offer the proven benefits of the techniques and their related technology, to a completely new sphere of operation. The concepts of automation therefore now include applications both "ON-LINE" and with "SAMPLE-CHANGER". The use of the sample changer is clear and really constitutes the operation of batched discrete analyses. This is typically applied to a total of 10 runs for instance, with multielement analysis normally overnight. This system still allows the user to individually prepare each sample according to his/her preferences and allows a large degree of flexibility with respect to sample variation. The possibility with this system for predeaeration of successive samples, greatly reduces the time per analysis. On the other hand, the possibility for true on-line automation is perhaps the most exciting, and does offer the most enterprising fields of operation. This concept allows remote switching of valves, sampling from flowing streams, sample pretreatment etc., all from the same basic programme that before offered segmented analysis, methods and electrodes mix. The system now becomes a very powerful tool indeed and some of the applications to date have illustrated the possibilities for these techniques to afford the user with results from media which under normal circumstances would have required perhaps

several discrete sample pretreatment stages using alternative
analytical techniques.

Current applications range from the analysis on-line of Copper
and Formaldehyde in an electroless Copper plating bath, determination of residual Nitric acid in 72% Sulphuric acid, Analysis
of Sulphur Dioxide, Iron (II) and Iron (total) in a power station
scrubber catalyst, to the determination of trace heavy metals
in desalinated water. These applications are tried and tested,
and even in cases of high ionic strength materials or systems
where the pH for example is extreme, all the most negative aspects
of the sample constitution can be turned to advantage. The most
clear from this point of view is that concerning the analysis
of desalinated water or even Seawater itself. In this case the
high ionic content of the Seawater is of benefit to the conductivity of the sample. On the other hand, sample pretreatment
for AAS would require exhaustive measures to remove all traces
of salinity in order to preserve the Carbon Furnace and remove
the possibility of memory effects. Seawater may be directly
placed in the Polarographic cell and, with perhaps the exception
of filtration, the analysis may be conducted directly on the
seawater sample; and this to the low and even sub ppb level.
The obvious advantage here is that there is no need to add or
extract with chemical agents alien to the sample itself, which
in turn leads to less chance of external contamination.

Discussion

It is hoped that the information contained in this presentation
will be of fresh inspiration to those familiar with the subjects
discussed. For those who are as yet unfamiliar with the topics
discussed it is hoped that this information will present an
introduction to a methodology which is highly developed and
time proven in the field of industrial analysis.

The developments described are mainly in two areas: those
with the hardware and to a certain extent new approaches with
regard to the software and data handling. The new hardware approaches
have already without any doubt had a very positive response,
both in the industrial sector as well as in the areas of more
academic interest. The need for a reliable Mercury electrode
system with an inherent safety system and reproducible drop
generation system seems in retrospect to have been a reason
for the lack of consideration of this technology in the past.
The added possibilities for complete automatic analytical proce-

dures has also found a very receptive audience. This is both
from the point of view of the auto-standard addition and calculation routines as well as the possibilities offered for Sample-
Changing and/or On-Line applications.

Automation may be as involved or as basic as the Analytical
situation requires. Items of hardware for Sample-Changer and
On-Line operation are common to both systems so that Instrumentation may be updated as and when required.

The days where a Polarographic method was always the last
resort and where it had to prove its efficacy several times
over are happily coming to a close. However it must always be
kept in mind that the increased sophistication in instrumental
design should not be used as a substitute for the detailed chemical
knowledge required to generate meaningful data from the equipment. In this respect Polarography is in the same league as
all other instrumental techniques.

M.R. Smyth and J.G. Vos
Electrochemistry, Sensors and Analysis
© Elsevier Science Publishers B.V., Amsterdam — Printed in The Netherlands

VOLTAMMETRIC DETERMINATION OF NITROUS ACID IN MIXTURES OF NITRIC AND SULFURIC
ACIDS BY OXIDATION OF NITRITE AT A CARBON PASTE ELECTRODE

J. ASPLUND
Nobel Chemicals AB, S-69 185 Karlskoga (Sweden)

SUMMARY
    A voltammetric method for determination of nitrous acid is presented. The de-
termination is based on oxidation of nitrite ions at a carbon paste electrode in
a 0.1 molar ammonium acetate solution.
    This voltammetric method can be used for determining nitrous acid in mixtures
of nitric and sulfuric acids. The sum of the strong acids, nitric, nitrous and
sulfuric acids, is determined by titration with sodium hydroxide. Nitric acid
is determined by electrometric titration with two polarized platinum electrodes
with Fe(II) in concentrated sulfuric acid.
    Results from the determination of nitric, nitrous and sulfuric acids in pro-
cess acids from the trinitrotoluene and nitrocellulose processes are presented.

INTRODUCTION

    Titrimetric procedures for the determination of nitrite in acid solution are

complicated by the instability and volatility of nitrous acid and its low rate

of reaction with many titrants. A simple method in which nitrite is determined

by coulometry at controlled potential has been developed (ref. 1). Nitrite is

oxidized directly to nitrate at a platinum electrode in pH 4.7 acetate buffer

solution. Possible loss of nitrous acid is not a factor in the analysis. Modifi-

cation of a platinum electrode by chemisorption of iodine has been found to im-

prove the reproducibility and to decrease the peak width in oxidation of nitrite

by linear scan voltammetry (ref. 2). Investigations on the use of both glassy

carbon and carbon paste electrodes for the oxidative determination of nitrite

have already been published (refs. 3,4).

    This work describes a simple, accurate method for determination of nitrous

acid in process acids from the trinitrotoluene and nitrocellulose processes. The

determination is based on oxidation of nitrite ions at a carbon paste electrode

in a 0.1 molar ammonium acetate solution. Nitric acid is determined by electro-

metric titration with Fe(II) in concentrated sulfuric acid and the sum of the

strong acids, namely nitrous, nitric and sulfuric acids, is titrated potentio-

metrically with sodium hydroxide.

EXPERIMENTAL

Apparatus

The voltammetric determinations were performed with a Metrohm Polarecord E
506 and a Metrohm EA 267 carbon paste electrode. All potentials were measured
and reported vs. the Ag/AgCl electrode (Metrohm EA 427). The current-potential
curves were recorded at a 10 mV s^{-1} scan rate.

A Metrohm Potentiograph E 536 with a Metrohm Polarizer E 585 (polarization
potential + 100 mV) and a Metrohm double platinum electrode EA 240 were used in
the electrometric titrations.

A Metrohm Titroprocessor E 636 was used in the potentiometric titrations. A
pH glass electrode (Metrohm EA 101) and a saturated calomel electrode (Metrohm
EA 404) were used with the titroprocessor.

Reagents

All chemicals were reagent grade or equivalent purity. Stock solution of nitrite, 250 mg $NaNO_2$/0.1 L, was prepared from a dried sample of E. Merck $NaNO_2$.
Micropipets were used to take aliquots of the nitrite stock solution.

A 0.2 molar Fe(II) solution was prepared dissolving 80 g ammonium iron(II)
sulfate hexahydrate, $(NH_4)_2Fe(SO_4)_2$ $6H_2O$ in 700 mL distilled water. The volume
is adjusted to 1000 mL with concentrated sulfuric acid. The Fe(II) solution is
protected against oxygen from the air by a nitrogen atmosphere or by a layer of
paraffin oil on top of the solution. The Fe(II) solution is standardized against
100 mg of dried potassium nitrate (KNO_3).

RESULTS AND DISCUSSION

Voltammetric determination of nitrite

Nitrous acid is dissociated more or less completely into nitrite and hydrogen ions in a water solution (pK_a = 3.2). Nitrite ions are oxidized at metal
electrodes, e.g. platinum, and carbon electrodes, e.g. carbon paste. In the
range from weak acid to neutral solution (pH 3 to 8.5) the overall oxidation
process is

$$NO_2^- + H_2O = 2 H^+ + NO_3^- + 2 e^- \tag{1}$$

With platinum as indicator electrode an inhibition of the oxidation process
by formation of the platinum oxide is observed.

Both the glassy carbon and the carbon paste electrodes show oxidation peaks
of similar shape, but the shape of the current-voltage curve for the oxidation
of nitrite is less drawn-out on the carbon paste electrode than on the glassy
carbon electrode. Due to the surface complications, the nitrite oxidation at
glassy carbon electrode is irreversible. As indicated by Fig. 1 a sharp peak is

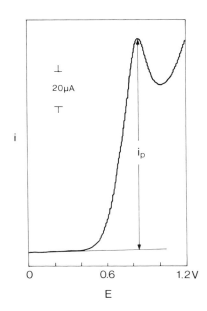

i

⊥
20μA
⊤

i_p

0   0.6   1.2 V

E

Fig. 1. Linear scan voltammogram from the determination of 0.1 mmolar nitrite
in a 0.1 molar ammonium acetate solution.

obtained with carbon paste as indicator electrode and even though oxidation
occurs at a relatively positive potential an unambigous baseline can be drawn.
The peak potential is at + 0.85 V in 0.1 molar ammonium acetate solution.

An electrode to be used for quantitative determinations by linear scan voltammetry would yield a peak current that was controlled by the rate of diffusion
of the analyte in the bulk solution.

In this case the peak current is proportional to the square root of the scan
rate and a plot of i_p vs $v^{1/2}$ should be a straight line. The effect of scan rate
on the peak current for the oxidation of nitrite at the carbon paste electrode
is summarized in Fig. 2. In the range 4 to 25 mV s^{-1} the peak current is proportional to $v^{1/2}$, which indicates that the current is limited by the diffusion of
nitrite to the electrode surface.

The voltammetric method based on oxidation of nitrite at a carbon paste electrode is specific for nitrite relative to nitrate since the latter ion is not
electroactive under the experimental conditions at hand. For example a 1000-fold
excess of nitrate does not alter the peak current. Dissolved oxygen does not
interfere, so sample deaeration is not required. In a 0.1 molar ammonium acetate
solution (pH 7) loss of nitrite due to the volatility and chemical instability
of nitrous acid was avoided because NO_2^- is the major species; no decrease of
nitrite concentration was observed over an one-hour period. The voltammetric

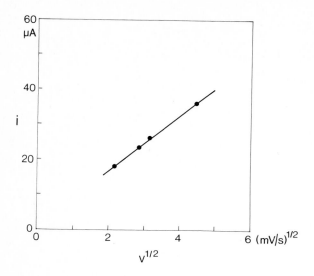

Fig. 2. Plot of the peak current vs the square root of the scan rate for the oxidation of 0.25 mmolar nitrite in 0.1 molar ammonium acetate at a carbon paste electrode.

peak current at the carbon paste electrode in the ammonium acetate solution was directly proportional to nitrite concentration.

Some results from the voltammetric determination of nitrous acid in process acids from the nitrocellulose process are summarized in Table 1.

Electrometric titration of nitrite

Standard Fe(II) solutions are often used in titrimetric determinations of nitrates. Depending on the conditions for the reaction nitrate ions react with Fe(II) ions either in the ratio 1:3 or 1:2.

In hot concentrated hydrochloric acid the following reaction takes place:

$$NO_3^- + 3\ Fe^{2+} + 4\ H^+ = NO + 3\ Fe^{3+} + 2\ H_2O \tag{2}$$

This reaction is too slow to be used for direct titrations. In practical applications excess Fe(II) is added and the reaction is driven to completion by boiling the solution. Excess Fe(II) is back titrated with Cr(III).

In concentrated sulfuric acid nitrate ions are reduced to nitrosylsulfuric acid:

$$NO_3^- + 2\ Fe^{2+} + 3\ H^+ + H_2SO_4 = HSO_3NO_2 + 2\ Fe^{3+} + 2\ H_2O \tag{3}$$

This reaction is sufficiently rapid to allow direct titration and determination of nitric acid in process acids is based on this reaction.

TABLE 1

Some typical results from determinations of nitrous acid in some acids from
the nitrocellulose process.

Sample	% HNO_2	Sample	%´ HNO_2
HS 144	0.32	LS 36	0.39
	0.31		0.35
	0.30		0.36
HS 147	0.33	LS 37	0.38
	0.31		0.37
	0.31		0.37

For the reaction to proceed stoichiometrically according to equation 3 the
temperature in the titration vessel must not rise above 40 ^{O}C. The water content
should also be the lowest possible, something that is achieved by keeping the
amount of sample small.

The end-point of the titration can be indicated either by the red complex
formed between nitrosylsulfuric acid and the first excess of Fe(II) ions or
instrumentally with two polarized indicator electrodes (ref. 5). In the instrumental method a potential is applied across the platinum electrodes, that are
immersed in the stirred solution and the current obtained during titration is
measured. By titration of nitrate with Fe(II) the current goes from zero to a
measurable value at the equivalence point, since the reduction of nitrate is irreversible and the Fe(II)/Fe(III) pair represents a reversible redox system.
The sudden change in current in the vicinity of the equivalence point is enough
for accurate location of the end point of the titration and facilitates automatic termination of the titration (Fig. 3).

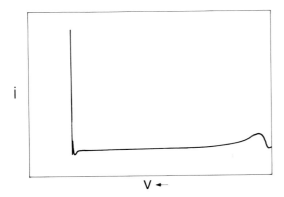

Fig. 3. The current as a function of the volume of added Fe(II) solution for
an electrometric titration with two polarized platinum electrodes of nitrate.

TABLE 2

Some typical results from determinations of nitrous, nitric and sulfuric acids in acids from the trinitrotoluene process with five nitrating apparatus.

Apparatus	% HNO_2	% HNO_3	% H_2SO_4	% H_2O	% Prod
A 1	–	9.94	60.28	29.78	–
2	0.51	7.15	73.60	18.74	3.47
3	2.31	10.51	79.64	7.54	10.76
4	1.06	9.93	85.08	3.93	12.85
5	0.10	12.05	88.40	-0.55	9.16
B 1	–	9.32	62.13	28.55	–
2	0.88	7.48	74.70	16.94	3.60
3	2.09	10.42	81.49	6.00	12.45
4	1.16	10.30	86.62	1.92	14.06
5	0.31	13.00	89.62	-2.93	16.76

Potentiometric titrations of strong acids

The total amount of strong acids, namely nitrous, nitric and sulfuric acids, in a process acid can of course be determined quite easily by visual titration with sodium hydroxide. Here, however the titration has been carried out potentiometrically with a Metrohm Titroprocessor E 636. With this instrument a complete evaluation of the composition of the process acid is made after input of the results from the previous determinations of nitrous and nitric acids.

In Table 2 some results from determinations of nitrous, nitric and sulfuric acids in process acids from the trinitrotoluene process are summarized.

REFERENCES

1 J.E. Harrar, Determination of nitrite by controlled -potential coulometry, Anal. Chem., 43 (1971) 143-145.
2 J.A. Cox and P.J. Kulesza, Oxidation and determination of nitrite at modified electrodes, J. Electroanal. Chem., 175 (1984) 105-118.
3 M. Sarwar and G. Willems, Voltammetric determination of nitrite: comparison of carbon paste and glassy carbon electrodes., J. Chem. Soc. Pak., 2 (1980) 123-125.
4 J.E. Newbery and M. Pilar Lopez de Haddad, Amperometric determination of nitrite by oxidation at a glassy carbon electrode, Analyst, 110 (1985) 81-82.
5 J. Asplund, J. Hazard. Mat., 9 (1984) 13-29.

M.R. Smyth and J.G. Vos
Electrochemistry, Sensors and Analysis
© Elsevier Science Publishers B.V., Amsterdam — Printed in The Netherlands

VOLTAMMETRIC DETERMINATION OF SOME HETEROCYCLIC MERCAPTANS

BRIAN LYNCH and MALCOLM R. SMYTH
School of Chemical Sciences, NIHE Dublin, Glasnevin, Dublin 9, Ireland.

ABSTRACT

The oxidative voltammetric behaviour of some heterocyclic mercaptans has been investigated using direct current (DC) polarography, differential pulse polarography (DPP), cyclic voltammetry (CV) and cathodic stripping voltammetry (CSV). These compounds all undergo a 1 electron mole^{-1} oxidation which is quasireversible in nature. The effect of pH and substituents has been studied, and DPP and CSV applied to the determination of selected mixtures.

INTRODUCTION

The aim of this work was to study the voltammetric behaviour of 2-mercapto-pyridine (I), 2-mercaptopyridine-1-oxide (II), 2-mercaptopyrimidine (III) and 2-mercapto-4-methylpyrimidine.HCl (IV), to observe substituent effects and to optimise conditions for their analysis in mixtures. The reaction of mercaptans at mercury electrodes has been well documented (1-6), especially the voltammetric behaviour of cysteine, cystine and glutathione (4-6). The most sensitive volta-mmetric technique for the analysis of mercaptans is cathodic stripping voltammetry (CSV) which involves (i) the oxidation of the mercaptan to form a mercury thiolate film on the electrode surface, followed by (ii) the reduction of the film in a cathodic scan. The high sensitivity is due to the preconcentration of the mercaptan (as a thiolate film) at the electrode surface (7-9). The CSV of thioamide drugs has also been studied (10).

METHODS

Apparatus

A Princeton Applied Research Model 303 SMDE was used in conjunction with a PAR Model 264 Polarographic Analyser and a Houston Instrument Model 2000 X-Y recorder for direct current (DC) polarography, differential pulse polarography (DPP) and differential pulse cathodic stripping voltammetry (DPCSV). For cyclic voltammetry (CV) the Model 303 was used in conjunction with a PAR Model 174 Polarographic Analyser and a Model 175 universal programmer. The cell was maintained at 25°C. A Ag/AgCl reference electrode and a Pt wire auxillary electrode were used. A stir-head bar and a PAR Model 305 stirrer were also used for DPCSV.

Chemicals

All mercaptans were obtained from Aldrich Chemical Co., except 2-mercaptopyrimidine which was obtained from Hiedel-De Haen AG. 1×10^{-2} and 1×10^{-3}M mercaptan standards were made up in methanol. A stock Britton-Robinson (BR) buffer was made up by mixing 0.04M acetic acid, boric acid and o-phosphoric acid in deionised water. To this stock solution 0.2M sodium hydroxide was added to bring the solutions to the required pH.

Procedures

The 1×10^{-2} and 1×10^{-3}M standards were diluted to various concentrations with BR buffer, and then transferred to the cell. Alternatively the mercaptans were added to the buffer in the cell using a micropipette via a port-hole in the support block. The solutions were purged with oxygen-free nitrogen for 8-10 min. The normal conditions for DC polarography and DPP were: drop time 1 s scan rate 2 mVs^{-1}, voltage range −1.0 to +0.3 V, pulse amplitude (for DPP) 50 mV. Cyclic voltammograms were recorded using a 'medium' hanging mercury drop, and the scan rate was varied from 20 mVs^{-1} to 200 mVs^{-1}. Conditions for CSV were: purge time 16 min., size of mercury drop medium, mod. amp 50 mV, deposition time 180 s, deposition potential +0.17 V for I and 0.07 V for the other compounds, stirring speed slow, equilibration time 30 s.

RESULTS AND DISCUSSION
Effect of pH

All compounds gave rise to a single oxidation wave using DC polarography. The variation of half-wave potential of this wave with pH is shown in Fig. 1. The polarographic pK_a values for the four compounds were calculated from the break in the curve from greater to lesser slope. The pK_a value for each compound corresponds to the dissociation

$$RSH + H_2O \rightleftharpoons RS^- + H_3O^+$$

In the region where the $E_{1/2}$ values are dependent on pH, the mercaptan exists mainly in its protonated form (RSH), while in the region where $E_{1/2}$ is independent of pH it exists as the anion RS^-.

A comparision of pK_a values obtained from ultraviolet (UV) spectrophotometry is given in Table 1.

Each compound showed little variation in diffusion current with pH. The current

due to 2-mercaptopyridine was slightly greater than that due to the other
compounds.

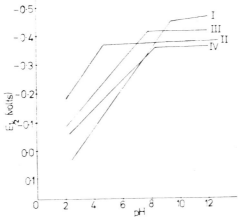

Fig. 1. Diagram showing the variation of half-wave potential with pH for each
compound.

TABLE 1

Comparison of pK_a values for heterocyclic mercaptans obtained from
polarographic and ultraviolet spectral studies.

Compound	Polarographic pK_a value	UV spectral pK_a value
I	9.5	9.75
II	4.6	4.40
III	7.6	7.25
IV	8.0	7.80

Electrode reaction

The electrode reaction responsible for the oxidation of these compounds is
given by

$$\text{(i)} \qquad Hg \rightleftharpoons Hg^+ + e^-$$
$$Hg^+ + RS^- \rightleftharpoons HgRS$$
$$\text{(ii)} \qquad 2HgRS \rightarrow Hg(RS)_2 \text{ film.}$$

The reversibility of the reaction was examined for each compound by plotting E_{DME} vs $\ln(i/(i_d-i))$ from the DC polarographic curves (where i is the current at potential E and i_d is the limiting current), for 0.1mM solutions at pH 6.0. For a reversible , one electron reaction, the slope of the graph should be equal to 0.026 mV @ 25ºC. Plots for 2-mercaptopyridine and 2-mercaptopyridine-1-oxide gave slopes of 0.314 and 0.324 respectively, while plots for 2-mercapto-pyrimidine and 2-mercapto-4-methylpyrimidine.HCl gave rise to slopes of 0.0286 and 0.0304 mV respectively. The reactions are therefore considered to be quasireversible, with the electrode reaction for the substituted pyrimidines being slightly more reversible than that for the substituted pyridines.

The anodic wave was found to be dependent on the diffusion of the mercaptan to the electrode surface. Plots of current vs. concentration were made for each substance using DPP. These curves show adsorption of the product at the electrode surface since the first peak reaches a maximum peak height and then decreases with futher increase in concentration. At the same time a more positive peak appears which increases linearly with concentration. This peak corresponds to multilayer formation. This is most obvious with 2-mercapto-pyridine. For the other compounds, the two peaks are not as well resolved. At certain concentrations, shifts of peak potentials were noticed which could be due to some change in the film structure or the formation of further layers.

Cyclic voltammograms of the compounds show enhanced reduction peaks, which indicates the presence of weakly adsorbed products (11). This is illustrated in Fig. 2 for 2-mercaptopyridine. Holding the electrode potential at +0.07 V for over two minutes and then scanning cathodically causes the formation of a much larger peak at more negative potentials, which is probably due to multilayers of adsorbed product being stripped off the electrode. At high concentrations (0.3mM) the anodic waves contain adsorption prewaves (which also indicate product adsorption), followed by the diffusion wave. These are not well enough resolved for accurate measurement of i_p. The reduction wave shows an adsorption spike on the first scan. Variation of peak current i_p with the square root of scan rate was found to be linear for the anodic wave of a 3×10^{-5}M solution of 2-mercaptopyridine, indicating again the dependence of the reaction on the diffusion of the mercaptan to the electrode surface.

Analytical Applications

Differential pulse cathodic stripping voltammetry (DPCSV) is the most sensitive of the techniques used for the analysis of these compounds. The electrode is held at a potential positive enough so that a mercury thiolate film is formed at the

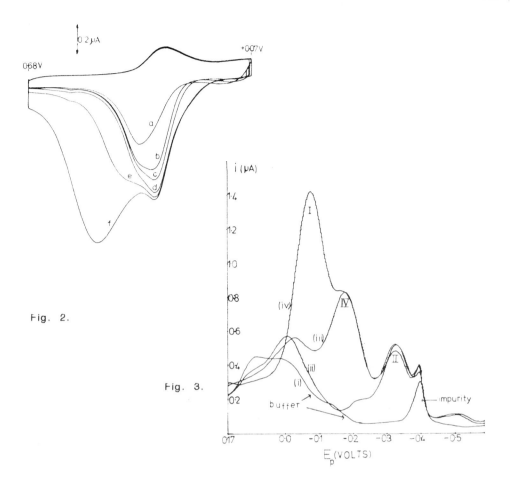

Fig. 2.

Fig. 3.

Fig. 2. Cyclic voltammograms of 4×10^{-5}M 2-mercaptopyridine in BR buffer pH 6.0; scan rate 100 mVs^{-1}, potential held at +0.07 V for (a) 0 s (b) 7 s (c) 10 s (d) 20 s (e) 45 s (f) 120 s.

Fig. 3. DPCSV analysis of a 2×10^{-7}M mixture of each compound at pH 4.0; (i) BR buffer pH 4.0 (ii) compound II (iii) compounds II and IV (iv) compounds I, II and IV.

electrode surface. The time for this deposition step can be varied. After 30
seconds equilibration a cathodic scan is applied resulting in the DPCSV
peak. Calibration curves were constructed for each compound and were found to
be linear in the range $1-7 \times 10^{-7}$M. A higher current was found for 2-mercapto-
pyridine , in agreement with the DC polarographic results reported above.   An
increase in deposition time results in an almost linear increase in peak height.
For a deposition time of 3 min., a 5×10^{-8}M solution of 2-mercapto-pyridine
gave a slight peak when compared with a blank scan of the buffer. However, for
a deposition time of 4 min., the peak was well resolved from the background
current.

   Equimolar mixtures of the compounds were examined using DPP and DPCSV to
test the techniques' abilities to resolve mixtures. The results were similar for
each method. With DPCSV, however, the peak resolution was slightly better
(see Fig. 3). Compounds I, II and III or IV gave rise to different peak potentials,
but III and IV were not resolved using either technique. 1×10^{-4}M concentrations
of each compound (I, II and IV) were mixed for DPP analysis, resulting in peak
potentials of -0.09 V, -0.35 V and -0.16 V respectively, with IV appearing as a
'hump' on the peak due to compound I (medium was BR buffer pH 4.0 and
pulse height 50 mV). For DPCSV, a 2×10^{-7}M mixture of compounds I, II and IV
was investigated (deposition time 3 min. at $+0.17$ V in BR buffer pH 4.0)
resulting in peak potentials of -0.07, -0.32 and -0.17 V respectively.

   The shifts in peak potentials are due to substituent effects. The major
difference is between 2-mercaptopyridine-1-oxide (II) and the other three
compounds. From Fig. 1 one can see that at low pH, compound II is easiest to
oxidise. At this pH the hydrogen is less strongly bound to the sulphur due to the
hydrogen bonding between the hydrogen and the oxygen atoms. This would also
explain its lower pK_a value. At higher pH's (above pH 10), where all compounds
exist in their anionic form, 2-mercaptopyridine is easiest to oxidise. A reason for
this could be that the pyrimidines (compounds III and IV) have 2 nitrogen atoms
compared with a single nitrogen atom and the two nitrogens in the ring have an
electron withdrawing effect (via resonance and inductance) stabilising the anion.
In compound II, the oxygen atom acts inductively as an electron withdrawer and
resonantly as an electron donator. At this high pH the inductive effect must be
the stronger of the two, since it is more difficult to oxidise compound II rather
than I.

ACKNOWLEDGEMENT

One of us (B.L.) wishes to thank the Institute of Chemistry of Ireland for financial support.

REFERENCES

1 F. Peter and R. Rosset , Anal. Chim. Acta, 79 (1975) 47
2 A.F. Kirvis and E. S. Gazda , Anal. Chem., 41 (1969) 212
3 D.A. Csejka, S.1. Nakos and E.W. DuBord, Anal. Chem., 47 (1975)   322
4 R.L. Birke and M. Mazorra, Anal. Chim. Acta, 18 (1980) 257
5 M. Youssefi and R.L. Birke, Anal. Chem., 49 (1979) 1380
6 M.1. Stanovich and A.J. Bard, J. Electroanal. Chem., 75 (1977) 487
7 1.M. Forence, J. Electroanal. Chem., 97 (1979) 219
8 J.1. Stock and R.E. Larson, Anal. Chim. Acta, 138 (1982) 371
9 E. Casassas, C. Arino and M. Esteban, Anal. Chim. Acta, 176 (1985) 133
10 I.E. Davidson and W. F. Smyth, Anal. Chem., 47 (1977) 1195
11 A.J. Bard and L.R. Faulkener, "Electrochemical Methods", John Wiley & Sons, 1980, p 530

M.R. Smyth and J.G. Vos   105
Electrochemistry, Sensors and Analysis
© Elsevier Science Publishers B.V., Amsterdam — Printed in The Netherlands

VOLTAMMETRIC STUDIES OF DIHYDROXAMIC ACIDS

J. A. AMBERSON

Department of Chemistry, Queens University, Belfast BT9 5AG

SUMMARY

Using differential pulse voltammetry (DPV), cyclic voltammetry (CV), and micro-coulometry (MC), five water soluble dihydroxamic acids were investigated. None of the acids were reducible, but each displayed three main oxidation peaks at a glassy carbon working electrode. pH 10 was chosen for quantitative work, and with DPV peak heights were found to be proportional to concentration within the range 2×10^{-5} to 1×10^{-3} mol l^{-1}. CV experiments indicated that the electrode reactions are irreversible; however peak heights showed linear dependence upon the square root of scan rate. MC experiments were unsuccessful, but the numbers of electrons and protons involved were estimated from the DPV experiments, enabling a two stage oxidation mechanism to be postulated.

INTRODUCTION

Monohydroxamic acids are well known as chelating agents of various metal ions; they have been investigated in the past mainly for their potential as analytical reagents (refs. 1,2).

Dihydroxamic acids contain two bidentate ligand groups and therefore are expected to be powerful chelating agents for metal ions. These acids are to be found in nature, mainly in micro-organisms where they are involved in the transport of iron (refs. 3,4).

Due to the importance of dihydroxamic acid—iron complexes in the biological transport of this metal, only the voltammetry of the naturally occurring complexes has been investigated in the past (ref. 5). In this present paper details are given of the voltammetric charateristics of simple chain aliphatic dihydroxamic acids of the general formula :

$$HO-NH-CO-(CH_2)_n-CO-NH-OH$$

The lower molecular weight, water soluble compounds (n=3 to n=7) only have so far been examined. These were prepared by J. Glennon and co-workers at Cork University (refs. 6,7).

In general these compounds exist in the keto form with hydrogen bonding occurring in the shorter chain, n=3 to n=5, molecules. They are weak acids and therefore can dissociate to form anionic species. Conclusions about the acid-base behaviour of the five compounds were drawn from the DPV experiments, as these were carried out over a wide pH range. This also allowed optimization for the analytical procedure .

EXPERIMENTAL

Reagents

Measurements were carried out using Prideaux buffers of constant ionic stength (ref. 8) as supporting electrolytes. Oxygen free nitrogen was used to deaerate the solutions. The glassy carbon electrode was polished with α-Al$_2$O$_3$ (BDH) between measurements. Only the dihydroxamic acids that did not diverge more than 3% from the theoretical percentage compositions, were used.

Apparatus

A Tacussel PRG 5 pulse polarograph coupled to a Hewlett-Packard 7035B X-Y recorder was employed for the DPV procedures. A pulse of 48ms duration was applied with 50mV amplitude, and a sampling period of 8ms. A scan rate of 4mVs^{-1} was selected for the pH studies, with 10mVs^{-1} preferred for the analytical procedures . Sensitivities used were normally between 5 and 25μA, full scale deflection.

The instrument used in the CV experiments was one built in the department. It consists of a potentiostat, or polarograph unit linked to a function generator (Hewlett-Packard 3310B), and a 100mV dual trace storage oscilloscope (Hewlett-Packard 1201B). The voltammograms were displayed on the oscilloscope screen and photographed.

For the MC experiments a Metrohm E-524 coulostat and E-525 integrator was used.

A three electrode cell was used in all experiments, consisting of a glassy carbon working electrode, a platinum wire auxilliary electrode and a mercury pool reference electrode.

RESULTS AND DISCUSSION

Differential pulse voltammetry

Measurements were made using 1x10^{-3} molar solutions, over the pH range 2-12. Both cathodic and anodic superimposed pulses were applied; the voltammograms obtained with anodic pulses were used for quantitative evaluation. None of the acids were reducible but each displayed three main oxidation peaks over the range 0 to +1.5 volts. For each acid, except n=3, two major peaks are present throughout the pH range, whereas the third peak is present only in acidic solutions. In the case of the n=3 compound, the first major oxidation peak was absent in acidic solution, but emerged in alkaline media. The third peak was generally found to be an adsorption peak.

Figure 1 illustrates the three main oxidation peaks; it shows the DPV curves for n=5 at pH 7. After this pH the third peak disappears, and the peak current of the first wave increases. In each case, at pH 10, the two peaks are well defined and suitable for analytical purposes.

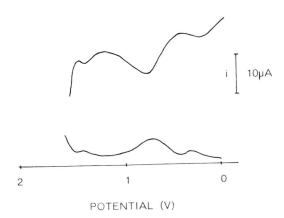

POTENTIAL (V)

Fig. 1. DPV traces obtained with 1×10^{-3} molar solution of n=5, pimelyl dihydroxamic acid at pH 7 (top: anodic superimposed pulses; bottom: cathodic superimposed pulses.)

To gain information on the acid-base behaviour, anodic peak potential was plotted against pH for each peak. Similar patterns were obtained for the first peaks of each acid, and for the second peaks; whereas peak three displayed variability.

In general peak one showed two breaks in the graph; the slopes of the various sections insinuated that in acidic solution at least one proton is involved in the electrode reaction, but above this pH no protons are involved. Similarly there are two breaks in the peak two graph, with no protons being involved in alkaline solution. The break points of these curves should give an indication of the pKa values of the acids.

Peak heights for the two main oxidation waves were found to be linearly dependent upon concentration, and calibration curves were drawn up over the ranges 2×10^{-5} to 1×10^{-3} mol l^{-1}. Figure 2 shows typical calibration curves for the n=3 acid.

108

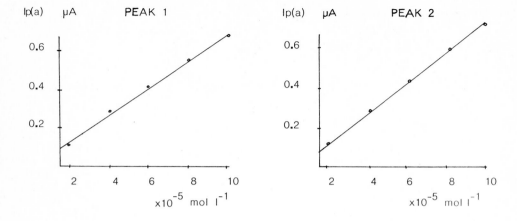

Fig. 2. DPV calibration curves for n=3, glutaryl dihydroxamic acid.

Cyclic voltammetry

CV experiments were carried out in pH 10 buffer, using scan rates of $0.02Vs^{-1}$ to $2.0Vs^{-1}$ over the potential range -0.2 to +1.8 volts. Again, two main oxidation peaks appeared on the anodic half cycle, with a third peak also present at faster scan rates. No peaks were present on the cathodic half cycle, indicating total irreversibility of the electrode reaction.

In general the peak current for peak 1 increases proportionally with the square root of the scan rate, in accordance with the Randles-Sevcik equation:

$$Ip = 2.69 \times 10^{-5} n^{3/2} A \, D^{1/2} Cv^{1/2}$$

whereas peaks 2 and 3 show linear correlation but not direct proportionality.

From these observations it was concluded that peak 1 could be due to an irreversible charge transfer with no chemical complications, but for peaks 2 and 3, an irreversible chemical reaction occurs following the initial irreversible charge transfer process.

Using CV it was again found that for the two main peaks, peak height is linearly dependent upon concentration, and calibration graphs were drawn up over similar concentration ranges to those of the DPV procedures.

Micro-coulometry

Unfortunately these experiments proved unsuccessful as the products formed during electrolysis quickly blocked the active surface of the glassy carbon electrode. A large area platinum electrode was tried in place of the glassy carbon, but this too became blocked before a measureable conversion could take place.

Oxidation mechanism

As the coulometric experiments proved unsuccessful the numbers of electrons involved were estimated from the measurements of the peak half-widths of the DPV experiments, using the equations:

$$W_{\frac{1}{2}} = 3.52RT/\beta nF$$

$$mV/pH = 59m/\beta n$$

where: $W_{\frac{1}{2}}$ = peak half-width; R = gas constant; T = temperature(K); n = number of electrons; F = Faraday constant; mV/pH = slope of peak potential vs pH graph; m = number of protons; β = transfer coefficient.

The value of β is merely taken as one which gives the most credible values of n. In these cases it was usually taken to be around 0.3.

The results indicate that the first oxidation is a two electron process; and the second is a one electron oxidation followed by a dimerization reaction, it can be explained by a similar mechanism to one proposed by Oliver and Waters (ref. 9) in 1971 for the chemical oxidation of monohydroxamic acids.

(i) <u>First oxidation.</u> $R-CO-NH-OH \longrightarrow R-CO-NO + 2e^- + 2H^+$

(ii) <u>Second oxidation.</u> $R-CO-NH-OH \longrightarrow R-CO-\overset{\bullet}{N}-OH + e^- + H^+$

Work continues with the investigation of alcohol soluble (n=8 to 14) acids, and it is hoped that after this, the electro-oxidation mechanism can be finalised.

ACKNOWLEDGEMENT
The co-operation of Dr. J. Glennon (Cork), S. I. Thompson (QUB) and Dr. G. Svehla (QUB) is gratefully acknowledged.

110

REFERENCES

1   Y.K. Agrawal, and R.D. Roshania, Bull. Soc. Chim. Belg., 89:3, 159-171 (1980).
2   Y.K. Agrawal, and S.A. Patel, Reviews in Analytical Chemistry, 4, 237-238
    (1980).
3   K.N. Raymond, and T.P. Tufano, The Biological Chemistry of Iron, H.B. Dunford
    D. Dolphin, K.N. Raymond, L. Sleiker, Eds.; D. Reidel Publishing Co.:
    Dordect, Holland, (1982) pp. 85-105.
4   J.B. Neilands, Ed. Microbial Iron Metabolism, Academic Press: New York (1974).
5   S.J. Barclay, B.H. Huynh, N.R. Raymond, Inorg. Chem. (1984), 23, 2011-2118.
6   D.A. Brown, R.A. Geraty, J.D. Glennon, and N. NiChoileain, Synthetic
    Communications, 15, (1985), 1159.
7   D.A. Brown, R.A. Geraty, N. NiChoileain, and J.D. Glennon, Inorg. Chem.
    (submitted).
8   C. Jordan, Microchem. J., 25, (1980), 492.
9   T.R. Oliver and W.A. Waters. J. Chem. Soc., 3, (1971) 677.

MECHANISTIC ASPECTS OF THE ELECTROCHEMICAL REDUCTION OF OXALATE

B.R. EGGINS and E.A. O'NEILL

Chemistry Department, University of Ulster at Jordanstown, Newtownabbey,
Co Antrim BT37 0QB (Northern Ireland).

SUMMARY
Voltammetry, chronoamperometry and controlled potential electrolysis have been used to study the electrochemical reduction of oxalate in aqueous solutions in the pH range 3 - 10. At pH 9 on both glassy carbon and mercury, two waves were observed, the first being a one electron wave on mercury but a two electron wave on carbon. The second was a four electron wave on both electrodes.

INTRODUCTION

In Britain, Goodridge and Lister (ref. 1) have optimised the reduction of oxalic acid on lead at low pH to give commercial yields of glyoxylic acid.

$$\begin{array}{c} CO_2H \\ | \\ CO_2H \end{array} \xrightarrow{\;+2e^- \;+\; 2H^+\;} \begin{array}{c} CHO \\ | \\ CO_2H \end{array} + \; H_2O$$

Picket and Yap (ref. 2) also studied this reaction under similar conditions.

In Germany, von Kaiser and Heitz (ref. 3) made a brief study on a range of electrodes mostly at pH1. They analysed their results by HPLC. The most relevant from our point of view was their result on nickel at pH9 in which they obtained a mixture of glyoxylate and glycollate (or malate and succinate) together with hydrogen evolution.

All these studies were at constant current.

EXPERIMENTAL

Voltammetry and chronoamperometry were carried out using a Bioanalytical CV-1A voltammeter and a Bryans 2900 x-y recorder.

All solutions were made up in 0.1 mol dm^{-3} tetramethylammonium chloride with 2 x 10^{-3} mol dm^{-3} oxalic acid. The pH was adjusted to the required value with tetramethylammonium hydroxide.

Controlled potential electrolysis was carried out using a Chemical Electronics 703A potentiostat with 0.02 mol dm^{-3} oxalic acid.

Product analysis was carried out by HPLC with an Altex Model 110A solvent metering pump, a Pye Unicam RU402P U.V. detector set at 210 nm and an Aminex HPX-87H ion exclusion column. The mobile phase consisted of 0.016 mol dm^{-3} H_2SO_4. Injection samples had a volume of 25µl and the flow rate of the mobile phase was 0.3 ml min^{-1}.

Chronoamperometry data were analysed by Computer program on a Commodore PET 4032 computer to give the semi-integral data.

RESULTS AND DISCUSSION

We began our studies with cyclic voltammetry on glassy carbon and on mercury over a range of pH values, but mostly at pH9 to tie in with our work on carbon dioxide (ref.4).

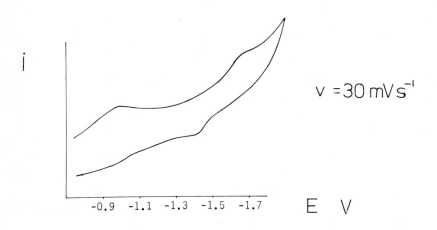

Fig. 1.   Cyclic Voltammogram of oxalate on glassy carbon at pH 9.0.

Two successive cathodic waves were observed (figure 1).

At pH9 on carbon these were at - 1.10V and - 1.70V vs S.C.E. On mercury they were at - 1.10V and - 1.80V. For both electrodes wave I did not shift with pH over the range pH 3 - 10. Wave II showed two sections, the first shifting anodically at 0.142V pH^{-1} on carbon but at 0.050V pH^{-1} on mercury up to pH 5.5 (pKa_2 for oxalic acid = 4.4); thereafter it remained fairly constant. (figures 2 (a) and (b)).

a

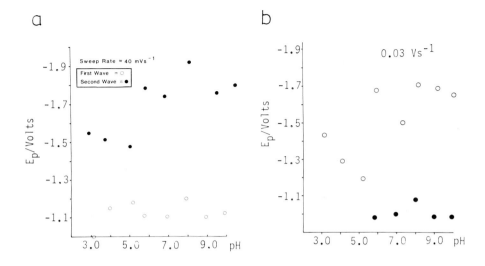

Fig. 2. Plot of Ep versus pH for oxalic acid on (a) mercury (b) Glassy carbon

The variation of current function with sweep rate indicated that on carbon the
mechanism involved an irreversible charge transfer with a following chemical
reaction. On mercury the behaviour indicated adsorption, which was also shown
by the pre-peak on the wave (figure 3) especially at rapid sweep rates.

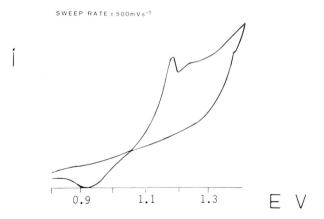

Fig. 3. Cyclic voltammogram of oxalate on mercury at pH 3.2

Adsorption would be expected on mercury as observed by Race (ref. 5). The use of some reversible behaviour is indicated by the anodic waves both on carbon and mercury.

From the Delahay equation approximate values for n for each wave were obtained as shown in Table I.

Table I. Apparent n values derived from voltammetric peak current data (Delahay equation) and from chronoamperometry (Cottrell equation), using the semi-integration method

Electrode	Delahay Equation		
	Wave I	Wave II	
	n_{app}	n_{app}	
Hg	0.87	2.65	
C	1.53	4.73	

Electrode	Chronoamperometry		
	Wave I	Wave II	
	n_{app}	n_{app}	
Hg	1.17	3.59	Cottrell
C	2.01	4.22	Cottrell
Hg	1.24	4.18	Semi-integration
C	2.17	5.01	Semi-integration

The low values for wave I on mercury were surprising, so we also studied these waves by chronoamperometry. The current-time data were analysed both by the Cottrell equation and by the semi-integration method of Oldham (ref. 6). Results are plotted in the form

$$i = i_0 - \beta m \qquad (1)$$

when m = semi-integral of current with time given by equation (2)

$$m = \frac{d^{-\frac{1}{2}} i}{dt^{-\frac{1}{2}}} = \frac{i_0}{\beta} - \frac{i_0 \exp(\beta^2 t) \mathrm{erfc}(\beta\sqrt{t})}{\beta} \qquad (2)$$

hence if $\beta = {}^k f/D^{\frac{1}{2}}$, n can be found

from $i = i_0 n FAC_0^b k_f \exp(\beta^2 t) \mathrm{erfc}(\beta\sqrt{t})$

A computer program was written to calculate m and evaluate n. Results are shown in Table I.

The results indicate that overall on mercury $n = 1$ for wave I and $n \cong 3$ for wave II. On carbon $n = 2$ for wave I and $n \cong 2 - 3$ for wave II.

The n = 2 result is expected for the conversion of oxalate to glyoxylate, and products can easily be postulated to explain the products from the second wave. However the n = 1 value for wave I on mercury is unexpected. To explain this, a dimerisation process may be postulated as in figure 4. The dihydroxytartaric acid is a known stable compound.

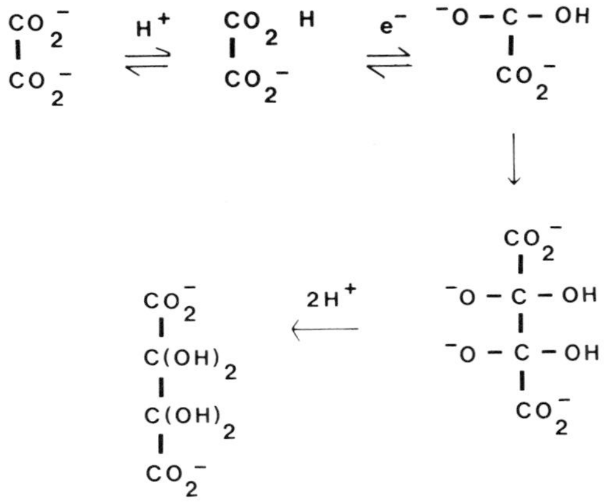

$$\begin{array}{c} CO_2^- \\ | \\ CO_2^- \end{array} \overset{H^+}{\rightleftharpoons} \begin{array}{c} CO_2H \\ | \\ CO_2^- \end{array} \overset{e^-}{\rightleftharpoons} \begin{array}{c} {}^-O-C-OH \\ | \\ CO_2^- \end{array}$$

$$\downarrow$$

$$\begin{array}{c} CO_2^- \\ | \\ C(OH)_2 \\ | \\ C(OH)_2 \\ | \\ CO_2^- \end{array} \overset{2H^+}{\longleftarrow} \begin{array}{c} CO_2^- \\ | \\ {}^-O-C-OH \\ | \\ {}^-O-C-OH \\ | \\ CO_2^- \end{array}$$

Fig. 4. Mechanism for the one-electron reduction of oxalate.

A series of transformations can be set up to explain all the values of n.

n	product
1	dihydroxytartaric acid
2	glyoxylate (or formate)
3	tartarate
4	glycollate or malate
5	succinate

This is shown in figure 5.

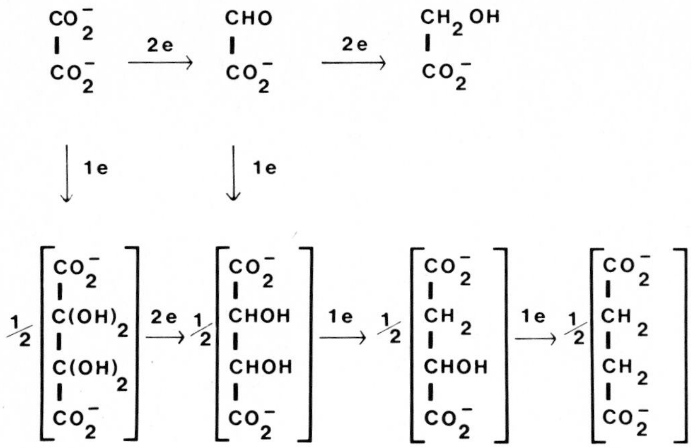

$$
\begin{array}{c}
\overset{CO_2^-}{\underset{CO_2^-}{|}} \xrightarrow{2e} \overset{CHO}{\underset{CO_2^-}{|}} \xrightarrow{2e} \overset{CH_2OH}{\underset{CO_2^-}{|}}
\end{array}
$$

$$
\downarrow 1e \qquad\qquad \downarrow 1e
$$

$$
\frac{1}{2}\begin{bmatrix} CO_2^- \\ | \\ C(OH)_2 \\ | \\ C(OH)_2 \\ | \\ CO_2^- \end{bmatrix} \xrightarrow{2e} \frac{1}{2}\begin{bmatrix} CO_2^- \\ | \\ CHOH \\ | \\ CHOH \\ | \\ CO_2^- \end{bmatrix} \xrightarrow{1e} \frac{1}{2}\begin{bmatrix} CO_2^- \\ | \\ CH_2 \\ | \\ CHOH \\ | \\ CO_2^- \end{bmatrix} \xrightarrow{1e} \frac{1}{2}\begin{bmatrix} CO_2^- \\ | \\ CH_2 \\ | \\ CH_2 \\ | \\ CO_2^- \end{bmatrix}
$$

Fig. 5. Overall reaction scheme for the reduction of oxalate.

Preliminary results for controlled potential electrolysis at -1.4V on mercury followed by HPLC analysis showed one main peak which appears to be formate rather than dihydroxytartarate, and minor peaks correspond to glyoxylate, glycollate and malate plus an unknown peak (figure 6).

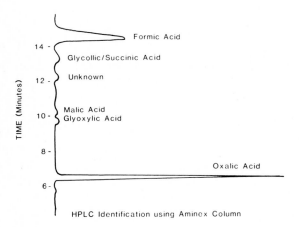

Fig. 6. Identification of products by HPLC from the reduction of oxalate at -1.4V on mercury at pH 9.0.

REFERENCES

1 F Goodridge and K Lister, Brit. Patent 1411371, 22 Oct (1975);
   F Goodridge, K Lister and R E Plimley, J.Appl.Electrochem.,10 (1980), 55.
2 D Pickett and K Yap, J.Appl.Electrochem.,4 (1974) 17.
3 Von H Kaiser and E Heitz, Ber. Bunsengeseltschaft, 77 (1973) 818.
4 E M Bennett, B R Eggins, J McNeill and E A McMullan, Anal. Proc., (1980) 356;
   B R Eggins and J McNeill, J.Electroanal. Chem.,148 (1983) 17.
5 W Race, J.Electroanal. Chem., 24 (1970) 315.
6 K B Oldham, J.Electroanal Chem.,145 (1983) 9.

M.R. Smyth and J.G. Vos
Electrochemistry, Sensors and Analysis
© Elsevier Science Publishers B.V., Amsterdam — Printed in The Netherlands

THE BEHAVIOUR OF OXYETHYLATED ALCOHOLS UNDER CONDITIONS OF TENSAMMETRY WITH ADSORPTIVE ACCUMULATION ON HMDE

Z. ŁUKASZEWSKI and M.K. PAWLAK

Institute of General Chemistry, Technical University of Poznań, 60-965 Poznań (Poland)

SUMMARY

The behaviour of oxyethylated alcohols 6-14, 10-10, 10-14, 18-14, 18-10 and 18-6 (the first number indicates the number of C atoms in the n-alkyl chain, and the second number indicates the average number of oxyethylene subunits) was investigated. Less hydrophobic surfactants (6-14, 10-10 and 10-14) behave similarly and form one single wide peak. Surfactants 18-14 and 18-10 form a single narrow peak which belongs to the monomer form of the surfactant. Surfactant 18-10, if its threshold concentration is achieved, also forms a second peak, corresponding to the pre-micellar form. In the case of 18-6 both peaks extend over the whole examined range of concentration. The relationships of the peak height to the preconcentration potential and to the concentration of surfactant were examined.

INTRODUCTION

Adsorptive accumulation of surfactants on the surface of HMDE prior to final measurement increases the possibilities for determining the trace level of concentration. In final measurements the Kalousek commutator technique (ref. 1), DP (refs. 2-4) and AC (refs. 4-6) modes can be used. When the latter technique is used to measure surfactants it is known as tensammetry with accumulation on HMDE, and is much more sensitive in comparison with classical tensammetry using DME. An additional advantage of tensammetry with accumulation on HMDE in comparison with its classical variety is the possibility to use the differentiating action of the preconcentration potential for the determination of components of mixtures (ref. 7).

The aim of this paper was investigation of the behaviour of oxyethylated alcohols (polyoxyethylene-n-alkylmonoethers) after their adsorptive accumulation on HMDE. These surfactants belong to the most popularly used group of nonionic surfactants, but the possibilities for determining their trace concentrations are still very limited. Some information about the behaviour of oxyethylated alcohols, mainly those having dodecyl chains, can be

obtained from papers concerning classical tensammetry (refs. 8-
-9) and differential capacity (ref. 10). However, the conditions
for making measurements in classical tensammetry and for differential capacity are substantially different in comparison with
tensammetry with accumulation on HMDE. The selection of surfactants in the present work aimed to investigate surfactants having
different ratios of hydrophobic and hydrophilic parts in their
molecular structure. The main focus was on polyoxyethylene-n-
-octadecylmonoethers, taking into account their practical importance.

EXPERIMENTAL

An OH-105 polarograph (Radelkis) with a voltage scan rate of
0.4 Vmin^{-1} was used. The applied amplitude of the alternating
voltage was 2 mV. Controlled-temperature HMDE equipment(Radiometer) having an additional molybdenum rod auxiliary electrode
was used. In the glass cell a small teflon beaker was placed   so
that the investigated solution was in contact with teflon.
    The following oxyethylated alcohols were used: 6-14, 10-10,
10-14, 18-14, 18-10 and 18-6. These surfactants were synthesized
at the Technical University of Poznań by oxyethylation of the
corresponding reagent grade n-alcohols. Thus they were polydis-
persal products, and together with the main products,they contain
free n-alcohol and poly(ethylene glycols).
    The sodium sulphate used for preparation of the base electro-
lyte was purified by double crystallization and heated at 600°C.
All solutions were prepared in water thrice-distilled from
quartz. Only freshly distilled water was used. Only glass and
quartz vessels were used. The supporting electrolyte in all the
studies was aqueous 0.5 M sodium sulphate. The investigated solutions were preliminarily thermostated at 25°C in measuring
flasks, and they were then introduced into the measuring cell
and underwent deaeration by the bubbling of oxygen-free nitrogen.
The solutions were simultaneously thermostated at 25°C. Just before preconcentration was started the stirrer was switched on,
the required preconcentration potential was set and a new mercury
drop was formed. The preconcentration time, which was 5 or 10
min., was measured from this moment. The registration of the tensammetric curve in the negative direction was made from immovable
solution, after a 1 min quiescent period.

RESULTS AND DISCUSSION

    Surfactants 6-14, 10-10 and 10-14 behave similarly. Each of
them forms one wide negative tensammetric peak. Examples of such
peaks are shown in Figure 1, which displays a series of peaks of
surfactant 10-14 obtained at different preconcentration potentials. In this Figure the decrease in the peak height is visible
with an increase of the negative value of the preconcentration
potential. A further increase in the negative value of preconcentration potential leads to the complete disappearance of the peak.

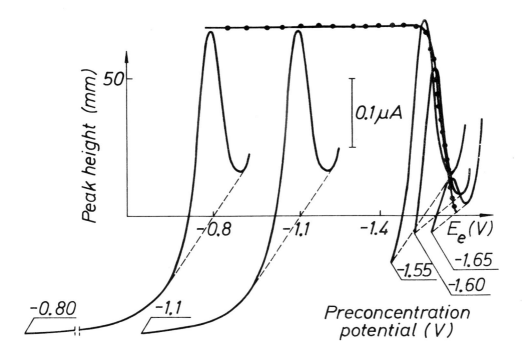

Fig. 1. The tensammetric curves for surfactant 10-14 obtained at different preconcentration potentials. The maxima of these peaks indicate the curve of dependence of peak height on preconcentration potential.
Concentration of surfactant: 0.20 mg l^{-1}; preconcentration time: 5 min.

The dependence shown in Fig. 1 really was based on a much higher number of measuring points than can be shown in the Figure. Very similar dependences were obtained for surfactants 6-14 and 10-10, whose dependences also decrease over similar ranges of preconcentration potential. At equal concentrations the heights of the peaks of these surfactants increase in the following sequence: 6-14, 10-10, 10-14, but the potentials of all these peaks are very similar. One can add that the peak of the monomer form of the surfactant, which could be described according to this papers numbering system as 12-8, is similar to the peaks of the investigated surfactants (ref. 9).

In contrast to the previous cases, surfactants 18-14, 18-10 and 18-6, which are more hydrophobic, form one or two narrow peaks (see Fig. 2). Basically surfactant 18-14 forms one negative peak; surfactant 18-6, two peaks; and surfactant 18-10, one or two peaks, depending on surfactant concentration.

122

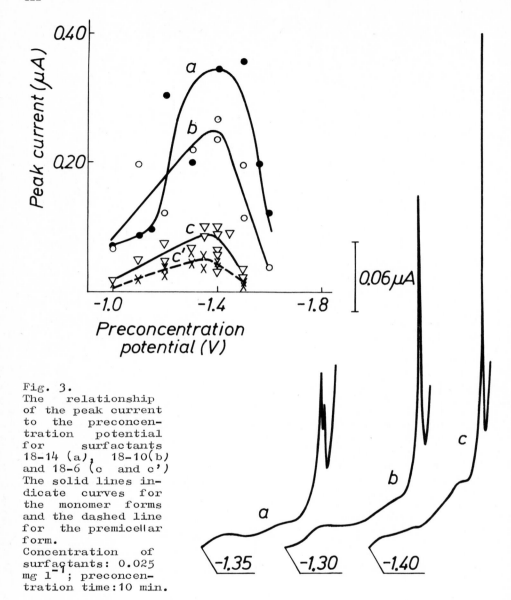

Fig. 3.
The relationship
of the peak current
to the preconcen-
tration potential
for surfactants
18–14 (a), 18–10(b)
and 18–6 (c and c')
The solid lines in-
dicate curves for
the monomer forms
and the dashed line
for the premicellar
form.
Concentration of
surfactants: 0.025
mg l^{-1}; preconcen-
tration time:10 min.

Fig. 2. Examples of tensammetric curves of surfactants: 18–6 (a),
18–10 (b) and 18–14 (c)
Concentration of surfactants: 0.025 mg l^{-1}; preconcentration
time: 10 min.

The less negative of these peaks seems to be caused by the monomer form of the surfactant; and the more negative peak is caused
by the associate premicellar form of surfactant.

Because the concentration of the examined solutions was far below the cmc values, these associate forms cannot be the micelles
of the surfactants. The tendency of the second peak to appear increased with the increase of hydrophobicity of the surfactant, in
the following sequence: 18-14, 18-10, 18-6; and likewise the tendency of the surfactants to association increased.

The dependence of the peak height on the preconcentration
potential for surfactants 18-14, 18-10 and 18-6 was examined.
The obtained results are shown in Fig. 3. Because of the formation of two peaks for surfactant 18-6 in Figure 3 this surfactant
is represented by two curves (a solid line for the monomer form

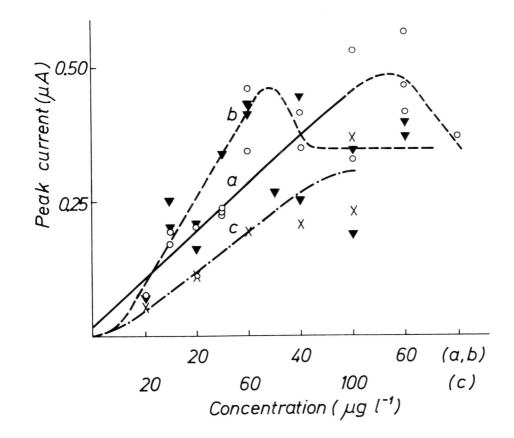

Fig. 4. The dependences of the peak current on concentration for
surfactants 18-10 (a) and 18-14 (b and c)
Preconcentration time: (a and b) 10 min, (c) 5 min; preconcentration potential: (a and b) -1.40 V, (c) -1.60 V vs. SCE.

and a dashed line for the premicelar form). The dependence of the peak height on the concentration of surfactant was also examined. In the case of surfactant 18-14 examination was made for preconcentration potentials of -1.40 V and -1.60 V vs. SCE and for preconcentration times of 5 and 10 minutes, respectively. In the case of surfactant 18-10 the experiments were only carried out for 5 minutes of preconcentration at a potential of -1.40 V vs. SCE. The results obtained are displayed in Fig. 4. Each of the curves in this Figure has a lengthy linear segment, which is useful for analytical purposes. The range of concentrations connected with this linear segment are shifted by the change of preconcentration time (see curves "b" and "c"). The saturation of the electrode surface with surfactant causes the stabilization of the peak height after achievement of a certain concentration of surfactant. A further decrease in the peak height is caused by the competition of the monomer and the premicellar associate forms on the electrode surface.

This work was supported by Research Program CPBP 01.17.

REFERENCES

1 Z.Kozarac, B.Cosović and M.Branica, J. Electroanal. Chem., 68(1976)75
2 R.Kalvoda, Anal. Chim. Acta, 138(1982)11
3 J.Wang, D.B.Lou, A.M.Farias and J.S.Mahmoud, Anal. Chem., 57(1985)158
4 E.Bednarkiewicz and Z.Kublik, Anal. Chim. Acta, 176(1985)133
5 H.Jehring and W.Stolle, Collect. Czech. Chem. Commun., 33(1968)1670
6 H.Batycka and Z.Łukaszewski, Anal. Chim. Acta, 162(1984)207
7 H.Batycka and Z.Łukaszewski, Anal. Chim. Acta, 162(1984)215
8 H.Jehring, U.Retter and A.Weiss, J. Electroanal. Chem., 52(1974)55
9 E.Müller and H.D.Dörfler, Z. Chem., 21(1981)28
10 W.S.Burykina, I.P.Kudina and M.A.Łoshkarev, Electrokhimya, 20(1984)857

M.R. Smyth and J.G. Vos
Electrochemistry, Sensors and Analysis
© Elsevier Science Publishers B.V., Amsterdam — Printed in The Netherlands

DIFFERENTIAL PULSE POLAROGRAPHIC DETERMINATION OF SOME OXYANIONS

Y.M. TEMERK, M.E. AHMED, M.M. KAMAL and Z.A. AHMED

Chemistry Department, Faculty of Science, Assiut University,
Assiut, Egypt.

SUMMARY
     The analytical application of differential pulse polarography
for the quantitative determination of TeO_3^{--}, TeO_4^{--}, AsO_2^- and VO_3^- has
been investigated. The dependence of the differential pulse peak
on various parameters was studied and the optimum conditions for
the analytical determination of these oxyanions were found. Results
of the analysis of mixtures of TeO_4^{--} - TeO_3^{--}, TeO_4^{--} - AsO_2^- and
TeO_4^{--} - VO_3^- are reported under the optimum conditions.

INTRODUCTION

     Analytical determination of inorganic anions is important for

assay purposes in commercial samples. In this context polarogra-

phic and voltammetric determination of some oxyanions have been

described (refs. 1-8). Detection limit for dc polarography usually

is ca. 10^{-4} M, but may be higher under non-optimum conditions. Since

differential pulse polarography should improve sensitivity at

trace level concentrations, the use of dpp techniques for deter-

mination of some oxyanions has been investigated (refs. 2-7). The

present work investigates the applicability of dpp for the quanti-

tative trace determination of TeO_3^{--}, TeO_4^{--}, AsO_2^- and VO_3^- under

the optimum conditions. The simultaneous determination of mixture

of TeO_4^{--} - TeO_3^- , TeO_4^{--} - AsO_2^- and TeO_4^{--} - VO_3^- was also investiga-

ted .

EXPERIMENTAL

Materials:

     All chemicals used were of analytical grade. 0.01 M stock solu-

tion of TeO_4^{--}, TeO_3^{--}, AsO_2^- and VO_3^- were prepared by dissolving

the accurate weight of the solid in twice-distilled deionized

water. The modified universal buffer series of Britton and

Robinson (pH 2-12) was used as supporting electrolyte and prepared

as given by Britton (ref. 9).

Instrumentation:

All polarograms were recorded with a Princeton Applied Research (PAR) Model 174A polarographic Analyser equipped with a PAR Model 172 A Drop Timer and Electrode Assembly. A thermostated Metrohm cell with a three electrode system was used.

RESULTS AND DISCUSSION

The dependence of the differential pulse peak on pH, pulse amplitude and drop time:

The differential pulse peaks of TeO_4^{--}, TeO_3^{--}, AsO_2^- and VO_3^- were investigated at various pH values. The peak height is clearly dependent on pH and the maximum response for TeO_3^{--} and TeO_4^{--} was found at pH 4.5 with acidic buffer solutions and pH 8.3 with alkaline solutions. With respect to AsO_2^- the differential pulse polarograms in acidic buffer solutions (< pH 5.2) indicate two consecutive peaks. However at pH 8.3 the reduction of AsO_2^- gave a concentration–sensitive peak height compared to media at other pH's. The reduction of VO_3^- displayed a maximum response peak at pH 4.5.

Differential pulse polarograms of TeO_3^{--}, TeO_4^{--}, AsO_2^- and VO_3^- recorded at various pulse amplitude indicate that the peak height increases as the pulse amplitude is increased. However at a pulse amplitude of 100 mV the differential pulse peaks of the investigated oxyanions gave the highest sensitivity and well defined peaks.

The dependence of the reduction peak of the investigated oxyanions on drop time shows that a drop time of 2 s gave the maximum response compared to that of 1 s. For the aforementioned results a drop time of 2 s was used in the present study.

Quantitative trace determination of TeO_3^{--}, TeO_4^{--}, AsO_2^- and VO_3^- by dpp.

The optimum conditions for the analytical determination of TeO_3^{--}, TeO_4^{--}, AsO_2^- and VO_3^- by dpp were found to be a drop time of 2 s, a scan rate of 2 mV s^{-1} and a pulse amplitude of 100 mV. Response of differential pulse peak height vs. concentration of each ion is linear at the optimum pH. The variation of i_p (μA) with the concentration of each oxyanion is represented by the straight line equation i_p = ac + b. The values of the regression coefficient are computed and assembled together with the straight line constants in Table 1. The validity of the method is supported by the constancy of the i_{dpp}/C values.

TABLE 1

Straight line constants for the calibration curves of the investigated oxanions under the optimum conditions.

Oxyanion	pH	Slope	intercept	S.D.* of the fit	r**
TeO_4^{--}	4.5	0.36 ± 0.012	0.021 ± 0.237	0.256	0.99
	8.3	0.29 ± 0.003	0.014 ± 0.017	0.025	0.98
TeO_3^{--}	8.3	0.20 ± 0.004	0.005 ± 0.025	0.036	0.99
AsO_3^{-}	8.3	0.36 ± 0.004	0.223 ± 0.113	0.143	0.97
VO_3^{-}	4.5	0.07 ± 0.003	-0.106 ± 0.065	0.089	0.98

S.D.* Standard deviation.
r** Regression coefficient.

Application of differential pulse technique for the analysis of some mixtures:

The larger difference in E_p for the reduction peaks of TeO_4^{--} and TeO_3^{--} at pH 8.3 and the pulse amplitude of the 100 mV permits the analysis of mixtures of TeO_4^{--} and TeO_3^{--}. Table 2 shows the results of the analysis of mixtures of TeO_4^{--} and TeO_3^{--}. The results agree well with the calibration curve constructed by this technique.

TABLE 2

Results of the analysis of mixtures of TeO_3^{--} - TeO_4^{--}.

Concentration taken, µM		Concentration observed, µM		Difference µM		Difference %	
TeO_3^{--}	TeO_4^{--}	TeO_3^{--}	TeO_4^{--}	TeO_3^{--}	TeO_4^{--}	TeO_3^{--}	TeO_4^{--}
0.96	0.96	0.83	1.05	-0.13	+0.09	13.5	9.4
2.37	2.37	1.95	2.55	-0.42	+0.18	17.7	7.6
4.47	4.47	4.10	5.00	-0.37	+0.13	8.2	2.9
7.10	7.10	6.65	7.55	-0.45	+0.45	6.3	6.3
9.43	9.43	9.10	9.80	-0.33	+0.37	3.5	3.9
11.76	11.76	10.90	12.30	-0.86	+0.54	7.3	4.6
14.08	14.08	14.50	14.25	+0.42	+0.17	3.0	1.2
16.39	16.39	17.00	16.70	+0.71	+0.31	4.3	1.9

A better separation of maximum peak height of AsO_2^{--} from TeO_4^{--} peak was obtained at pH 8.3. Response of height(pulse peak)as a function of concentration of both ions of mixture is linear. The observed concentrations are based on the comparison of the measured i_p with the previously discussed calibration curves for each anion alone.

The difference in E_p for TeO_4^{--} - VO_3^- pair is ~0.4 V which permits simultaneous determination of these oxyanions. The results for the analysis of mixtures of TeO_4^{--} and VO_3^- indicates that the concentration range is 9-30 μM under the optimum conditions.

REFERENCES

1 R.M. Issa, B.A. Abd-El Nabey; Z. Anal. Chemie, 231 (1967) 339.
2 F.T. Henry, T.O. Kirch and T.M. Thorpe; Anal. Chem., 51 (1978) 215.
3 W. Holak; Anal. Chem.; 52 (1982) 2189.
4 J.N. Gaur, S.K. Bhargava, S.K. Sharma; Ind. Chem. Soc., 57(9) (1980) 898.
5 J.C. Rat, D. Burnel, M.F. Hutin; F.Kolla; Ann. Falsif. Expert. Chim. Toxicol, 75 (808) (1982) 265.
6 J.P. Arnold and R.M. Johnson; Talanta, 16, (1969) 1191.
7   D.J. Myers and J. Osteryoung; Anal. Chem., 45 (1973) 267.
8 J.J. Lingane and L. Meites; J. Am. Chem. Soc., 69 (1948) 1021.
9 H.T.S. Britton, Hydrogen Ions, 4th Edition, Chapman and Hall (1952) 313.

ANALYTICAL POTENTIOMETRY

M.R. Smyth and J.G. Vos
Electrochemistry, Sensors and Analysis
© Elsevier Science Publishers B.V., Amsterdam — Printed in The Netherlands

THEORY AND APPLICATION OF POTENTIOMETRIC METHODS OF ANALYSIS

G. SVEHLA

Department of Chemistry, Queen's University, Belfast BT9 5AG

SUMMARY

The most important developments in potentiometric methods of analysis have been connected to the introduction of new ion-selective indicator electrodes. These are classified and two important classes (electrodes incorporating neutral molecular carriers and field effect transistors) are discussed in more detail. The response is treated in terms of the Nernst and Nikolskii equations and membrane potentials. Construction and geometry, selectivity, sensitivity, response time, stability and temperature dependence are treated briefly. Some points on the accuracy and precision are emphasized.

INTRODUCTION

The origins of potentiometry can be traced back to 1841, when Poggendorf (ref. 1) first described a circuit suitable for the measurement of cell voltages. This circuit, simplified by DuBois-Reymond in 1867 (ref. 2) has been in general use until the advent of modern pH-meters in the first half of this century. The correlation between ion concentrations and electromotive forces was first described by Nernst in 1888 (refs. 3,4). The first potentiometric titration was carried out by Behrend in 1893 (ref. 5), who used a mercury indicator electrode for the determination of halides. Soon after Peters (1898, ref. 6) introduced the term redox potential, Crotogino (1900, ref. 7) described the first potentiometric redox titration. Cremer in 1906 (ref. 8) discovered that certain glass membranes are pH-sensitive; the first glass electrode (which was also the first ion-selective membrane electrode) was constructed by the famous Fritz Haber (with Klemensiewicz, 1909, ref. 9) who later turned his interest towards the synthesis of ammonia. Abegg, Auerbach and Luther compiled the first comprehensive table of electrode potentials (1911, ref. 10). Potentiometry, as an analytical method, gained popularity in the 1920-s, mainly as a result of the activities of Müller (ref. 11) and Kolthoff and Furman (ref. 12), who published the first monographs on the subject. In 1937 Nikolskii alone (ref. 13) and also with Tolmacheva (refs. 14,15) when explaining the so called "alkaline error" of glass electrodes, described an equation in which a selectivity coefficient first appeared and where the idea of sodium-selective glass membranes were first mentioned. The development of new ion-selective electrode systems started in the 1960-s and has been gaining momentum ever since. These modern developments have been well summarised in monographs (refs. 16-20) and reviews (refs. 21-25).

ION-SELECTIVE ELECTRODES

The operation of ion-selective electrodes (ISE-s) is based upon an ion-selective membrane which, in the ideal case, is permeable only to the analyte ion, but non-permeable to all other ions (and molecules), including the analyte counter-ion. If two solutions, containing the analyte ion in different concentrations are separated by such a membrane, diffusion will start but comes to a halt soon, leading to an equilibrium (Donnan, 1911, ref. 26). If the activities of the ions on the left and right side of the membrane are a_l and a_r respectively, a potential difference will arise across the membrane, which can be measured by immersing two reference electrodes into these solutions. The potential difference can be expressed by the Nernst equation as:

$$E = (RT/zF)\ln(a_l/a_r) \tag{1}$$

The solution on the right may be encapsulated with the reference electrode into one electrode body; in this ISE the activity a_r is kept constant, and so the potential measureable by such an electrode, if immersed into a solution of a_i activity becomes:

$$E = E^O + (RT/z_iF)\ln(a_i) = E^O + (2.303RT/z_iF)\log(a_i) \tag{2}$$

where z_i is the ionic charge and E^O is the standard potential, which includes the logarithm of the ion activity on the right. The value of the pre-logarithmic factor for room temperature (25°C) is 0.059 V/dec; thus a 10-fold change in the activity of a monovalent analyte ion will result in a 59 mV change in the electrode potential.

The above equations are valid for ideal membranes only. If the membrane is permeable by other ions as well, their activities will also influence the response. The potential of such electrodes can be described by the modified Nikolskii--equation:

$$E = E^O + S\log[a_i + \sum K_{ij}(a_j^{z_i/z_j})] \tag{3}$$

here a_j is the activity of an interfering ion with z_j charge and K_{ij} is the so called selectivity coefficient (for the analyte ion i with respect to ion j). The lower
the value of K_{ij}, the more selective is the electrode towards the analyte ion.
The pre-logarithmic factor S is usually lower than 0.059/z.

Fig. 1 shows the response curves of ISE-s as predicted from equation 3. The
range of higher concentrations with linear response can be utilised for analytical
purposes. It can be seen that the limit of determination is the better, the lower
the selectivity coefficient and the lower the activity of the interfering ion.

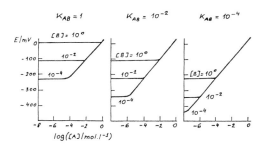

A: Analite ion, B: Interfering ion

Fig. 1. Response curves of ISE-s.

CLASSIFICATION OF ION-SELECTIVE ELECTRODES

According to the nature of the ion-selective membrane, ISE-s may be classified
in the following way:

```
Solid-state electrodes
        with homogeneous membranes:
            glass, crystal, nonporous materials
        with heterogeneous membranes:
            inert matrix (PVC, silicone rubber, parchment,
            hydrophobised graphite) + precipitate
Electrodes with liquid membranes
Electrodes with semi-solid membranes
        support: PVC, cellulose acetate
        ion-exchanger:
            (bulky) cations (for anions)
            (bulky) anions (for cations)
            neutral molecular carriers (ionophores)
Special electrodes
    Gas sensors
    Enzyme electrodes
    MOSFET-s, CHEMFET-s and ISFET-s
```

Among these, electrodes with neutral molecular carriers and ion-selective field effect transistors (ISFETS) seem to be in the focus of attention of researchers; they therefore merit a more detailed discussion.

ELECTRODES WITH NEUTRAL MOLECULAR CARRIERS

Some uncharged molecules with lipophilic character are capable of transporting ions across a hydrophobic membrane. Their selectivity towards a given ion may be enhanched by the addition of a secondary complexing agent. The incorporation of these into a membrane yields highly selective and stable electrodes. In certain cases plasticisers have to be included as well, which often results in lower cell resistance and minimises drift. Outstanding examples for such sensors are the potassium-selective Valinomycin-based electrode (ref. 27) and the hydrogen-selective n-dodecylamine-based electrode (ref. 28); both were first introduced by Simon and co-workers.

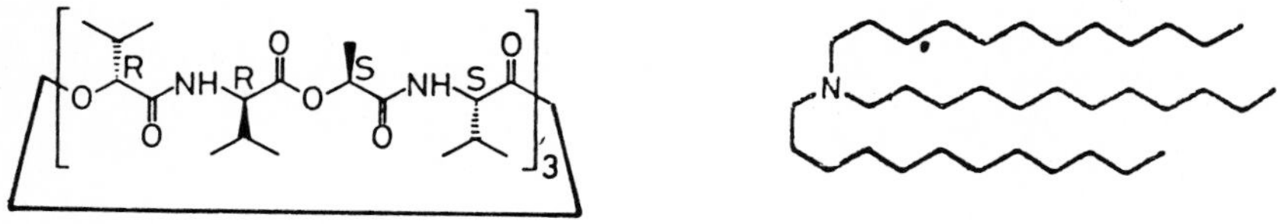

Valinomycin (K^+-selective)

n-dodecylamine (H^+-selective)

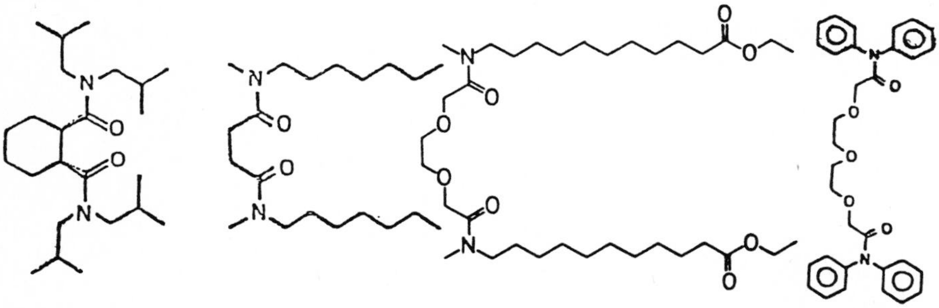

(Li^+-selective)   (Mg^{2+}-selective)   (Ca^{2+}-selective)   (Ba^{2+}-selective)

Fig. 2. Some selective ionophores.

According to their structure, one can distinguish between cyclic and acyclic ionophores. Both types possess an internal cavity, into which the analyte ion is trapped. The size of the cavity determines the maximal size of the ion which can be complexed, while the chemical nature of the substituents to be found in the inside of the cavity decides whether the ionophore will complex a cation or an anion. Cation-exchanger ionophores usually contain etheric, carbonyl or phenolic oxygens in their cavities (Fig. 2, see also ref. 19).

Some alkyl calixaryl esters, synthesised by McKervey and co-workers in Cork (ref. 29) are selective ion-exchangers. In Belfast we were successful in building sodium- and caesium-selective electrodes by incorporating these into a membrane (see paper PS3 of these Proceedings).

ACCURACY OF POTENTIOMETRIC MEASUREMENTS

At this point it is important to discuss the role of the electrode resistance in the accuracy of potentiometric measurements (Fig. 3).

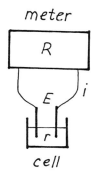

In the figure E represents the electromotive force to be measured, r is the resistance of the cell (which, in this case, is practically equal to the resistance of the ISE), and R is the input resistance of the meter. In order to measure the signal, a finite i current has to flow in the circuit:

$$i = E/(r + R)$$

Fig. 3. A potentiometric circuit.

This current causes a $\Delta E = ir$ potential drop across the cell, which represents the error in the reading.

If we wish to measure the electromotive force with a maximum error of say 0.1 per cent ($\Delta E/E \leq 0.001$), from the above equations it follows that the condition $R \geq 999r$ must be fulfilled. In other words, <u>the input resistance of the meter must be at least a thousand times greater than the resistance of the ISE.</u> The resistance of an ISE is usually in the order of 10^6 ohm, this means that the input resistance of the meter must be at least 10^9 ohm. In fact, modern pH-meters are usually designed with an input resistance greater than 10^{11} - 10^{12} ohm, to cater for the rather high resistance of the (H^+-selecti-ve) glass electrode. To achieve this high input resistance, appropriate devices (vacuum tubes or field-effect trasistors) have to be applied in the first amplifica-tion stage. Because of these high resistances, the signal is very weak and has to be protected by using shielded cables.

ION-SELECTIVE FIELD EFFECT TRANSISTORS (ISFET-S)

The conductivity between the source and drain of field-effect transistors (FET-s) depend mainly on the potential of the gate (ref. 30). If used at the input stage of a pH-meter, the gate potential is regulated by the incoming signal, which in turn is dependent on the potential of the ISE-membrane.

Experience has shown that the ion-selective membrane can be incorporated into the gate of the FET, and the whole device, known as an ISFET (sometimes CHEMFET) can be used directly as a probe with ion-selective properties. An ISFET is therefore a combination of an ISE and an amplifier, from which the signal can be transmitted without the difficulties mentioned above.

ISFET-s are on the whole still in the development stage. Opinions about their practical viability are mixed. Some say they were oversold by forceful marketing at a stage, when teething problems were still to be sorted out. A few of these (like proper encapsulation and insulation, as well as unwillingness of industry to start mass production) still exist. Others are enthusiastic without qualification. Some excellent examples (like the four-function device for the simultaneous monitoring of K^+, H^+, Ca^{2+} and Na^+, ref. 31) have been reported.

RESPONSE TIME

A very important practical characteristic of an ISE is the speed by which it follows variations in the analyte concentration. Some electrodes respond almost simultaneously, while others slowly. In principle, it is the rate of attaining the (heterogeneous) chemical equilibrium of the ion-exchange process, which has the greatest effect on the response time.

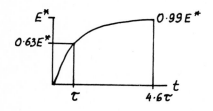

In Fig. 4 the measured potential (E) is shown as the function of time (t). The response curve can be described with the equation:

$$E = E^*[1 - \exp(t/\tau)]$$

Fig. 4. Response time.

where E^* is the final (equilibrium) potential of the ISE and τ is the time constant. Note that at $t = \tau$; $E = 0.63E^*$, in other words, the time constant equals
the time needed for the potential to rise to 63% of its final value. Note also
that 95% of the final response is reached in $4.3\,\tau$, while to acquire 99% response
a time elapse of $4.6\,\tau$ is needed.

If the time constant is less than 100 ms, the electrode can be called instantaneous. Depending on the purpose of application, time constants up to several
hundred seconds may be acceptable. If the time constant runs to several thou-
sands of seconds, the practical value of the electrode is limited.

TEMPERATURE EFFECTS

When using ISE-s, it must be borne in mind that their response depends
on temperature. The response equations (like eq. 1 to 3), though contain the
temperature as a variable, do not describe the temperature dependence adequate-
ly, because the E^o standard potential is also temperature dependent. The variation of the latter with temperature can be expressed through the Gibbs-Helmholtz
equation (ref. 32), which, for practical purposes can take the form:

$$E^o_T = E^o_{298} + (T\text{-}298)(\partial E/\partial T)_p$$

here E^o_T is the standard potential at (absolute) temperature T, E^o_{298} is the
(tabulated) value of the standard potential at room temperature, and $(\partial E/\partial T)_p$
is the temperature coefficient of the electrode at constant pressure. This equation is valid as long as the enthalphy of the electrode reaction remains constant
within the applied range of temperature. The temperature coefficient can be
determined by measuring the electrode potential at different temperatures.
It must be emphasized, that the potential of ordinary reference electrodes (calomel, silver/silver-chloride) varies strongly with temperature (because of the
variation in the solubility of the KCl electrolyte) and should be taken into account, when calculating the temperature coefficient (ref. 18, p. 129). The standard
potential of the hydrogen electrode is equal to zero (by definition) at any temperature.

The variation of the Nernstian pre-logarithmic factor must also be kept
in mind when evaluating potentiometric data. Its value for 25°C is $0.059/z$,
while at 37°C (body temperature) it takes the value of $0.0615/z$ V/dec.

DRIFT

Measurements with ISE-s are often subjected to drift. If the electrode is used in a batch operation, measuring samples individually, this drift can be eliminated by frequent re-calibration between samples. A simple calculator programme is available to correct such results (ref. 33). If however the electrode is used continuously (e. g. continuous flow or flow injection monitoring, extended intravenal or intravascular application) drift can cause serious errors and it is advisable to test such electrodes before application.

There are various types of drift. Parallel drift means that the slope of the calibration curve remains unchanged, only the intercept [of the E vs log(a$_i$) curve with the E - axis] moves. More serious is a concentration-dependent drift which results in changes in the slope of the calibration curve; the line often being piovoted at a medium concentration point. Random drift shows no regular trend.

Among the several reasons which may cause drift, a few are worth to be listed. Frequently it is the reference electrode which is to blame: a steady diffusion of solvent into the reference solution (e. g. concentrated KCl) can cause changes in the reference electrode potential. If such a danger exist, it is advisable to use such a solution in the liquid junction, which matches the ionic strength of the sample (like using saline for measurements in blood). Chemical changes caused by the sample in the membrane can lead to high cell resistances and hence to drift. For the same reason, ISE-s must be stored in a dilute solution of the ion for which they are specific. Variations in temperature and pressure may also cause drift.

PRECISION OF ANALYTICAL POTENTIOMETRY

Let us finally examine the precision (reproducibility) of analytical potentiometric measurements. This is important if one wishes to obtain a realistic picture on the practical value of the analytical result. As the correlation between analytical signal (electrode potential) and result (activity or concentration) is logarithmic, the effect of the precision of the measurement (carried out with s_E standard deviation) on the precision of the result (s_a) can only be expressed through the law of propagation of errors (ref. 34). In a simple case the response equation can be Nernstian, taking the form:

$$E = E^O + (0.059/z)\log(a_i) = E^O + (0.025/z)\ln(a_i) \qquad (4)$$

from which the a$_i$ ion activity can be expressed as:

$$a_i = \exp[40z(E-E^O)] \tag{5}$$

Applying the law of propagation of errors, we can write:

$$s^2_a = [(\partial a/\partial E)s_E]^2 \tag{6}$$

where the $(\partial a/\partial E)$ partial derivative takes the form:

$$(\partial a/\partial E) = (40z)\exp[40z(E-E^O)] \tag{7}$$

The relative standard deviation (coefficient of variation) of the activity can be expressed by combining equations 4 to 7 as:

$$s_a/a_i = 40zs_E \tag{8}$$

In practical terms, eq. 8 means that if the standard deviation of the potential measurement is 1 mV, the relative standard deviation of the ion activity value is 4 per cent for monovalent, 8 per cent for divalent ions etc. At routine potentiometric measurements this is a realistic figure. If the potential measurement is carried out with a better precision (say $s_E = 0.1$ mV, after taking special precautions by controlling temperature and environment), these percentages may become lower. For this reason, direct analytical potentiometry is not really suitable to determine main components of samples; if that is the task, it is better to apply potentiometric titrations (using the same ISE), when the end-point can usually be determined with a relative standard deviation of 0.5 per cent or less.

REFERENCES

1. J. C. Poggendorf: Annal. Phys. Chem. 54(1841) 161.
2. E. DuBois-Reymond: Arch. Anat. Physiol. 1867, 417.
3. W. Nernst: Z. physik. Chem. 2(1888) 613.
4. W. Nernst: Z. physik. Chem. 4(1889) 129.
5. R. Behrend: Z. physik. Chem. 11(1893) 466.
6. R. Peters: Z. physik. Chem. 26(1898) 193.
7. F. Crotogino: Z. anorg. Chem. 24(1900) 225.
8. M. Cremer: Z. Biol. 47(1906) 562.
9. F. Haber and Z. Klemensiewicz: Z. physik. Chem. 67(1909) 385.
10. R. Abegg, F. Auerbach and R. Luther: "Messungen elektromotorischer Krafte galvanischer Ketten" Abhandl. d. Bunsenges. Nr. 5, Knapp, 1911.
11. E. Müller: "Die elektrometrische Massanalyse" T. Steinkopf, 1923.
12. I. M. Kolthoff and N. H. Furman: "Potentiometric Titrations" Wiley, 1931.
13. B. P. Nikolskii: Zh. Fiz. Khim. SSSR 10(1937) 495.
14. B. P. Nikolskii and T. A. Tolmacheva: Zh. Fiz. Khim. SSSR 10(1937) 504.

140

15. B. P. Nikolskii and T. A. Tolmacheva: Zh. Fiz. Khim. SSSR 10(1937) 513.
16. H. S. Rossotti: "Chemical Applications of Potentiometry" Van Nostrand, 1969.
17. G. J. Moody and J. R. D. Thomas: "Selective Ion-Sensitive Electrodes", Merrow, 1971.
18. G. Svehla: "Automatic Potentiometric Titrations" Pergamon, 1978.
19. W. E. Morf: "The Principles of Ion-Selective Electrodes and of Membrane Transport" Elsevier, 1981.
20. D. Amman, W. E. Morf, P. Anker, P. C. Meier, E. Pretsch and W. Simon: "Neutral Carrier Based Ion Selective Electrodes", Pergamon, 1983.
21. E. P. Serjeant: "Potentiometry and Potentiometric Titrations" Wiley, 1984.
22. G. H. Fricke: Anal. Chem. 52(1980) 259R.
23. M. E. Meyerhoff and Y. M. Fraticelli: Anal. Chem. 54(1982) 27R.
24. M. A. Arnold and M. E. Meyerhoff: Anal. Chem. 56(1984) 20R.
25. M. A. Arnold and R. L. Solsky: Anal. Chem. 58(1986) 84R.
26. F. G. Donnan: Z. Elektrochem. 17(1911) 572.
27. L. A. R. Pioda, V. Stankova and W. Simon: Anal. Lett. 2(1969) 665.
28. D. Amman, F. Lanter, R. A. Steiner, P. Schultess, V. Shijo and W. Simon: Anal. Chem. 53(1981) 2267.
29. A. M. McKervey, E. M. Seward, G. Ferguson, B. Ruhl and S. J. Harris: J. Chem. Soc. Chem. Commun. (1985) 388.
30. H. V. Malmstadt, C. G. Enke and S. R. Crouch: "Electronics and Instrumentation for Scientists" Benjamin, 1981. P. 168 et f.
31. A. Sibbald, P. O. Whalley and A. K. Covington: Anal. Chim. Acta 159(1984) 47.
32. P. W. Atkins: "Physical Chemistry", 2nd Ed., Oxford UP 1982, p. 166 et f.
33. G. Svehla and E. L. Dickson: Anal. Chim. Acta 136(1982) 369.
34. G. Svehla in W. F. Smyth (Ed): "Electroanalysis in Hygiene, Environmental, Clinical and Pharmaceutical Chemistry", Elsevier, 1980, p. 21.

M.R. Smyth and J.G. Vos
Electrochemistry, Sensors and Analysis
© Elsevier Science Publishers B.V., Amsterdam — Printed in The Netherlands

ANALYTICAL POTENTIOMETRY IN THE STUDY OF MODEL SIDEROPHORE REAGENTS

J.D. GLENNON

Department of Chemistry, University College, Cork, Ireland.

SUMMARY

Siderophores are naturally occurring sequestering agents for ferric ion and are involved in the microbial transport of Fe(III). Synthetic models of these agents, as well as the siderophores themselves, have found many uses in medicine, biochemistry and analytical chemistry. The coordinating functional group with the strong affinity for Fe(III) is often hydroxamate in nature. The strongly chelating aliphatic dihydroxamic acids, $HOHNCO(CH_2)_n$-$CONHOH$, (n = 3-14) are synthetic models of naturally occurring dihydroxamic acids and are the subject of the present investigations by analytical potentiometry. The reagents form unusual but strong dimeric complexes which represent the predominant and biologically active complex species. The results also indicate the high analytical selectivity of the reagents for Fe(III). Studies on the analytical applications and on the coordination chemistry of these reagents are underway.

INTRODUCTION

The expanding and overlapping areas of inorganic biochemistry and bio-technology highlight the need for the development and application of suitable techniques for the study and analysis of biological ligands and complexes (Ref. 1). A knowledge of total metal concentration is often no longer the main goal of the analytical method but rather what is required is a knowledge of the number and nature of the complex species present.

One such area which presents a challenge and opportunity involves the study of metal sequestering agents. The insolubility of ferric hydroxide at physiological pH and the essential nature of iron for microbial growth apparently engendered the production of a wide range of powerful sequestering agents for ferric ion - the siderophores (Ref. 2). These ligands are involved in the microbial transport of Fe(III) into the cell and have found many uses in medicine, biochemistry and chemistry. The functional groups in these sidero-phores are the hydroxamate and catecholate bidentate chelating groups as found in ferrichromes, ferrioxamines and enterobactin. Advantage has been taken of these coordinating groups in analytical applications such as gravimetric and extraction-spectrophotometric determinations (Ref. 3). The optimisation of these analytical procedures for metal analysis is facilitated by the knowledge of the species present in aqueous solution.

Recent research in our laboratories has led to the development of a con-venient method of synthesis of strongly chelating dihydroxamic acids,

HOHNCO$(CH_2)_n$-CONHOH, (n = 3-14) which were previously not readily available. (Ref. 4).

Rhodotorulic Acid

Dihydroxamic Acid Reagent (n = 3-14)

In this report the results from the potentiometric analysis on two reagents (n = 3 and 5) are reviewed in detail as illustrative examples of the determination of complex species in aqueous solution by analytical potentiometry. These investigations involve the direct measurement of proton liberation on Fe(III) complexation using the glass electrode and computer-assisted data refinement. The work also probes the nature and analytical selectivity of the interaction of Fe(III) with the selected examples of the strongly chelating dihydroxamic acids.

EXPERIMENTAL METHODS

Materials

Glutaryl (n = 3) and pimelyl (n = 5) dihydroxamic acids were synthesised according to previously published procedures (Ref. 4) and were of analytical grade. Stock solutions were prepared from distilled and deionized water and stored under an inert atmosphere. The base used was carbonate-free 1.010 M NaOH, standardised against weighted amounts of dried potassium hydrogen phthalate. A stock solution of ferric nitrate, 0.515 M in 0.02 M HNO_3, was subsequently prepared from Analar Fe$(NO_3)_3 \cdot 9H_2O$. The stock acid and salt solution was 0.024 M in 0.15 M $NaNO_3$. The metal and acid solutions were standardised using atomic absorption spectrometry and potentiometric titration respectively.

Potentiometric Titrations

Potentiometric titrations were performed using a Radiometer Automatic Titration apparatus consisting of a digital PHM84 pH meter, autoburette ABU80, titrator TTA80 and an automatic recorder REC61 servograph. The 50 ml titration vessel was thermostatically controlled at 25 ± 0.05°C and the electrode pair

consisted of a Radiometer G2040 C glass electrode and a K4040 reference
electrode. A computer interface link to a BBC microcomputer was constructed,
allowing automatic data collection at every 0.1 pH unit. A schematic outline
of the apparatus is shown in Figure 1.

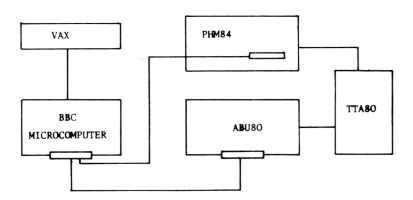

Fig. 1. Schematic Diagram of Computer-Titration Apparatus Link-up.

A series of titration curves are recorded with metal and ligand concentrations varied at constant mineral acid concentration (Ref. 5). An excess
of ligand over metal is used to ensure full coordination and to prevent metal
precipitation. For the Fe(III) interaction with pimelyl dihydroxamic acid,
curves were obtained for C_L = 3.452 x 10^{-3} M, 5.460 x 10^{-3} M and 7.720 x 10^{-3} M
with C_{Fe} = 1.029 x 10^{-3} M and for C_{Fe} = 5.148 x 10^{-4} M, 1.029 x 10^{-3} M and
1.389 x 10^{-3} M with C_L = 5.460 x 10^{-3} M. A similar procedure was followed for
the Fe(III) glutaryl dihydroxamic acid system (Ref. 6).

Determination of Species and Stability Constants

The equilibrium reactions occurring between the metal ion M, proton H and
anionic hydroxamate L can be represented by the general reaction equation

$$pM + qH + rL \rightleftharpoons M_pH_qL_r$$

where p, q and r are the stoichiometric quantities of M, H and L respectively.
The stability of the species formed is given by the stoichiometric
equilibrium constant, β_{pqr}, expressed in terms of concentrations at constant
ionic strength, temperature and pressure.

$$\beta_{pqr} = \frac{[M_pH_qL_r]}{m^p h^q l^r}$$

where m, h and l are the concentrations of free metal ion, hydrogen ion and hydroxamate ligand respectively. The relations to obtain the unbound metal ion concentration and ligand L concentration at any specified pH values have been described (Ref. 5). The titration data was fed to the program PLOT-3, yielding values for the unbound portions of metal and the proton liberation term $\delta H^+/\delta C_M$ together with $\delta H^+/\delta C_L$ and the free ligand concentrations. The program GUESS-3 was then used to set up a matrix of the terms $m^p h^q l^r$ for each proposed species at each selected pH value. This matrix then served as the input data for the program LEASK-4 which uses an iterative least-squares minimization procedure to calculate the stability constants β_{pqr} and the final species distribution.

RESULTS AND DISCUSSION

Information on the nature and stability of the complex species in aqueous solution is important in understanding the biological activity of dihydroxamic acids and in optimising analytical procedures for the utilisation of these reagents in metal analysis. The molar proton liberation on metal coordination $\delta H^+/\delta C_M$ is directly determined at the glass electrode and provides important information on the on-set and progress of complexation. A plot of $\delta H^+/\delta C_M$ for the Fe(III)-glutaryl dihydroxamic acid system is shown in Figure 2.

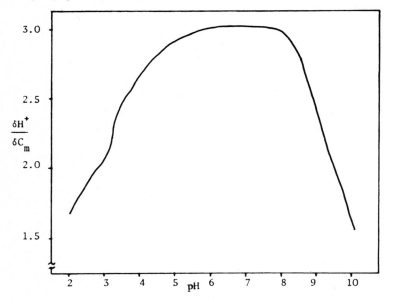

Fig. 2. Molar proton liberation $\delta H^+/\delta C_M$ for the Fe(III)-glutaryl dihydroxamic acid system as a function of pH.

The ligand pK_a values are $pK_1 = 9.75$ and $pK_2 = 8.67$ (Ref. 7) implying that a maximum of two protons can be liberated per coordinated ligand in neutral and acidic solutions. The molar proton liberation is well advanced even at pH = 2 indicating that strong complexation occurs in acidic solution. The rapidly increasing proton liberation from pH = 2 to pH = 5 is consistent with increasing complexation of free Fe(III) and with an increasing ligand number. Some levelling off of the proton liberation at a value of 2 occurs at pH = 2.7 consistent with the formation of an FeL complex species. However, the most informative feature of the proton liberation data, is the almost constant value of 3 obtained in the pH range 5-8 indicating the presence of a major complex species involving 1.5 ligands per mole of Fe(III). This stoichiometry is well supported by a plot of $\bar{n}$ versus pH (Ref. 6). Further support for a major complex species, Fe_2L_3, comes from the largely invariant visible absorption characteristics of the orange-red solution with λ_{max} = 425 nm over this pH range (Ref. 7). This pattern is also true of the Fe(III)-pimelyl dihydroxamic acid system. The predominance of the Fe_2L_3 complex species is very clearly shown in the preliminary species distribution diagram obtained for this system and presented in Figure 3 (Ref. 8).

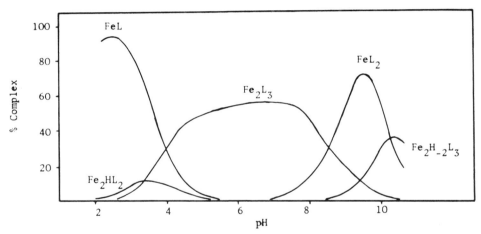

Fig. 3.   Species distribution diagram for the Fe(III)-pimelyl dihydroxamic acid system as a function of pH.

From the species distribution it can be seen that the initial complex is an FeL species which at its maximum at pH 2.7 represents greater than 90% of the total Fe(III). By pH = 4 this complex is overtaken by the Fe_2L_3 species (log β_{pqr} = 41.06), which is the major complex species up to pH = 8.5.

This strong octahedral and dimeric coordination of Fe(III) by these synthetic dihydroxamic acid reagents mimics that of the naturally occurring dihydroxamate siderophores (Ref. 9). The proposed structure of the Fe_2L_3 complex is shown in Figure 4.

146

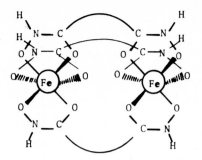

Fig. 4.   Proposed structure of the major Fe_2L_3 complex species.

Further studies are underway to investigate the coordination chemistry of these unusual dimeric complexes. The high affinity of these reagents for Fe(III) is also under investigation in analytical applications involving instrumental analysis.

ACKNOWLEDGEMENTS

The researchers who have contributed to this work include in no small way Professor D.A. Brown and Dr. R. Geraty at the collaborating bioinorganic laboratory at the Chemistry Department of University College Dublin. I would also like to thank the postgraduate researchers at University College Cork, Ms. N. Ni Choileain and Mr R. McMahon.

REFERENCES

1.   G.H. Morrison, Analytical Chemistry 55 (1983) 2017.
2.   J.B. Neilands, in Microbial Iron Metabolism, J.B. Neilands (Ed.), Academic Press, New York (1974) pp. 3-34.
3.   J.K. Agrawal and S.A. Patel, Reviews in Analytical Chemistry, 4 (1980) 237-278.
4.   D.A. Brown, R.A. Geraty, J.D. Glennon and N. Ni Choileain, Synthetic Communications 15 (13) (1985) 1159-1164.
5.   B. Sarkar and T.P. Kruck, Can. J. Chem. 51 (1973) 3541.
6.   N. Ni Choileain and J.D. Glennon, Rev. Port. Quim., 27 (1985) 287-288.
7.   D.A. Brown, R.A. Geraty, J.D. Glennon and N. Ni Choileain, Complexation Behaviour of Dihydroxamic Acids with Metal Ions, submitted for publication.
8.   R. McMahon, N. Ni Choileain and J.D. Glennon, Potentiometric Analysis of Strong Fe(III) Complexation by Pimelyl Dihydroxamic Acid, in preparation.
9.   S.J. Barclay, B.H. Huynh and K.N. Raymond, Inorganic Chemistry 23 (1984) 2011.

M.R. Smyth and J.G. Vos   147
Electrochemistry, Sensors and Analysis
© Elsevier Science Publishers B.V., Amsterdam — Printed in The Netherlands

POTENTIOMETRIC DETERMINATIONS OF AMINO ACIDS USING ENZYMATIC AND BACTERIAL
ELECTRODES

B.J. VINCKE, J-C. VIRE and G.J. PATRIARCHE
Free University of Brussels (U.L.B.), Institute of Pharmacy, Campus Plaine
205/6, Boulevard du Triomphe, B-1050 Brussels, (Belgium).

SUMMARY
    Biosensors using enzymes and bacteria associated to an ammonia gas sensing
electrode are described for the potentiometric determination of amino acids,
based on a deamination reaction. The parameters involved in the optimization
of the electrode response, such as the nature of the buffer, pH, temperature,
but also microbiological factors, are discussed in details for each electrode.
Performances of the electrodes and interferences which may occur are presented.
Opportunities of using bacteria to produce specific unstable enzymes are
pointed out.

INTRODUCTION

    Since several years, electrochemical techniques have been associated to

enzymatic processes in order to obtain high performing analytical methods

owing to the selectivity of enzyme reactions and the sensitivity of electro-

chemical sensors (refs. 1,2).

    However, if more than two thousand enzymes are now well identified, most of

them are unstable or not commercially available. In addition, the use of

enzyme electrodes implies that the catalyst will be physically or chemically

immobilized.

    More recently, bacterial and tissular electrodes were proposed, which solve

these problems and offer several advantages, such as a wider choice of enzymes,

a longer lifetime, a better tolerance towards physico-chemical parameters

(ref. 2). This will be illustrated by the determination of two amino acids:

asparagine and tryptophane.

EXPERIMENTAL

    An ammonia gas sensing electrode (Universal Sensors Inc., type 019733582 or

Tacussel p NH_3-1) is used for the construction of the enzyme or bacterial

electrodes. Potentials are monitored with a Tacussel Minisis 6000 millivolt-

meter in conjunction with a Göertz Servogor 120 recorder. Measurements are

performed in a thermostatic cell (± 0.2°C).

All chemicals are of analytical or pharmacopoeia pure grade (Difco and Merck)
The strains used are collection strains: Serratia marcescens ATCC 13880
(asparagine) and Escherichia coli B/It 7-A ATCC 27553 (tryptophane).
Tryptophanase is a Sigma Chemical Co. product.

RESULTS AND DISCUSSION

The ammonia gas sensing electrode is modified by setting the enzyme or the
bacterial membrane against the hydrophobic membrane. The catalytic layer is
protected by a dialysis or a cellulose acetate membrane, the latter being
preferred in order to avoid a dialysis phenomenon which prevents the elimination of water and ammonia from the catalytic layer and increases the response
time, the base potential and the stabilisation time.

The bacterial membrane is prepared as follows:the stock strain, conserved
at 4°C on Heart Infusion Agar slants, is cultured at 37°C in the liquid growth
medium. The third subculture is filtered on a microporous cellulose acetate
filter. This one, coated with the bacteria, is dried 30 minutes at 37°C and
the electrode membrane is cut   out.

Asparagine

Asparagine has been determined according to the following reaction:

$$\text{L-asparagine} \; + \; H_2O \; \xrightarrow[\text{L-asparaginase}]{\text{bacterial}} \; \text{L-aspartic acid} \; + \; NH_3$$

Two growth media were tested: the first, said "poor medium" contains yeast
extract and tryptone. The second, or "rich medium", is the Brain Heart
Infusion medium (B.H.I. Difco) (refs. 3,4).

Studying the influence of the growth medium on the electrode response, we
have pointed out that the better response was obtained when the substrate,
asparagine, was added to the growth medium (1.5 g per 100 ml). This
procedure induces or increases the synthesis of asparaginase by the bacteria
(ref. 5).

As shown by the figure 1, the brain heart infusion medium, corresponding to
the curve B, increases the proliferation of the cells, giving rise to a better
sensitivity; but a wider linearity range is obtained with the first medium
(curve A) which will be preferred for analytical purposes.

Asparaginase is very unstable and enzyme electrodes equipped with a liquid
membrane of purified asparaginase lose the half of their activity after three
or four days. Chemically bounded enzyme electrodes are still more stable
(refs. 6,7).

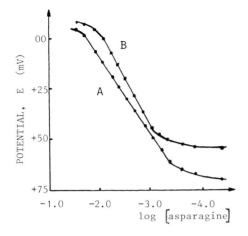

Fig. 1. Influence of the growth medium.

Temperature: 30.0°C :

0.05 M tris-HCl buffer, pH 7.8

A) first medium

B) B.H.I. medium.

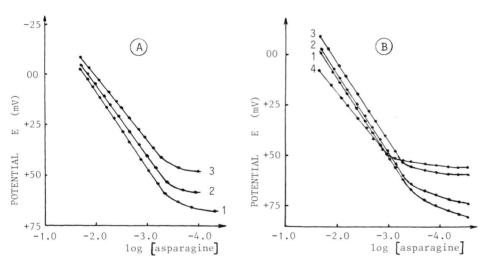

Fig. 2. Stability of the bacterial electrode.
A) conserved in the first medium at 4°C; 1) 1st and 2nd days; 2) 3th day;
   3) 5th day.
B) conserved in the first medium at 30°C under shaking; 1) 1st day; 2) 16th
   day; 3) 22nd day; 4) 27th day.

The enzyme activity of bacterial electrode also decreases if they are conserved in their growth medium at 4°C or in the buffer at room temperature (Fig. 2 A). The slope and the linearity range decrease. In such conditions, the growth of the cells is inhibited. But if the electrode is conserved in its growth medium at 30°C under slow shaking, proliferation of the bacteria is maintained at the surface of the electrode and its stability is enhanced up to more than three weeks (Fig. 2 B). During the fourth week, the slope decreases slowly as the linearity range and the response time increases.

TABLE 1

Influence of the nature of the buffer, its concentration and of the pH on the slope and the linearity range of the electrode response.

Buffer and concentration	pH	Slope (mV/dec)	Linearity range (mole . 1^{-1})
Tris - HCl 0.1 M	8.5	35	$2 . 10^{-2}$ to $8 . 10^{-4}$
Phosphate 0.1 M	8.5	48	$3 . 10^{-2}$ to $3 . 10^{-3}$
Tris - HCl 0.1 M	7.8	36	$2 . 10^{-2}$ to $6 . 10^{-4}$
Phosphate 0.1 M	7.8	50	$4 . 10^{-2}$ to $2 . 10^{-3}$
Tris - HCl 0.05 M	7.8	38	$2 . 10^{-2}$ to $4 . 10^{-4}$
Tris - HCl 0.1 M	7.4	37	$2 . 10^{-2}$ to $6 . 10^{-4}$

The physico-chemical parameters are also of great importance (Table 1). If a phosphate buffer increases the sensitivity, the tris-hydrochloride buffer allows to obtain a wider linearity range. A variation of pH from 7.4 to 8.5 does not strongly affect the slope and the linearity range, allowing to work with bacterial electrodes in a wider pH range than with enzyme electrodes, for which this parameter is more critical (ref. 6).

The best response time is obtained at 30°C but, again, the maximum is not so sharp than that observed with an enzyme electrode.

Although microbial electrodes have sometimes a longer response time than enzyme electrodes, due to the thickness of the bacterial membrane, this parameter does not exceed 4 to 9 minutes, the longer time being observed with the lower concentrations.

From this study, the best operating conditions of the asparagine electrode can be summarized as follow: 0.05 M tris-hydrochloride buffer at pH 7.8, 30°C, conservation of the electrode in the growth medium at 30°C under shaking. The characteristics of the electrode are then: a slope of 38 mV per decade, a linearity range from 2.10^{-2} to 4.10^{-4} M, a response time of 4 to 9 minutes and a lifetime of about 26 days.

Serratia marcescens contains several enzymes able to deaminate other amino acids, thus some interferences may occur. From a study covering 18 amino acids and urea, it appears that only L-glutamine and L-histidine give a relative response of 20 %, the others having a relative response lower than 1 %.

Tryptophane

Tryptophane can be determined using the tryptophanase synthetized by Escherichia coli. This enzyme catalyses the following reaction:

$$\text{L-tryptophane} + H_2O \xrightarrow[\substack{\text{or bacterial} \\ \text{tryptophanase}}]{\text{tryptophanase}} \text{indole} + \text{pyruvic acid} + NH_3$$

The bacterial electrode is constructed following the same procedure as for asparagine, the cells being cultured in a "poor medium" with indole as inducer (674 B/It-7a medium) (ref. 4). A broth medium containing yeast extract and bactotryptone was also tested but the sensitivity of the electrode was strongly decreased, due to a lower enzyme activity.

The characteristics of the bacterial electrode were compared with those obtained for the enzyme electrode covered with a crude extract of tryptophanase. 30 mg of the enzyme and 0.5 mg of pyridoxal phosphate as cofactor are put on a cellophane membrane and fixed against the hydrophobic membrane of the electrode. The electrode is then immersed during one hour in the buffer at 37°C in order to dissolve the enzyme. Physical immobilization in agar gel and chemical immobilization with glutaraldehyde on a cellulose acetate membrane were also tested but the stability of such electrodes is not better than that of solubilized enzyme. This will be due to the impossibility for the immobilized enzyme to react with its cofactor, the pyridoxal phosphate. This cofactor of tryptophanase must always be added to the solution but at a higher concentration for the enzyme electrode (20 mg for 100 ml) than for the bacterial electrode (2.5 mg for 100 ml), the bacteria containing it partially.

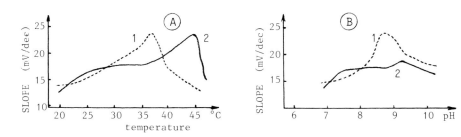

Fig. 3. Effect of temperature (A) and of pH (B) on the slope of the electrodes
        0.1 M phosphate buffer.
(A) pH 8.3 : 1) enzyme electrode ; 2) bacterial electrode.
(B) T = 30°C ; 1) enzyme electrode ; 2) bacterial electrode.

In order to maintain the best linearity range, the working temperature must stay below 35°C for both electrodes. While the sensitivity of the bacterial electrode is not affected by a change of temperature between 25° and 35°C, the slope of the enzyme electrode decreases with this parameter. As a compromise, we have choosen a working temperature of 30°C for both electrodes (Fig. 3 A).

The effect of pH is also critical for the enzyme electrode where the maximum slope appears at pH 8.7 but with a restricted linearity range. This effect is not so pronounced with the bacterial electrode and a pH of 8.3 was selected (Fig. 3 B).

The tris-hydrochloride buffer employed for asparagine, but also for urea (ref. 8) cannot be used in this case, the electrode response being inhibited. A 0.1 M phosphate buffer gives the best results.

The bacterial electrode is also conserved in its growth medium at 37°C in order to regenerate the enzyme synthesis. The enzyme electrode is conserved at 4°C in the working buffer.

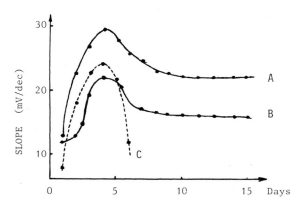

Fig. 4. Stability of enzyme and bacterial electrodes.
        0.1 M phosphate buffer, pH 8.3 ; T = 30°C
A : bacterial electrode regenerated in "poor medium".
B : bacterial electrode regenerated in broth medium .
C : solubilized enzyme electrode.

Figure 4 shows that both electrodes exhibit a maximum of sensitivity at the fourth day. This may be probably attributed to the complex part taken by pyridoxal phosphate in the steric modification of the enzyme (ref. 9). The slope of the bacterial electrode is then stabilized, giving a lifetime of three weeks. The sensitivity of the enzyme electrode falls rapidly after the maximum and this sensor cannot be used more than four days. Immobilization of the enzyme does not increase the stability.

Several compounds may interfere, due to the fact that tryptophanase catalyses other reactions, the substrates being tryptophane, cysteine and serine (refs. 10,11). Molecules which are structurally related to tryptophane also interfere: tryptamine or 5-hydroxytryptophane. In addition, Escherichia coli has also other deaminating enzymes, able to react with asparagine, glutamine and aspartic acid. These interferences also occur with the enzyme electrode because it is a crude extract not well purified.

Table 2 compares the characteristics of both sensors.

TABLE 2

Comparison of the characteristics of enzyme and bacterial electrodes for tryptophane determination.

Criteria	Bacterial electrode	Enzyme electrode (solubilized)
Linearity range (mole . 1^{-1})	1.10^{-2} to 3.10^{-4}	1.10^{-2} to 5.10^{-4}
Slope (mV/dec)	22	18
Response time (minutes)	10 to 15	10 to 15
Stability	3 weeks	4 days
Cofactor concentration	2.5 mg/100 ml	20 mg/100 ml
Interferences	Asparagine Glutamine Aspartic acid	Asparagine Glutamine Aspartic acid
Stability of stock material	Several years on H.I.A. slants at 4°C.	Less than 2 months (lyophilized form)

CONCLUSIONS

This study shows that there are many similarities between enzyme and bacterial electrodes. The linearity range, the slope and the response time are quite similar, but the stability of the bacterial electrode is enhanced, owing to the regeneration of the living cells during the conservation of the electrode. This possibility is of great interest when the enzyme is unstable or cannot be immobilized. The selectivity, which is often better for the enzyme, depends on the purification of the catalyst or on the metabolism of the bacteria. In addition, the enzyme is not always available or may be unstable, even in its lyophilized form, whereas the strain is available and may be conserved at 4°C during several years.

REFERENCES

1 G.G. Guilbault, Enzyme Electrodes in Analytical Chemistry, in Wilson and Wilson's (Ed.), Comprehensive Analytical Chemistry, vol VIII, Elsevier, Amsterdam, 1977, p. 1.
2 G.G. Guilbault, Analytical Uses of Immobilized Enzymes, M. Dekker Inc., New York, 1984.
3 B.J. Vincké, M.J. Devleeschouwer, G.J. Patriarche, Anal. Letters, 16(B9), (1983) 673.
4 American Type Culture Collection, Catalogue of Strains I, 15th ed., 12301 Parklawn Drive, Rockville, Ma 20852, U.S.A., 1982.
5 J. Brisou, Techniques d'Enzymologie Bactérienne, Masson, Paris, 1971.
6 G.G. Guilbault, E. Hrabankova, Anal. Chim. Acta, 56 (1971) 285.

7   D.A. Ferguson, J.W. Boyd, Jr., A.W. Phillips, Anal. Biochem., 62 (1974) 81.

8   P. Eliard, A. Laudet, J-C. Viré, G.J. Patriarche, Analusis, 10 (1982) 182.

9   Y. Morino, E.E. Snell, J.Biol. Chem., 242 (1967) 5591.

10   W.A. Newton, Y. Morino, E.E. Snell, J. Biol. Chem., 240 (1965) 1211.

11   W.A. Newton, E.E. Snell, Proc. Nat. Acad. Sci., 48 (1962) 1431.

M.R. Smyth and J.G. Vos
Electrochemistry, Sensors and Analysis
© Elsevier Science Publishers B.V., Amsterdam — Printed in The Netherlands

NEUTRAL CARRIER-BASED ION-SELECTIVE ELECTRODES

D. DIAMOND

Department of Chemistry, Queen's University, Belfast BT9 5AG

SUMMARY

The performance of sodium and caesium ion-selective electrodes based on a new series of ionophores, alkyl calixaryl acetates, is presented. Response improvement due to the inclusion of tetra(p-chlorophenyl)borate in the membrane is demonstrated. Selectivity coefficients for group I and II cations have been determined and compared with similar electrodes.

INTRODUCTION

Neutral carriers are uncharged, lipophilic molecules capable of selectively transporting metal ions across a hydrophobic membrane. The outstanding selectivity of such molecules for certain cations has lead to their incorporation as sensing agents in several very successful sensors, like the Valinomycin (K^+, refs. 1 and 5) and Tri-n-dodecylamine (H^+, ref. 10) electrodes. Interest in this area of electrochemistry continues to grow, with over 160 references quoted in a recent biannual review (ref. 2).

ELECTRODE RESPONSE

The potential arising at the membrane/sample interface is ideally described by the Nernst-equation:

$$E = E^O + S \log a_i \tag{1}$$

where E = the measured potential in V

E^O = a standard potential dependent on the entire cell measuring system and temperature (but independent of the a_i ionic activity)

S = the prelogarithmic factor, giving the slope of the E vs $\log a_i$ curve. For true Nernstian response its value is $2.303RT/z_i F$ (where R = the gas constant, T = the absolute temperature, z_i = the charge of the ion and F = the Faraday constant).

This equation predicts a response of 59.2 mV to a decade change in the activity of a monovalent primary ion (a_i) at room temperature.

However, other ions may penetrate the membrane phase and generate an additional response. This behaviour is fully accounted for by the Nikolski equation:

$$E = E^O + Slog[a_i + K_{ij}^{pot} (a_j)^{z_i/z_j}] \qquad (2)$$

here a_j = the activity of the interfering ion and K_{ij}^{pot} = the potentiometrically derived selectivity coefficient.

The selectivity coefficient thus accounts for the contribution of various interfering ions (j) to the observed potential (E). For low concentrations of such ions ($a_j \rightarrow 0$), or almost total specificity for the primary ion ($K_{ij}^{pot} \rightarrow 0$), a Nernstian type response to a_i will be observed.

IONOPHORE STRUCTURE

The compounds used in this research are structurally similar to certain crown ethers and cavitands noted for their ion complexing behaviour. They possess a central cavity with inward facing phenolic and carbonyl oxygen atoms capable of complexing with suitably sized cations, particularly alkali metal ions. Outward facing aryl and alkyl groups ensure solubility of cation complexes in the hydrophobic membrane phase.

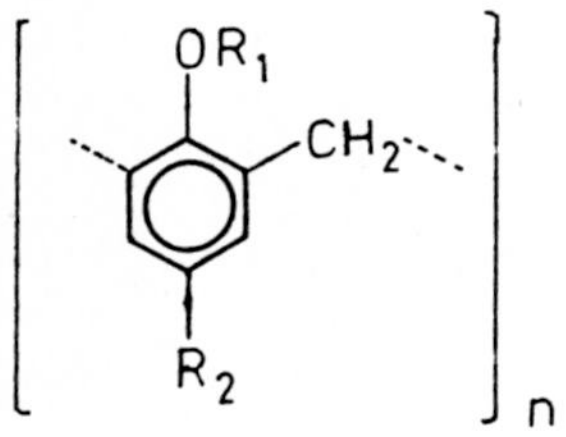

n	R_1	R_2	Ligand
4	CH_2CO_2Et	But	1(b)
4	CH_2CO_2Me	But	1(c)
6	CH_2CO_2Et	But	2(b)
6	CH_2CO_2Et	H	2(d)

Fig. 1. Structure of alkyl calixaryl acetates.

These four ligands were the most promising of a series synthesised at University College Cork. Phase transfer measurements (ref. 4) indicated that the smaller tetramers were most efficient at complexing Na^+ ions, whereas the hexamers were best for Cs^+.

EXPERIMENTAL

To make up unmodified electrodes, the compounds were dissolved in trichloro-benzene (TCB) [1(b) and 2(d) 1% m/m; 1(c) and 2(b) 0.5% m/m; the reduced concentrations of 1(c) and 2(b) being due to their lower solubility in TCB]. Silver-/silver chloride internal reference electrodes were constructed from fine silver wire (Johnson Matthey $\emptyset$ = 0.004") by chloridising for 10 minutes in 10% sodium hypochlorite solution. A glass pipette was utilised as the electrode body. With the internal reference electrode in place, the thicker end was sealed and the internal reference solution (0.1 M $NaNO_3$) injected. The membrane solution was then carefully added from a syringe needle to fill the final 1 mm of the pipette tip. After construction, electrodes were left in distilled water for at least two hours to allow the membrane/water interface to stabilise. All solutions were made up as chlorides in doubly distilled water without ionic buffers. Potential measurements were taken with a METROHM 632 pH-meter using a fibre wick staurated calomel external reference electrode. The effect of K^+ leakage from the reference was checked using nitrate solutions and a Ag/AgCl external refe-rence, and found to be insignificant.

Modified electrodes were made up in a similar way, with the inclusion of tetra-p-chloro-phenylborate (TPCB) in the ratio of 1:4 by mass to the exchanger.

The effect of pH on the electrode response was investigated using the mixed solution method (ref. 3, pp 14-16). Buffers of constant ionic strength to cover the pH-range 2-12 were made up as described by Jordan (ref. 9), using LiCl/LiOH in place of KCl/KOH. Each buffer was diluted to give an approximate lithium ion concentration of 0.01 M. The primary ion concentration was kept constant at 10^{-4} M. At these concentrations, the lithium ions were not found to interfere with the electrode response.

DISCUSSION

Electrodes based on 1(b) and 1(c) clearly do not display the expected Na^+-specificity and are generally unstable when in use (fig. 2). Although 2(b) and 2(d) are Cs^+-selective, their slopes are sub-Nernstian. High background noise levels and interference on the signal due to magnetic stirring or local movement indicate a high membrane resistance.

This behaviour has, in the past, been explained in terms of low exchange current densities. Although carriers may be ion-specific, they tend to form weak complexes, resulting in poor cation extraction efficiency (ref. 8, p. 18). Con-sequently, the electrode response may be dominated by the extraction capability of the solvent, which will generally favour larger cations.

158

The addition of TPCB in carefully controlled amounts can improve interfacial kinetics and result in lower electrode resistance (ref. 8, p. 18). A good excess of carriers should be maintained to prevent dominance in electrode response by the exchanger, which is generally better for larger cations and has in fact formed the basis of electrodes selective for such cations (ref. 6, p. 214).

In addition, the inclusion of permanent anion sites within the membrane reduces anion interference at higher activities leading to a wider Nernstian response range (ref. 8, p. 19).

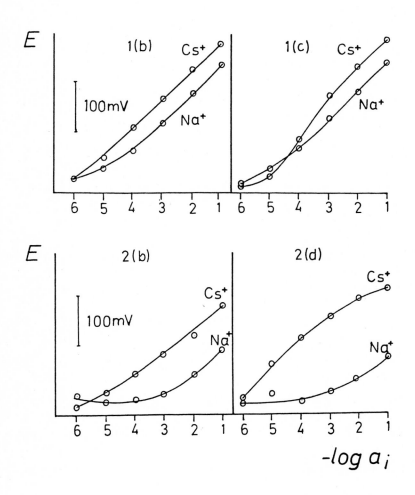

Fig. 2. Response of unmodified electrodes.

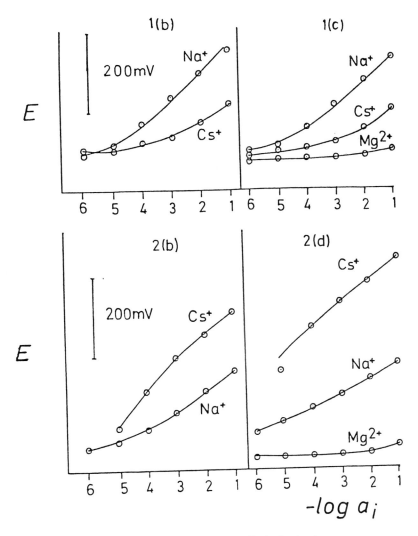

Fig. 3. Response of modified electrodes.

A striking change in the response of tetramer based electrodes is evident, with the expected Na^+-selectivity being manifested (fig. 3). In addition, the response slope and stability of the hexamer-based electrodes have improved. A more comprehensive list of selectivity coefficients is given below.

Table 1. $\log(K_{ij}^{pot})$ figures for each electrode ($a_i = 10^{-2}$)

For calculation of the selectivity coefficients, Na^+ was taken to be the primary ion for 1(b) and 1(c), and Cs^+ for 2(b) and 2(d).

Primary ion:	$i = Na^+$			$i = Cs^+$	
Ligand:	1(b)	1(c)	Na^{+*}	2(b)	2(d)
Interfering ion, j =					
Cs^+	-1.44	-1.66	-2.5	0	0
Rb^+	-2.54	-2.37	-2.4	-0.88	-1.27
K^+	-1.82	-2.00	-2.4	-0.91	-2.47
Na^+	0	0	0	-2.23	-3.87
Li^+	-2.64	-2.52	+0.4	-4.16	-5.41
Ba^{2+}	-4.9	-4.7	-1.4	-4.1	<-6.0
Ca^{2+}	-5.7	-5.3	-0.5	-5.5	<-6.0
Mg^{2+}	<-6.0	-5.9	-2.8	-6.0	<-6.0

*figures for ligand 5 from ref. 8, table 6.

The selectivity pattern obtained for each ligand is in good agreement with phase-transfer efficiency figures, published by McKervey et al.(ref. 4), suggesting that the inclusion of the exchanger is necessary for the carriers to function efficiently. Data for the best Na^+-ligand given in a recent review (ref. 8) is included for comparison. While selectivities for Cs^+, Rb^+ and K^+ are of the same order for this carrier and 1(c), the latter does not suffer from interference from Li^+ and has virtually no response to group II. ions.

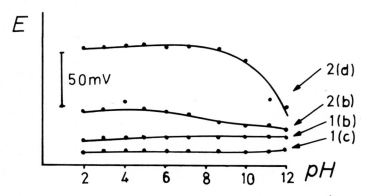

Fig. 4. pH-dependence of electrode response, $a_i = 10^{-4}$.

As indicated in figure 4, all electrodes (and especially the Na^+-electrodes) showed good stability over the pH-range examined. With negligible response up to pH=2, selectivity for primary over hydrogen ions must be 100-fold or better ($K_{ij}^{pot} \leqslant 10^{-2}$).

CONCLUSIONS

All four compounds have been shown to work well as potentiometric sensors. The modified electrodes show good selectivity for the primary ion, virtually no response to divalent cations and are stable over a wide pH-range. Response times are short (90% response in less than 3 s), and lifetimes long (at least 1 month stored in distilled water). Electrodes using 1(c) as sensor have better selectivity for Na^+ over K^+, Li^+, H^+ and Ca^{2+} than others currently used for blood/urine sodium monitoring [see refs. 7 p. 49 and 8 (table 6, ligands 6 and 7)].

The Cs^+-ligand 2(d) has better selectivity than the single example quoted by Amman et al.(ref. 8, p. 56), and could possibly form the basis of caesium determinations in the future.

ACKNOWLEDGEMENTS

The co-operation of Professor A. M. McKervey (UCC), Miss E. M. Seward (UCC) and Dr. G. Svehla (QUB) is gratefully acknowledged.

REFERENCES

1. L. A. R. Pioda, V. Standova and W. Simon: Anal. Lett. 2(1969), 665.
2. M. A. Arnold and R. L. Solsky: Anal. Chem. 58(1986) 84R.
3. G. J. Moody and J. D. R. Thomas: "Selective Ion-Sensitive Electrodes", Merrow Technical Library, 1971.
4. A. M. McKervey, E. M. Seward, G. Ferguson, B. Ruhl and S. J. Harris: J. Chem. Soc. Chem. Commun. (1985), 388.
5. H. F. Osswald, R. Asper, W. Dimai and W. Simon: Clin. Chem. 25(1979), 39.
6. W. E. Morf: "The Principles of Ion-Selective Electrodes and of Membrane Transport" (Vol. 2. in the series: "Studies in Analytical Chemistry"), Elsevier, Amsterdam, 1981.
7. P. C. Meier, D. Amman, W. E. Morf and W. Simon: in J. Koryta (Ed): "Medical and Biological Applications of Electrochemical Devices", Chapter 2. Wiley (1980).
8. D. Amman, W. E. Morf, P. Anker, P. C. Meier, E. Pretsch and W. Simon: "Neutral Carrier Based Ion Selective Electrodes" [Vol. 5. of the series: J. R. D. Thomas (Ed): "Ion-Selective Electrode Reviews"], Pergamon (1983).
9. C. Jordan: Microchem. Journal 25(1980), 492.
10. D. Amman, F. Lanter, R. A. Steiner, P. Schultess, Y. Shijo and W. Simon: Anal. Chem. 53(1981), 2267.

M.R. Smyth and J.G. Vos
Electrochemistry, Sensors and Analysis
© Elsevier Science Publishers B.V., Amsterdam — Printed in The Netherlands

POTENTIOMETRIC MICROELECTRODES BASED ON SURFACE ACTIVATED CARBON FIBERS

M. JOSOWICZ, K. POTJE, H.-D. LIESS[1] and J.O. BESENHARD, A. KURTZE[2]

[1]Faculty of Electrical Engineering, Universität der Bundeswehr München,
D-8014 Neubiberg (West Germany)

[2]Department of Inorganic Chemistry, Westfälische Wilhelms-Universität Münster,
D-4400 Münster (West Germany)

SUMMARY
   Mechanically highly stable and electronically well conducting carbon fibers
can be transferred by electrochemical activation into host matrices, which are
able to accomodate electroactive sensing material either in pores or in hydrophilic surface films of chemically modified carbonaceous material. The preliminary results obtained with modified carbon fibers to a silver/silver chloride
reference electrode are the subject of this paper.

INTRODUCTION
   A large number of ion selective electrodes with solid internal contact not
containing any internal reference system have been reported. The most known
type of these microelectrodes is the "Selectrode", developed by Ružička (ref.
1). These electrodes are large in dimension and have a high electrical resistance varying from 10 kΩ to 10 MΩ. It is possible to design microelectrodes
for biomedical applications of very small dimensions and low electrical resistance in which the electronic conductor is a carbon fiber. Highly oriented
carbon fibers ("graphite fibers") are characterized by extreme stiffness, exceeding that of steel. The typical diameter of commercial fibers is ca. 8 µm;
this diameter may be reduced to less than 1 µm by air oxidation (Fig. 1) in a
temperature gradient (ref. 2). It is known that the modification of carbon
fibers can be achieved by using intercalation methods or intensive oxidative
surface treatment. The problem is to adjust the pretreatment procedure in such
a way that the mechanical stability will not suffer. It is the aim of the pretreatment to attack not only the "edge" but also to open the "basal plane"
carbon atoms, which provides an expanded electroactive surface. High current
density anodic oxidation in aqueous electrolytes is the preferred method to
produce thick layers of oxygen containing functional groups on carbon fibers.
Repetitive oxidation/reduction current pulses promote the formation of uni-
form layers of "oxides" (ref. 3). These "oxides" have interesting properties:

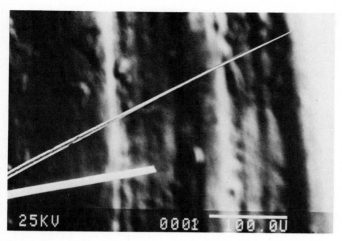

Fig. 1. SEM micrograph of thermally pointed and original carbon fibers.

(i) in spite of the high degree of functionalisation the oxides are still electronic conductors - this is because the anodic oxidation is "automatically attenuated" - in areas where the electronic conductivity decreases due to overoxidation;

(ii) the oxides are hydrophilic and microporous and can be soaked with any polar liquid;

(iii) the oxides have surprisingly high cation exchange capacity (ca. 2 meq/g C) (ref. 4).

The combination of these three properties allows the insertion of the sensing material not only on the surface of the carbon fiber but also inside the carbon fiber structure. The advantage of carbon fiber microelectrodes over conventional potentiometric glass capillary microsensors (ref. 5) is the noticable lower internal resistance. Therefore, these low-impedance carbon fiber based microelectrodes might be especially useful when fast processes have to be monitored.

The hydrophilic and microporous surface zone of carbon fibers can accomodate the reactive components of a typical reference electrode, e.g. Ag/AgCl/ KCl(aq). The preparation and performance of this electrode will be shown.

EXPERIMENTAL

For this study, single carbon fibers (Celion GY 70, Celanese Corp.) of 8 μm diameter were placed within platinum plates to form a good electrical contact. The electrochemical preparation of these electrodes was done by galvanostatic or potentiostatic methods using a potentiostat WENKING ST 72. A platinum plate, placed perpendicular to the carbon fiber, was used as the auxiliary electrode.

For galvanostatic preconditioning the following pulse train current was applied for 24 minutes in 1 M Na_2SO_4 using a WAVETEK MODEL 142 function generator: pulse height: 16.5 μA/g C; pulse duration: 60 ms; dc bias: - 2.75 μA/g C; period: 90 ms.

For the potentiostatic preconditioning the following voltage pulse train was applied for 3 minutes in 0.01 M H_2SO_4 using the same function generator (WAVETEK): pulse height: 2.6 V; pulse duration: 70 ms; dc bias: - 0.3 V; period: 80 ms. After these preparations, the carbon fiber was placed in deionised H_2O for 2 h and stored in 0.1 M $AgNO_3$ for about 20 h at room temperature in the dark. Retention of silver in the fiber was achieved by repeated dipping in a solution of 7% PVC powder and 7.5% p-nitrophenyloctylether in tetrahydrofuran. The amount of silver was estimated by cyclic voltammogram.

After pretreatment of the fiber exchanged silver was converted to metallic silver by applying a constant current of - 0.18 mA/μg C for 40 s in 3 M KCl and subsequently partly oxidized to silver chloride in 0.01 M HCl saturated with AgCl at a constant current of 1.6 μA/μg C for 40 s.

To complete the reference system an Agar-Agar matrix was used, which was produced from 1% Agar-Agar solution in 3 M KCl saturated with AgCl and silanized with Prosil-28 (4%).

RESULTS AND DISCUSSION

In our work we have treated the carbon fiber as a conducting matrix and investigated the effect of electrochemical modification of the basic carbon fiber structure for its suitability as a reference microelectrode. As expected the potentiostatic preparation of the carbon fiber yielded carbon "oxides", characterized by cation exchange properties. The cyclic voltammogram of potentiostatically pretreated carbon fiber, exchanged with Ag^+, washed and coated with PVC as described above, is shown in Fig. 2.

Peak (1) corresponds to the reduction of still adherent Ag^+ which, of course, forms insoluble AgCl in the KCl electrolyte. Peak (2) is the reduction of Ag^+ in the oxide layer and peak (3) is the irreversible reduction of these surface oxides. The metallic silver on a carbon fiber, almost free of electrochemically carbon "oxide" groups is reoxidised to silver chloride in peak (4). During the second cathodic sweep shown in Fig. 1b, the rereduction of this AgCl to metallic Ag occurs in peak (5) and peak (6) is the same as peak (4) but slightly decreased. Peaks (5) and (6) are reversible. They verify the reference electrode character of the single modified carbon fiber.

In further study the carbon fiber was pretreated by applying the galvanostatic oxidation/reduction pulse method. After Ag^+ from $AgNO_3$ solution was

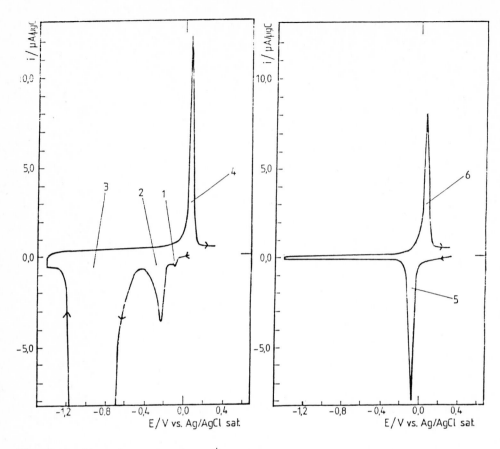

Fig. 2. Cyclic voltammogram of Ag^+ charged carbon fiber in saturated KCl; scan rate: 25 mV/s vs Ag/AgCl(sat.).

incorporated into the carbon fiber structure the reduction to metallic silver followed by oxidation to AgCl was carried out by a constant current as previously described.

The electrode was stored in 3 M KCl solution saturated with AgCl between measurements. The equilibrium potential of the carbon fiber silver/silver chloride electrode was obtained by measuring the potential difference of this electrode E1 and the standard reference electrode Ag/AgCl/3 M KCl (Metrohm) E2=211.5 mV at 20 °C (vs NHE) in 3 M KCl. The potential difference E1-E2 was in the range of -1 to -3 mV. The measurement of exchange current in the close vicinity of the equilibrium potential forms an linear plot shown in Fig. 3. The transfer of the current which flows across the carbon fiber/solution interface is high, so that no significant distribution of equilibrium occurs and, therefore, classifies this electrode as a nonpolarized electrode.

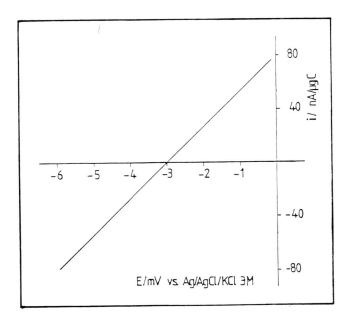

Fig. 3. Micropolarization test for electrode reversibility in 3 M KCl vs Ag/AgCl/3 M KCl; scan rate: 2 mV/s.

ACKNOWLEDGEMENTS

The authors express their thanks for the financial support of the Deutsche Forschungsgemeinschaft.

REFERENCES

1   J. Ružička and C.G. Lamm, Anal. Chim. Acta, 54 (1971) 1.
2   J.O. Besenhard, A. Kurtze, R.F. Sauter, M. Josowicz, H.-D. Liess, K. Potje,
    Proc. Carbon '86, Baden-Baden, Fed. Rep. of Germany, June 30 - July 4,
    Deutsche Keramische Gesellschaft, Bad Honnef, 1986.
3   E. Theodoridou, A.D. Jannakoudakis, J.O. Besenhard, R.F. Sauter,
    Synth. Met., 14 (1986) 125.
4   E. Theodoridou, A.D. Jannakoudakis, R.F. Sauter, J.O. Besenhard,
    Proc. Carbon '86, Baden-Baden, Fed. Rep. of Germany, June 30 - July 4,
    Deutsche Keramische Gesellschaft, Bad Honnef, 1986.
5   H. Freiser, Ion selective electrodes in analytical chemistry, Vol. 1 and
    Vol. 2, Plenum Press, New York, 1978 and 1980.

M.R. Smyth and J.G. Vos

Electrochemistry, Sensors and Analysis

© Elsevier Science Publishers B.V., Amsterdam — Printed in The Netherlands

USE OF POTENTIOMETRIC AND POLAROGRAPHIC METHODS IN THE STUDY OF
N N' N" N"' TETRAKIS (2-HYDROXYPROPYL) ETHYLENE DIAMINE (EDTPA) AS A CHELATOR
IN ANAEROBIC TYPE ADHESIVES

M. C. BRENNAN AND R. G. LEONARD

Loctite (Ireland) Limited, Whitestown Industrial Estate, Tallaght, Dublin 24,
Ireland.

ABSTRACT

Potentiometric titration was used to quantitatively determine the
solubility of chelators in anaerobic adhesives and this result was confirmed
by stability experiments utilising the Arrhenius relationship. Polarography
has shown that EDTPA forms a predominant 1:1 complex with copper(II)which is
useful in estimating the quantity of chelator to add to such products in
order to chelate the expected level of contaminating metal. Similar studies
could be applied to other metals.

INTRODUCTION

The well known and accepted chelator EDTA is used extensively as a
complexone and its practical value lies in its ability to form stable water
soluble complexes with metal ions. In certain organic chemical industries,
however, such a chelator is ineffective because of its poor solubility,
particularly in methacrylate/acrylate based anaerobic type sealants and
adhesives. This problem has prompted a careful investigation to find a
suitable substitute. The chelator, when formulated into such adhesives, must
meet the following criteria: (i) be soluble; (ii) chelate the contaminant
metals; (iii) deactivate the metals; (iv) not interfere with the chemistry
of the product (refs. 1-3).

The presence of metals in such products may derive from the manufacturing
process or from raw materials, and trace levels may have pronounced adverse
effects on the final product quality. Such problems may be avoided if the
concentration of the free metal is reduced to a level so that their influence
is minimised or they are effectively chelated so as to render them harmless.

In this investigation, four potential chelators; EDTA Na_4 (ethylenediamine tetraacetic acid tetrasodium salt), HEDTA Na_3 (N-hydroxyethylene-
diamine triacetic acid trisodium salt), EDTPA (tetrakis-(2-hydroxypropyl
ethylenediamine) and DETPA Na_5 (diethylenetriamine pentaacetic acid penta

sodium salt), were carefully considered on the basis of their ability to deactivate copper in a typical sealant.

Potentiometric titrations were used to determine the solubility of the chelators in this matrix,and differential pulse polarography was used in a study to confirm that complexation of the copper (II) was occurring and to investigate the stoichiometry of the reaction (refs. 4-7).

METHODS
Potentiometric Titrations

Potentiometric titrations were carried out at a platinum electrode using a Metrohm E536 potentiometer in conjunction with an E535 dosimat and a calomel reference electrode.

Ten gram samples of a typical anaerobic sealant containing the appropriate chelator at the 0.05% (w/w) level were diluted with 40 cm³ of an ethanol/water (90/10) mixture. To this solution was added 2 drops of phenolphthalein solution and an aliquot of 1M Na OH to make the solution basic. Then HCl (1M) was added until the pH of the solution was approximately neutral. Dichloroacetic acid buffer (10 cm³) [2M chloroacetic acid adjusted to pH 3 with 1M sodium acetate] was then added and the solution titrated with ferric chloride (2 x 10^{-3}M). The concentrations of the chelator in ppm in the adhesive were then calculated from:

$$\text{ppm Chelator} = \frac{(Ts - Tb) \times M(Fe) \times 1000}{\text{Weight of Sample}}$$

Ts = Sample titre, Tb = blank titre
M(Fe)= Molarity Fe Cl_3

The determination of shelf-life of a typical anaerobic formulation

The test for determining the shelf-life of anaerobic adhesives is based on the increase in viscosity of the adhesive with increasing temperature. By using the Arrhenius relationship (equation 1), the life of the adhesive may be predicted with reasonable accuracy. The assumptions are made that, in the heterogeneous chemical mixture, viscosity increases result from diffusion - influenced chemical reactions, that the activation energies of these reactions are relatively independent of temperature and that the rate of viscosity change with temperature follows this mathematical relationship.

$$\ln k = \ln A - \frac{Ea}{RT} \qquad \text{(Equation 1)}$$

k = specific reaction rate
A = frequency factor constant
Ea = activation energy
T = absolute temperature
R = gas constant

Samples were stored at room temperature, 77°F and at two elevated temperatures t_1 = 120°F and t_2 = 160°F. Viscosities were measured periodically using the HAAKE "Falling Ball Viscometer". A maximum viscosity was selected as a "cut off" point (Vc) in predicting stability. This was usually the point of minimum acceptable extrusion rate. The rate of increase in viscosity(k)at each temperature was plotted against $\frac{1}{T}$, and the straight line extrapolated for t_1 and t_2 readings to the reciprocal of 77°F. The rate was then estimated and the time to reach the "cut off" viscosity at 77°F calculated. This time is the "estimated shelf life".

Polarographic Measurements

Polarographic measurements were carried out in the differential pulse mode using a PAR Model 384 Polarographic Analyser in conjunction with a PAR Model 303 SMDE. A series of copper-EDTPA mixtures were prepared in 0.1M sodium acetate buffer (adjusted to pH 6.6 with glacial acetic acid) and scanned from +0.100 to -0.225 vs Ag/Ag Cl. Each solution was degassed for 100 seconds, the drop time was 0.5 seconds, the pulse height was 50 mV and the scan rate was 8 mVs^{-1}.

RESULTS AND DISCUSSION
Potentiometry

The results in Table 1 show the concentration of chelators (in ppm) determined by potentiometry remaining in the sealant formulation after various sample treatments.

Sample/Treatment	EDTA	DETPA	HEDTA	EDTPA
1. Added ppm	636	739	936	600
2. Stirred 0.5 hours - sampled immediately	636	739	826	610
3. After 0.5 hours standing supernatant liquid taken	19	0	110	602
4. As above after 4 days standing	27	30	175	610
5. After 5 days standing and centrifuging at 3000 rpm for 20 minutes	42	25	180	596

Table 1.   Effect of various sample treatments on concentration of chelators (in ppm) in a typical adhesive formulation.

Acceptable precision was only obtainable (with this method) above 40 ppm and concentrations lower than this gave very large standard deviations between several titrations.

These results demonstrate that EDTA and DETPA concentrations in this anaerobic formulation are very low on the basis of solubility. Therefore, the DETPA was not considered further.

Shelf-life Studies

A study was then made of the effectiveness of HEDTA, EDTPA and EDTA in removing the interference of Cu(II) ions from a typical adhesive formulation. The results of this study are shown in Tables 2 and 3.

ppm Chelator added	HEDTA		EDTPA		EDTA	
	Fresh	3 days standing	Fresh	3 days standing	Fresh	3 days standing
500	550 days+	550 days+	600 days+	600 days+	600 days+	600 days+
400	550 "	550 "	600 "	600 "	600 "	600 "
300	550 "	550 "	600 "	600 "	600 "	600 "
200	550 "	550 "	600 "	600 "	600 "	600 "
100	550 "	550 "	600 "	600 "	600 "	600 "
50	550 "	550 "	600 "	600 "	600 "	600 "

Table 2.   Results of Stability Studies Without copper of HEDTA, EDTPA and EDTA.

ppm Chelator added	HEDTA		EDTPA		EDTA	
	Fresh	3 days standing	Fresh	3 days standing	Fresh	3 days standing
500	550 days+	550 days+	600 days+	600 days+	600 days+	85 days+
400	550 "	400 "	600 "	600 "	600 "	85 "
300	550 "	180 "	600 "	600 "	600 "	80 "
200	165 "	100 "	480 "	400 "	600 "	40 "
100	61 "	29 "	325 "	205 "	600 "	40 "
50	14 "	20 "	85 "	60 "	600 "	40 "

Table 3. Results of Stability Studies <u>With</u> Copper of HEDTA, EDTPA and EDTA.

In Table 2, the stability studies indicated that all formulations were similar and do not seem to interfere with the chemistry of the cure system. However, in Table 3, the formulations were 'spiked' with 0.5 ppm copper(II) when fresh, and after three days standing at room temperature. (NOTE: in the case of the solutions aged for three days only, the supernatant liquid was tested). Both the HEDTA and EDTA did not deactivate the copper to any appreciable extent.

The EDTPA was satisfactory at 230-300 ppm, maintaining shelf-life similar to "no added" copper study. The fact that HEDTA and EDTA did not chelate the copper on the aged sealant confirmed the poor solubility of these chelators as shown by potentiometric titration.

Polarographic Study of EDTPA and Copper

The N N' N" N''' tetrakis-(-2 hydroxypropyl) ethylenediamine (EDTPA) has so far shown that it maintains good stability at low concentrations and may be considered as a potential chelator. The aim of the polarographic study was to determine the number of chelate ligands that are attached to each copper(II) ion using Job's method of "continuous variations". This method involves a measurement of the peak potential (Ep) for a series of solutions of copper(II) and EDTPA prepared according to Job's method (ref. 8). The advantage of this technique is that both the complexed and uncomplexed copper ion can be plotted on an X-Y recorder (Figure 1).

174

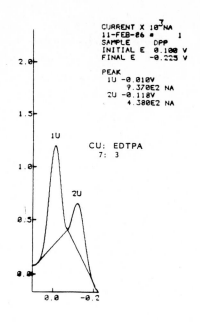

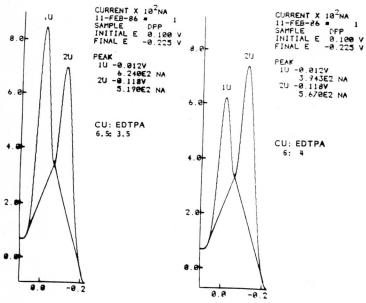

Fig. 1. Selected Polarograms from Cu-EDTPA Study.

A plot of the mole ratio against the peak currents for each of the
complexed peaks is shown in Figure 2   The intercepting point of maximum peak
potential for the copper - EDTPA complex shows that a stable complex is
formed having a molar ratio of 1:1.   This indicates the existence of a 1:1
copper - EDTPA complex, possibly with an octahedral structure.

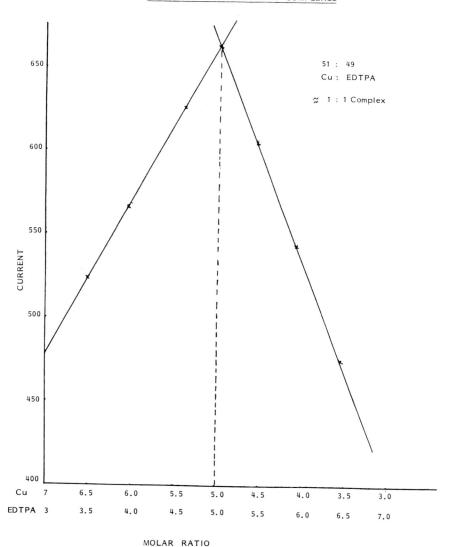

Fig. 2.   Job's plot for complexation of Cu(II) and EDTPA in adhesive
formulation.

CONCLUSION

The EDTPA was shown to be the most soluble of the four chelators considered by potentiometry and no detectable chemical interference in the anaerobic adhesive was observed. It was shown by the shelf-life studies to be able to deactivate the "spiked" copper in the same adhesive. Polarography has shown that the chelator forms a 1:1 complex with copper(II)ion and the peak potential for the complex shifted to a more negative potential.

REFERENCES

1. Industrial Adhesives and Sealants. Edited by B. S. Jackson, Hutchinson Book Company.
2. Adhesives Materials, Their Properties and Usages by Irving Katz, Foster Publishing Company.
3. Adhesive Bonding, Charles V. Cage, McGraw-Hill Book Company.
4. The Sequestration of Metals [Theoretical Considerations and Practical Applications], Robert L. Smith.
5. Metal Chelation - Principles and Applicaitons, Colin F. Bell.
6. Inorganic Complex Compounds - R. Kent Murmann.
7. Modern Approach to Inorganic Chemistry - Bell and Lott, Second Edition.
8. Ann. Chim. (Paris) 9, 113 (1928); 6, 97 (1935) Job P.

BIOELECTROCHEMISTRY

ELECTRON TRANSPORT MECHANISMS AND ENERGY TRANSDUCTION IN MITOCHONDRIA

F. H. MALPRESS

Department of Biochemistry, Queen's University, Belfast, N. Ireland

SUMMARY

Mitochondrial energy transduction may be seen as a threefold process. The release of energy by electron flow from more electronegative to more electropositive redox couples, the conversion of this electromotive force into a protonmotive force, and the use of the protonmotive force to cause activating conformational changes in the membrane-bound enzyme which mediates ATP synthesis. The first two stages are considered here, and a coulombic alternative is given to the orthodox chemiosmotic presentation of the p.m.f.

The subject under consideration is a problem of bioenergetics, grounded in electrochemistry. Various aspects have been admirably treated in a number of recent reviews (refs. 1-5), from which it will readily be seen that there are many points of conflicting interpretation and that the evidence does not always lead to agreement. To bring coherence to the totality of molecular events that accompany the gross, membrane-located reaction:

$$NADH + H^+ + \tfrac{1}{2}O_2 + 3P_i + 3ADP \longrightarrow NAD^+ + 3ATP + 4H_2O \qquad (1)$$

requires an understanding of multiple integrated mechanisms of great intricacy. In contrast, an overall thermodynamic assessment of the reaction is simple. We can resolve eqn. (1) into two parts, an oxidative reaction, producing energy:

$$NADH + H^+ + \tfrac{1}{2}O_2 \longrightarrow NAD^+ + H_2O + energy \qquad (2)$$

and a chemical synthesis, requiring energy:

$$3ADP + 3P_i + energy \longrightarrow 3ATP + 3H_2O \qquad (3)$$

The oxidation links two redox couples ($NADH/NAD^+$, H_2O/O) for which the E_m values are approximately -0.3V and +0.8V, a potential gap of 1.1eV. Since we know that 1eV has an energy equivalent of 1.6×10^{-22}kJ, and that two electrons are transferred for every molecule of pyridine nucleotide oxidized:

$$NADH \longrightarrow NAD^+ + H^+ + 2e^- \qquad (4)$$

180

we may calculate the energy released by the oxidation of 1gmol NADH when molecular oxygen is the ultimate acceptor of reducing equivalents: it is 211kJ. An estimate of the free energy requirement for the synthesis of 1gmol ATP (eqn.3) under in vivo conditions, is about 60kJ. With efficient conversion we might anticipate the synthesis of three molecules of ATP for every NADH molecule oxidized; this expectation is realised experimentally.

The problem of mechanism therefore is how to convert chemical bond energy presented in the form NADH, into the chemical bond energy of three terminal pyrophosphate bonds of ATP (eqn.1). Much early work designed to detect this transfer through the agency of chemical intermediates proved abortive and will not be considered here, for all recent evidence suggests that we have to regard electrical forces at the membrane as the true mediating influence between e^--flow and synthesis. Figure 1 shows, in skeletal form, the pattern of electron transfer between the redox couples of the mitochondrial inner membrane ranged in their sequence of increasing electrical potential. It illustrates several points of interest: (i) the components of the electron transport chain can be conveniently sub-fractionated into four complexes, three of which have a potential span commensurate with the synthesis of one ATP molecule for the passage of two electrons; these syntheses have been demonstrated (refs.6-8). Complex II (not shown in Fig.1) is non-phosphorylating, it brings reducing equivalents from succinate into the main scheme at the point of coenzyme Q. Its mechanism resembles that of Complex I (Hydrogen transfer through a flavin nucleotide, e^--transfer via non-haem iron groups) but it shows no proton-pump activity (see (v)). (ii) H-atom transfer is mediated by nucleotide or quinonoid molecules as a $2e^-$ movement, though as seen in Fig.1 the oxidation of the hydroquinone, QH_2, as the substrate of coenzyme Q cytochrome c reductase is a complicated process involving the re-cycling of one electron through the cytochrome b doublet (ref.9) (iii) simple electron

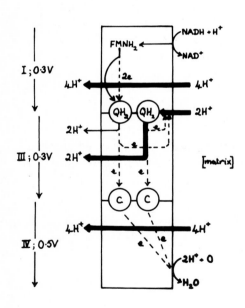

Fig. 1. Electron (- - -), H^+ (———) and H^+-pump (heavy line) pathways for the phosphorylating complexes of the respiratory chain. Individual carriers are not identified except FMN, coenzyme Q and cytochrome c.

transfer takes place between metal-protein centres as a $1e^-$ movement:

$$M_1^{2+} + M_2^{3+} \longrightarrow M_1^{3+} + M_2^{2+} \tag{5}$$

The metal is usually iron present in the haem porphyrin-ring structure (the cytochromes) or in labile thiopeptide association as non-haem iron, where it is liganded to cysteine or inorganic sulphur (nFe.nS), but in Complex IV (cytochrome oxidase) we find both haem iron (cytochromes a, a_3) and copper acting cooperatively. (iv) The movement of an electron from one metal centre to another must be an optimised activity if the transfer process is to be efficient. The most likely mechanism is by an e^--jump between nearly equipotential electronic orbitals of a donor and acceptor atom. It is a movement to be reckoned in terms of probabilities and will be largely dependent on the length of the jump. Williams (ref. 1) has suggested 15Å as a realistic estimate. Molecular proximity and the proper active site orientations of successive redox pairs thus become factors of first importance and offer a good practical reason for the location of electron transport chains within the confines of a membrane structure. Under a variety of influences: local electrical potential effects (ref. 1), micro-environmental changes following lateral diffusion (ref. 10) electrostatic forces between the apoproteins of reacting species (ref. 11), we may envisage electronic energy orbitals of a donor and an acceptor molecule adjusting to a point of common value where the free energy for an electron jump is momentarily reduced to near zero (ref. 12). The completed jump, which is the redox reaction, will destroy this predisposing condition. (v) In both Complex I and Complex III there is a stage at which H-atom transfer leads to the liberation of protons:

$$\text{Complex I.} \quad FMNH_2 + Fe_2^{3+}.S_2 \longrightarrow FMN + Fe_2^{2+}.S_2 + 2H^+ \tag{6}$$

$$\text{Complex III.} \quad QH_2 + (b^{3+})(c^{3+}) \longrightarrow Q + (b^{2+})(c^{2+}) + 2H^+ \tag{7}$$

At these sites therefore it is quite possible that energy transfer could become a property of the movement of protons, derived from substrate molecules and increasingly energized by e^--transfer through the sequence of haem or non-haem iron centres. In Complex I, for example, the potential gap between the flavin nucleotide and coenzyme Q is covered by 6-7 electron jumps, all of the form given in eqn. 5. Judged purely on the basis of paper redox reactions these substrate-H changes ($H \rightarrow H^+ + e^-$) are compensating within the system as a whole and summate to eqn. 1. The question arises whether they are suitable, or sufficient to become the transient vehicle of the energy exchange. Williams (ref.2) has offered proposals along these lines: 'It is the exchange of electrons between heavy metals and organic molecules carrying bound hydrogen which releases the proton into the membrane phase before its limited diffusion and passage through the ATP-synthase channel'. But although this may be a proper description for the derivation of protons from substrate-bound, or apoprotein hydrogen, a consideration of much greater significance is introduced by the correlated activity of all the ATP-synthesizing complexes as proton pumps. This switch of energy from an electromotive force to a protonmotive mode, not through the release of bound-H within the membrane, but by the movement of ionic-H^+ across the membrane, is now very widely accepted as the resolving transition in the energy transduction mechanism (Fig.1). It follows that proton movements must assume an importance equal to that of e^--flow in any proposed mechanism.

The evidence for proton pump activity is conclusive. It comes from experiments using the entire respiratory chain; from studies on single complexes separately observed within the membrane by the use of inhibitors and added electron 'sinks' of suitable E_m value (ref.13); or again, by the isolation of an individual complex and its incorporation in an artificial liposomal vesicle (ref.14). In all these cases the energization of the system by small pulses of an appropriate substrate in the presence of a permeant ion (which serves to prevent change in the membrane potential of the system), led to a demonstrable movement of protons into the bulk phase, which could be stoichiometrically related to the size of the pulse, that is to electron flow. There is as yet no final agreement on these $H^+/2e^-$ ratios, but the extensive studies from Lehninger's and from Wikstrom's laboratories (refs.15,16) exemplify the efforts being made and the difficulties encountered in this field of research. The demonstration that the pH of the mitochondrial matrix increases on energization (refs.17,18) confirms the transmembrane character of the pumps and raises an immediate problem: since ATP-synthase is placed at the matrical side of the membrane the utilization of any protonic force generated by the translocation of protons from the matrix to the bulk phase will require the return of protons through the membrane to the synthase concurrently and in equal measure. To probe further into these proton counterflows almost certainly confronts us with complex conformational changes in specific elements at the respective sites, and the opening and shutting of H^+-bonding peptide channels sensitive to such changes, on all of which our knowledge is rudimentary and speculative.

It is clear however that after proton translocation two conditions of charge separation could be present: H^+/e^- separation within the membrane and H^+/OH^- separation across the membrane. To these must be added a third, until recently completely disregarded, $H^+/fixed\ charge^-$ separation at the membrane surface (ref.19). Our judgement on which of these three is the significant development will determine our view on the nature of the protonmotive force in synthesis, for, at this point, charge separation has become the repository of our energy resource, and the proton the common factor within our choice. This choice

differentiates the four important contemporary views schematized in Fig.2;
they are radically distinct, especially so in regard to their electrical
characterization.

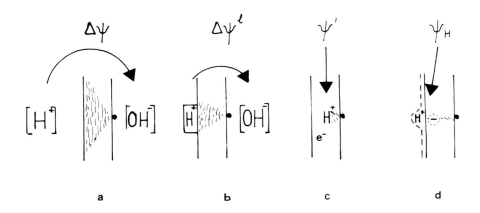

Fig.2. Mitochondrial energization. Four versions of the charge separation
and the electrical contribution to the protonmotive force: a, chemiosmotic
(transmembrane, delocalized, H^+/OH^-); b, chemiosmotic (transmembrane, localized, H^+/OH^-); c, intramembrane (activated proton, H^+/e^-); d, coulombic
(surface, highly localized, $H^+/fixed$ charge$^-$). Hatched areas: proton catchment;
•, ATP-synthase; – – – – – , limit of diffuse double layer.

A useful initial comparison may be made in terms of the thermodynamic
parameters which identify the functional force in each case:

a. chemiosmotic (delocalized), $\quad \Delta G = n\left[F\Delta\Psi - 2.3RT\Delta pH\right]$

b. chemiosmotic (localized), $\quad \Delta G = n\left[F\Delta\Psi^\ell - 2.3RT\Delta pH^\ell\right]$

c. intramembrane, $\quad \Delta G = n\left[F\Psi' + \Delta(\Delta G^\circ)\right]$

d. coulombic, $\quad \Delta G = nF\Psi_H$

ΔG is the energy required for the synthesis of 1gmol-ATP, n the H^+/P ratio
and F equals $96kJmol^{-1}V^{-1}$. The electrical forces arising from energization
are: delocalized ($\Delta\Psi$) and localized ($\Delta\Psi^\ell$) transmembrane potentials, an
intramembrane potential (Ψ') and a surface electrostatic potential (Ψ_H)
respectively.

A very confident consensus in favour of the chemiosmotic form 'a' was evident up to 1980. Since that time more attention has been paid to the weaknesses of this hypothesis and to the possible merits of alternative formulations of the protonmotive force. Much of the credit for this liberalization of ideas must be given to a review article written by Kell (ref.20), outlining an 'electrodic view' of electron transport phosphorylation. It collated and amplified a great deal of evidence which had been accumulating in a rather piecemeal and ineffective way, against the idea of a bulk-phase electrochemical force and in favour of a similar force localized at, or near, the membrane surface (ref.21). The new concept was also chemiosmotic (Fig.2, type b) and embraced a number of variant presentations which differed in the reasons offered for localization; for example, surface fixed charge attractions (refs.21-23) or spaced coupling units operating as proton microcircuits (ref.24), though this latter view has recently been held in question (ref.25).

Less precision can be given to the forces involved in type 'c', in which e^--transfer leads to an intramembranal activation. In this model a specific role is given to molecular energy changes, $\Delta(\Delta G^\circ)$, which might supplement the protonic force in promoting synthesis; a purely conformational, or proton-independent 'collision' hypothesis of energy transduction (ref.26) would presumably depend entirely on this molecular term.

The coulombic hypothesis (Fig.2, type d) brings a new solution to the problem. Its distinguishing characteristics are the acceptance of the experimentally proven increase in negative fixed charge formation, which e^--flow produces on the inner membrane surface (ref.27), as a primary event of mitochondrial energy transduction; and the conviction that this activity cannot be divorced from the second primary event: proton translocation (refs.19, 28-32).

The writer had earlier noted the importance of fixed charge formation but had seen its function as the surface localization of an electrochemical force (refs.21,22). For two very persuasive reasons the possibility of an electrostatic force had been discounted: first, the literature at the time, and to this day, was full of references to the actual measurement of a transmembrane electrochemical force of full synthetic capability (ca.220mV); second, the increase ($\delta\Psi_0$) in the surface potential when mitochondria were energized was always of the order 5 - 10mV. Since it would be this increase that would determine the field effect upon any proton in the aqueous phase, it seemed quite inadequate to generate any useful protonic force through charge-charge interaction.

Nevertheless, the appearance of negative fixed charges and positively charged protons at the same time and, in all likelihood, in very close proximity, since both are the result of e^--flow through the phosphorylating complexes, presented tantalizing possibilities for energy conversion. Three considerations led ultimately to a hypothesis based on electrostatic forces which, in the writer's view, offers a tenable and distinctive alternative interpretation of proton-motive activity. First, all measurements of an electrochemical force, and indeed all the basic evidence upon which such a force appears to rest, come from grossly artificial experiments. This contention has been treated elsewhere at length (ref.28) and the arguments will not be repeated here. The conclusion, however, is that an enhanced electrochemical force active in synthesis, whether considered as localized or delocalized, simply does not exist. It is a 'phantom' parameter, a consequence under the conditions customarily used for its detection and measurement, of the substitution of permeant cation binding for proton binding at the fixed charge sites. The experimental procedures disrupt the natural event and provide no basis for deductions about <u>in vivo</u> mechanism. The connection between the 'phantom' and the coulombic energy parameters ($\Delta\tilde{\mu}_H = f\Psi_H$) has also been discussed (refs.27,28). Secondly (Fig.3), all measurements of $\delta\Psi_0$ refer to the whole inner membrane surface. This could represent the resultant potential of a very small energized area ($\delta\Psi_0 = 220$mV), surrounded by a much larger neutral area upon which e^--flow produces no effect. If this is so the experimental data indicate that only about 2-3 per cent of the surface should be regarded as truly energiz-able. Thirdly, a coulombic force of 220mV between unit-ary positive and unitary negative point charges in a medium of permittivity 20, will be present when the distance between the charges is about 3Å. This is a mere fraction of the full extent of the diffuse double layer in the presence of such a surface force.

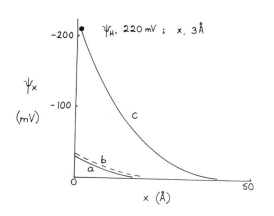

Fig.3. Potential profiles for the diffuse double layer [$\Psi_x = \Psi_0 \exp(-x/\lambda)$ where λ is the Debye length. $\Psi_0 \simeq \zeta$-potential].
a, non-energized ($\Psi_0 \simeq -35$mV)
b, energized (whole membrane, $\Psi_0 \simeq -40$mV)
c, energized (fixed charge sites, $\Psi_0 \simeq -220$mV)

A great advantage of the coulombic approach is that by giving an estimate of fixed charge formation it allows a quantitative treatment of the protonmotive force on a molecular and time-structured basis. The procedure for these calcul-ations has been published (ref.30) and supplemented with a later critical emendation (ref.32). It gives a picture of the kinetics of the force far removed

186

from the generalised protonic impulsion which characterizes the chemiosmotic hypothesis or its variants. For the latter the transmembrane force is a cumulative force of protons maintaining variable high charge ($\Delta\Psi$) and osmotic (ΔpH) differentials across the membrane at all times; after translocation the protons become a 'population'. In complete contrast the coulombic hypothesis finds that every proton is unique; it is translocated, held and then utilized by its paired fixed charge at one phosphorylating site, from which it never escapes. Fixed charges are continually regenerated through a 'fixed charge cycle' comprising, in a minimal model, three interchanging forms: a (neutral), b (charged) and c (H^+-associated). The H^+/fixed charge$^-$ ratio is one (ref.32). A problem arises in that we do not yet know the charge-carrying capacity of the fixed charge molecules at the phosphorylating sites. This, however, affects only the quantitative and not the qualitative appraisal of the mechanism. In Fig.4 single and triply charged forms (b^{1-} and b^{3-}) have been used to illustrate time relationships. The 'life' of a translocated proton must be of an order commensurate with the duration of the fixed charge which it neutralises. For the b^{3-} form this 'blink' of manifest charge is about 18ms; it is succeeded at any given site by a period lasting about 350ms when no charge is present (forms a and c). If we therefore consider one energized zone with its three phosphorylating sites, a translocated proton might be detected for no more than 15 per cent of any time span.

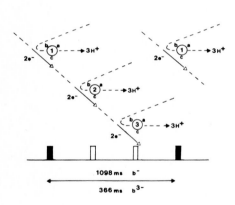

Fig.4. Time-related events in one energized zone during one repeat period (H^+/2e$^-$ = 3). ① ② ③, fixed charge cycles at Complexes I, III and IV respectively. See text for further explanation.

These relationships may be deduced from the blink-pattern shown as rectangles at the bottom of Fig.4, where the 'blinks' at site 1 (Complex I) have been filled to emphasize the repeat period at one site. For a putative b^{1-} form each blink of the pattern would involve three revolutions of the fixed charge cycles; the repeat period would be three times longer, but there would be three times as many energized zones (refs.30,32).

Whatever mechanism of energy transduction is thought correct - chemiosmotic or coulombic - the same number of protons will be used for the same amount of synthesis. It is therefore interesting to note that Westerhoff and Chen (ref. 33) have recently given a theoretical treatment which attempts to explain some of the anomalies encountered in chemiosmotic theory on the assumption that, for a localized version of the hypothesis, the mean number of protons present at any time at any site, is small. In the writer's view, while this conclusion flirts with the truth, the authors' argument is vitiated by its prior acceptance of the protonmotive force as electrochemical. This has been the perpetuated assumption in transduction studies since Mitchell first propounded his hypothesis in 1961, but it is certainly not the only interpretation the evidence allows.

In 1966 (ref.34) Mitchell gave a summary of four basic postulates for the chemiosmotic hypothesis. These seemed to match and to be satisfied by the observed experimental data of mitochondrial energization. The postulates were therefore highly prejudicial in favour of the chemiosmotic view. Unfortunately the omission of one further experimental datum, unrecognised in 1966 - the generation of negative fixed charges by electron transport chain activity - has now undermined this concurrence of chemiosmotic requirements and experimental actualities; the authority of the hypothesis is weakened accordingly.

The prediction that the basic postulates might be used '--- as the target for critical experiments designed to show that the chemiosmotic hypothesis may be untenable' (ref.34), has been realised now for many years through fixed charge evaluations (refs.27,28). Any hypothesis has now to face five, not four, basic postulates, or at least show good reason why the fifth postulate should be ignored. Until this is done the chemiosmoticists' repeated claims to authority in the field of energy transduction (e.g.ref.35) must be judged uncritical at best and very possibly entirely fallacious.

REFERENCES

1   R.J.P. Williams, The multifarious couplings of energy transduction, Biochim. Biophys. Acta, 505 (1978) 1-44.
2   R.J.P. Williams, The symbiosis of metal and protein functions, Eur. J. Biochem., 150 (1985) 231- 248.
3   P. Mitchell, The correlation of chemical and osmotic forces in biochemistry, J. Biochem., 97 (1985) 1-18.
4   M. Wikstrom, K. Krab and M. Saraste, Proton-translocating cytochrome complexes, Ann. Rev. Biochem., 50 (1981) 623-655.
5   Y. Hatefi, The mitochondrial electron transport and oxidative phosphorylation system, Ann. Rev. Biochem. 54 (1985) 1015-1069.
6   C.I. Ragan and E. Racker, The reconstitution of the first site of energy conservation, J. Biol. Chem. 248 (1973) 2563-2569.
7   E. Racker and A. Kandrach, Reconstitution of the third segment of oxidative phosphorylation, J. Biol. Chem. 248 (1973) 5841-5847.
8   A. Alexandre, F. Galiazzo and A.L. Lehninger, On the location of the H^+-extruding steps in site 2 of the mitochondrial electron transport chain, J. Biol. Chem. 255 (1980) 10721-10730.

188

9   P. Mitchell and J. Moyle, The role of ubiquinone and plastoquinone in chemiosmotic coupling between electron transfer and proton translocation, in G. Lenaz (Ed.), Coenzyme Q, John Wiley, New York, 1985, pp145-163.

10 C.R. Hackenbrock, Lateral diffusion and electron transfer in the mitochondrial inner membrane, Trends Biochem. Sci., 6 (1981) 151-154.

11 D.C. Rees, Electrostatic influence on energetics of electron transfer reactions, Proc. natl. Acad. Sci. U.S.A., 82 (1985) 3082-3085.

12 R.A. Marcus and N. Sutin, Electron transfers in chemistry and biology, Biochim. Biophys. Acta, 811 (1985) 265-322.

13 A. Villalobo, A. Alexandre and A.L. Lehninger, H^+ stoichiometry of sites 1 + 2 of the respiratory chain of normal and tumor mitochondria, Arch. Biochem. Biophys., 233 (1984) 417-427.

14 K. Krab and M. Wikstrom, Proton-translocating cytochrome c oxidase in artificial phospholipid vesicles, Biochim. Biophys. Acta, 504 (1978) 200-214.

15 A.L. Lehninger, B. Reynafarje, R.W. Hendler and R.I. Shrager, The H^+/O ratio of proton translocation linked to the oxidation of succinate by mitochondria, FEBS. Lett. 192 (1985) 173-178.

16 K. Krab, J. Soos and M. Wikstrom, The H^+/O ratio of proton translocation linked to the oxidation of succinate by mitochondria, FEBS Lett. 178 (1984) 187-192.

17 P. Mitchell and J. Moyle, Estimation of membrane potential and pH difference across the cristae membrane of rat liver mitochondria, Eur. J. Biochem., 7 (1969) 471-484.

18 M. Wikstrom, Pumping of protons from the mitochondrial matrix by cytochrome oxidase, Nature, Lond., 308 (1984) 558-560.

19 F.H. Malpress, Energetic consensus, Nature, Lond., 289 (1981) p335.

20 D.B. Kell, On the functional proton current pathway of electron transport phosphorylation. An electrodic view, Biochim. Biophys Acta, 549 (1979)55-99.

21 G.P.R. Archbold, C.L. Farrington, S.A. Lappin, A.M. McKay and F.H. Malpress, Proton translocation: the ρ-zone interpretation, Abstr. 11th. FEBS Meeting, Copenhagen, (1977) A4-10 p653.

22 G.P.R. Archbold, C.L. Farrington, S.A. Lappin, A.M. McKay and F.H. Malpress, Oxygen-pulse curves in rat liver mitochondrial suspensions, Biochem. J. 180 (1979) 161-174.

23 T.H. Haines, Anionic lipid headgroups as a proton-conducting pathway along the surface of membranes, Proc. natl. Acad. Sci. U.S.A., 80 (1983) 160-164.

24 H.V. Westerhoff, B.A. Melandri, G. Venturoli, G.F. Azzone and D.B. Kell, A minimal hypothesis for membrane-linked free-energy transduction, Biochim. Biophys. Acta, 768 (1984) 257-292.

25 J.F. Nagle and R.A. Dilley, Models of localized energy coupling, J. Bioenerg. Biomembr. 18 (1986) 55-64.

26 E.C. Slater, J.A. Berden and M.A. Herweijer, A hypothesis for the mechanism of respiratory chain phosphorylation not involving the electrochemical gradient of protons as obligatory intermediate, Biochim. Biophys. Acta, 811 (1985) 217-231.

27 F.H. Malpress, Some implications of fixed-charge formation during electron transport chain activity, Biochem. Soc. Trans., 12 (1984) 399-401.

28 F.H. Malpress, A coulombic hypothesis of mitochondrial oxidative phosphorylation, J. theor. Biol. 109 (1984) 501-521.

29 F.H. malpress, Local energized proton hypotheses of mitochondrial oxidative phosphorylation, J. theor. Biol. 109 (1984) 523-532.

30 F.H. Malpress, The coulombic hypothesis of mitochondrial energy transduction: an attempt to quantify relationships in the energized zones, J. theor. Biol. 111 (1984) 397-405.

31 F.H. Malpress, The coulombic hypothesis of mitochondrial energy transduction, Abstr. 13th. Internat. Congress Biochem. Amsterdam, 1985 p757.

32 F.H. Malpress, A coulombic hypothesis of mitochondrial energy transduction, Printed privately, Belfast (1985) 8pp.

33 H.V. Westerhoff and Y. Chen, Stochastic free energy transduction, Proc.
    natl. Acad. Sci. U.S.A. 82 (1985) 3222-3226.
34 P. Mitchell, Chemiosmotic coupling in oxidative and photosynthetic phosphor-
    ylation, Glynn Research Ltd. Bodmin, England. 1966.
35 P. Mitchell, Keilin's respiratory chain concept and its chemiosmotic
    consequences, Science, 206 (1979) 1148-1159.

ELECTROCHEMICAL ANALYSIS OF NUCLEIC ACIDS

J.-M. SÉQUARIS

Institute of Applied Physical Chemistry, Nuclear Research Center (KFA), Jülich,
P.O. Box 1913, D-5170 Jülich (Federal Republic of Germany)

SUMMARY
     Fundamental aspects of electrochemical analysis of nucleic acids <u>in vitro</u>
are reviewed in case of using sweep voltammetry at a stationary mercury electrode. This method provides quantitative information on hydrodynamic and
interfacial properties of DNA. Diffusion coefficient and size of conformational
changes of DNA subjected to various genotoxic agents (ultrasound and δ-irra-
diation, chemical methylation, Pt compounds interaction) can now be determined with the same technique.

INTRODUCTION

     Since pioneering work by Palećek (ref. 1), applications of electrochemical
methods to investigate nucleic acids structures have demonstrated a sensitivity
competing with classic spectroscopic methods (ultraviolet, cicular dichroism
spectroscopies) and hydrodynamic measurements (viscosity, sedimentation). However, a large choice of experimental parameters inherent in electrochemical
methods can discourage potential users. Furthermore, quantitative results which
may state qualitative information are sometimes lacking. Therefore it is attempted in this paper to review some fundamental aspects of the application
of the electrochemistry in nucleic acid research.

     The key role played by nucleic acids and especially deoxyribonucleic acid
(DNA) in delivering genetic information in living cells has incited investigation and elucidation for their physical and chemical structures. Thus X-ray
data have established that the DNA molecule is a double helix. The backbone
of each helix consists of repeating units of deoxyribose sugar and phophate.
The backbones are linked by hydrogen bonds between pairs of four kinds of
nucleic base; adenine (A) = thymine (T), guanine (G) ≡ cytosine (C). However,
various possible fluctuations from this solid rigid structure have been
evoked in solution under physiological conditions. Indeed, such effects as
temperature, electrolyte concentration, ligand interactions, and also more
complex interactions with proteins, may induce conformational changes of DNA
structure. Thus localized or transient conformational changes make it difficult to answer two basic questions: what is the organization of DNA in
chromosomes and how does DNA expression take place (ref. 2)? A second aspect

of DNA structural investigations considers the effect of genotoxic agents e.g. environmental chemicals, radiations and viruses with possible mutagenic and carcinogenic consequences. The understanding of their mechanisms of action at the molecular level would help to improve genetic safety. The approach to these different problems by electrochemistry is based on the measurement of accessibility of DNA residues to an electrode surface. Thus the adsorption of electrochemical markers of DNA by such nucleic bases is generally followed through their oxidation or reduction reactions. Recently the application of surface-enhanced Raman scattering (SERS) spectroscopy also permits their vibrational spectra to be recorded (ref. 3). Indeed conformational states of DNA can be characterized electrochemically by an accessibility of nucleic bases minimum for the native DNA in solution or its solid structure model B-DNA. In this structure the sugar/phosphate backbone forming the outer surface of the helix hinders sterically the contact of the surface electrode with the inner nucleic bases. Whereas, in the case of a single stranded (ss) DNA obtained by denaturation (heating, extreme pH values) the electrochemical signals are maximum. Unstacking of nucleic bases and rupture of hydrogen bonds render the nucleic bases residues totally free to interactions (ref. 1). It must be noted that the non-accessibility of nucleic bases in native DNA is an ideal case. Indeed,various solution and interfacial conditions may enhance fluctuations in helical configuration stability and increase the adsorption of nucleic bases. However, the two conformational states of reference, denatured and native, generally serve to quantify the conformational changes of DNA subjected to genotoxic agents.

ELECTROCHEMICAL SIGNALS

Depending on the nature of signals from adsorbed nucleic acids, various experimental conditions can be chosen (ref. 4). The nature of the signal, faradaic or non-faradaic, depends on whether or not an electron exchange has taken place. However, limited space in this paper does not permit a comprehensive review of all the used techniques to be given. Therefore in the case of faradaic oxidation signals,useful information is given by Brabec's papers (ref. 5).

In the present work, applications have been restricted to the utilization of the mercury electrode. Furthermore, results obtained with the hanging mercury drop electrode (HMDE), a stationary electrode, are only considered in order to standardize the method and to obtain quantitative results. Thus in case of sweep voltammetry the electrochemical response of DNA I (current intensity) can be thus generalized:

$$I = \text{function } (k, c, Fr., D^{1/2}, t_s^{1/2}) \tag{1}$$

where k includes experimental parameters (potential sweep rate, mercury surface area, nature of signals), c: total concentration of electrochemical markers of DNA, Fr: fraction of accessible markers in DNA (macromolecular parameter), D: diffusion coefficient of DNA and t_s: adsorption time at starting potential E_s. As adsorption of DNA proceeds irreversibly at the electrode surface, two types of voltammetric responses I can be thus discerned. For adsorption time t_s lower than the time required by saturation t_m of the surface $t_s < t_m$, I depends on $D^{1/2}$ and c x Fr. i.e. on hydrodynamic properties of DNA and accessibility of electrochemical markers; while for adsorption time $t_s \geq t_m$, I only depends on c x Fr. These two types of responses are experimentally obtained by plotting I against the square root of t_s. Thus for $t_s^{1/2} < t_m^{1/2}$, I varies linearly while for $t_s^{1/2} \geq t_m^{1/2}$, I is constant.

The nature of the voltammetric responses depends on pH. From pH 4 to 7 voltammetric signals correspond to the reduction of accessible adenine and cytosine residues in DNA. The electrode processes, governed by interfacial protonation of nucleic bases residues, are irreversible (ref. 6). The strongly adsorbed reduction products act

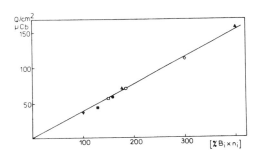

Fig. 1: Voltammetric charge Q responses for single stranded nucleic acids. E_s -0.4 V; 0.5 M McIlvaine buffer, pH 5.6; t_s 40 s; c 50 μg ml^{-1}; v 1 V·s^{-1}.
▽ poly C U (50/50); ■ denatured Micrococcus Lysodeikticus DNS (28 % AT); ▢ poly ACGU (24/26/22/28); ● denatured Calf Thymus DNA (56 % AT); △ poly AU (45/55); O poly ACU (32/27/41); ◇ poly AC (50/50); ▲ poly A at pH 5.9. Bi reducible nucleic bases, n_i electrons number (see also text)

194

as catalysts for hydrogen evolution. Kinetics of reduction of adsorbed DNA, such as the blocking film properties of reduced biopolymer, have been investigated elsewhere (ref. 7) with a potentiostatic double step-sweep method. In Fig. 1, the time integral of reduction peak in single sweep voltammetry, or charge Q, for various single-stranded nucleic acids is plotted as a function of percentage of reducible nucleic bases (B_i: adenine, cytosine) multiplied by the number of electrons involved in their reduction (ref. 8). With the assumption that the number of electrons is 4 and 2 respectively for adenine and cytosine moieties (ref. 9), it is possible at pH 5.6 to get a linear relationship. Furthermore, if a reduced monolayer for single-stranded polynucleotides such as denatured DNA is considered, an average surface concentration for mononucleotide units of 4×10^{-10} mole cm^{-2} can be calculated, which corresponds to a surface area of 42 $\overset{o}{A}{}^2$ per nucleotide. This value is confirmed by other investigations in the case of mono- and dinucleotides (refs. 10-11). The rather compact interfacial structure for nucleotide residues favours a perpendicular orientation of nucleic bases to electrode surface with the phosphate/sugar backbone oriented towards the solution. This specific adsorption of nucleic bases residues under low-charged interfacial conditions has also been observed by SERS spectroscopy (ref. 3).

From pH 7 to pH 9, voltammetric signals of low intensity correspond to interfacial double layer capacitance changes. At pH 8 (Fig. 2) the sweep voltammetric signals are of a non-faradic nature and are connected with adsorption and reorientation phenomena of components of adsorbed DNA at HMDE polarized cathodically. Depending on the native or denatured states of DNA in solution, specific desorption peaks characteristic of adsorbed double stranded, ds,

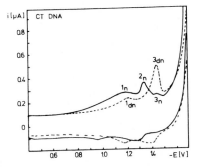

Fig. 2: Sweep voltammograms with nonfaradaic peaks of adsorbed denatured (--- 1_{dn}, 3_{dn}) and native (—— 1_n, 2_n, 3_n) CT DNA, c 75 μg ml^{-1}; E_s -0.4 V; t_s 120 s; 0.5 M McIlvaine buffer, pH 8; v 0.2 V·s^{-1}; T 25 °C

(peak 2) respectively and single stranded, ss (peak 3) forms can be shown
on the voltammetric responses. Although the nature of the more cathodic peak
3 has been assigned to a desorption of nucleic bases residues freely accessible after strand separation, the identity and particularly the conditions
of appearance of the peak 2 are questionable (refs. 12-15). However, from
the peak potential and current intensity dependence on the potential sweep
rate, ionic strength and pH, peak 2 can be assigned, in agreement with other
authors, to a rather slow process of desorption of nucleic bases moieties
from the double-stranded, ds, DNA structure. The broad peak 1 for native DNA
must involve various interfacial phenomena such as orientation of nucleic
bases and desorption of sugar/phosphate backbone.

HYDRODYNAMIC RESULTS
    The dependence of the reduction response of ss and ds DNA at pH 5.6 on
$t_s^{1/2}$ (ref. 6) has been used to determine diffusion properties of native and
modified DNA by genotoxic agents (ref. 13). Thus the diffusion coefficient
of DNA can be calculated from

$$\Gamma_t = 1.128 \cdot D^{1/2} \cdot c \cdot t \cdot s^{1/2} \tag{2}$$

$$D = 0.786 \cdot \Gamma_s^2 \cdot c^{-2} t_s^{-1} \tag{3}$$

At time of saturation $t_s = t_m$, $\Gamma m = 4 \ 10^{-10}$ mole$\cdot$cm^{-2}

$$D = 1.257 \ 10^{-19} \cdot c^{-2} t_m^{-1} \tag{4}$$

(4) is a numerical relation which gives D (cm$^2 \cdot$ s^{-1}) with c (mononucleotides·
cm^{-3}), t_m (s).
    In case of ds DNA this relation can only be applied at adsorption poten-
tial E_s -1.2 V (against SCE) when all the nucleic bases from DNA are equally
accessible (see below "Interfacial Results"). In Fig. 3, changes in diffusion
coefficient of DNA caused by ultrasonic and γ -ionizing radiations _in vitro_
have been measured. These variations have been plotted against main damages
as double-strand breaks (DSB) after ultrasonication and single-strand breaks
(SSB) after γ -irradiation. Thus the decrease in molecular weight of native
DNA subjected to ultrasound increases the diffusion coefficient more greatly
than the rise of flexibility due to the accumulation of SSB for low γ -irra-
diation doses (0 → 12 KRad). The direct quantitative estimation of D of
DNA by voltammetry must also be stressed in comparison with indirect methods

196

such as sedimentation techniques. For rapid
analytical purposes another method based on
alternative current voltammetry may be preferred. This method is based on the measurement of the time dependence of the electrical
properties on the mercury surface by the adsorption of nucleic acids (capacity current).
Thus relative variations of D_r for damaged
DNA have been used to test the reactivity
of genotoxic alkylating chemicals (ref. 16).

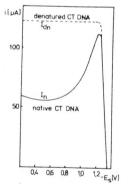

INTERFACIAL RESULTS

Fundamental

The knowledge of the adsorption behaviour of DNA at the electrode surface is
of crucial importance for the interpretation
of voltammetric results and their analytical
applications. Furthermore the charged mercury surface, whose double layer properties
are well characterized, is a suitable model
by simulating electrical and hydrophilic properties encountered in biological surfaces.
Indeed compaction of DNA by proteins into
chromosomes or transient opening of the helical structure by an enzymatic system to express genetic information are typical interfacial phenomena. Use of stationary electrodes
as HMDE thus permits the investigation of the
potential dependences (E_s) of contributions
from surface charges and interfacial structure
of water on conformation state of adsorbed DNA.
Under saturation conditions, $t_s \geqslant t_m$, the potential dependences of I are reported in Fig.
4 and 5. At pH 5.6, faradaic responses for ss
DNA are independent of E_s. While for ds DNA
voltammetric responses show minimal values
from E_s around the potential of zero charge

Fig. 3: Dependence of diffusion coefficients of sonicated 0 and γ-ir-
radiated + CT DNA on strand breaks
induced in native DNA (Mw 10 x 10^6)
per 10^6 daltons (see also text)

Fig. 4: Dependence of sweep voltammetric faradaic response on adsorption potential E_s at full coverage for native (—) and dena-
tured (---) CT DNA. c 75 $\mu g \cdot ml^{-1}$;
t_s 150 s; 0.5 M McIlvaine buffer,
pH 5.6; v 10 V$\cdot$s^{-1}

values at more cathodic potential values, E_s -1.2 V. At pH 8 (Fig. 5) a
comparable dependence for non-faradaic peak 3 characteristic of ss DNA
can also be observed with native (———)
and denatured DNA (---). Furthermore,
in case of native DNA, the opposite
potential dependence of peak 3 (ss
DNA) and peak 2 (ds DNA) also demonstrates an interdependence between
the two structures under interfacial
conditions. Thus it can be stated
that under adsorption conditions
around E_s -1.2 V, at highly negatively charged surface of about -15
$\mu C/cm^2$, the extent of accessibility
of nucleic base residues is close
for ss and ds DNA. Furthermore,
identical peak 3 in sweep voltammetric responses at pH 8 from E_s -1.2
V confirms that ds DNA undergoes a
conformational change into a single-
stranded structure. This observation,
first made by Valenta and Nürnberg
(ref. 17), has been confirmed under
various medium conditions (refs.
7, 13-15, 18,19) and at a carbon
electrode surface (ref. 5). The
effects of the surface electrical

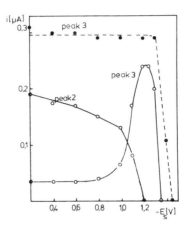

Fig. 5: Dependence of non faradaic
sweep voltammetric peaks for native
(———, 2_n, 3_n) and denatured (---, 3_{dn})
CT DNA on adsorption potential E_s at
full coverage. Other experimental conditions in Fig. 2

charges and the associated electric field gradient on the polyelectrolyte properties of DNA (ref. 15) on the permanent dipoles of nucleic bases can be
evoked to explain interfacial reorganization also accompanied by water release.
Indeed, modification of hydration layer of DNA stabilized by cations is a
prerequisite for access to nucleic base residues. It must be noted that the
disappearance of signals at more negative adsorption potentials E_s than -1.2 V
at pH 8 (Fig. 5) is due to the non-direct adsorption of DNA in the presence of
a surface layer of high-polarized water dipoles (hydrophilic surface). At
pH 5.6 (Fig. 4) the non-adsorption potential region of DNA is preceded
by a reduction potential region as has been shown by the potentiostatic
double step-sweep method (ref. 7).

Although the interfacial reorganization of the ds DNA into a rather compact ss structure (T: $4 \cdot 10^{-10}$ mole cm^{-2}) is an interesting simulation of possible biological interfacial interactions, the presence of this process at negative potentials must be not overlooked for sweep voltammetric responses from more positive

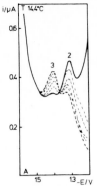
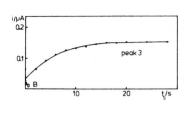

Fig. 6: Variation of the non-faradaic sweep voltammetric response of adsorbed CT DNA after potentiostatic double step polarization from E_c - 0.4 V to E_s -1.2 V at full coverage. A) —— : E_s -0.4 V, t_c 60 s; E_s - 1.2 V, t_s 0 s; -·-·-: E_c - 0.4 V, t_c 60 s; E_s -1.2 V, t_s (0 → 20 s). B) Time dependence of non faradaic peak 3_n during t_s at E_s -1.2 V.
c 260 $\mu g \cdot ml^{-1}$; 0.4 M NaCl, phosphate buffer pH 8.2; v 0.2 $V \cdot s^{-1}$, T 14.4 $^{\circ}C$

E_s. Indeed, in the case of stationary electrodes, possible effects of the history of the whole voltammetric experiment must be taken into consideration. Thus kinetic parameters of interfacial reorganization of preadsorbed native DNA at more positive potential E_s -0.4 V have been investigated by a potential jump to the "destabilizing" potential region of -1.2 V. The potentiostatic double-step jump (ref. 7) from -0.4 V to -1.2 V is followed by a voltammetric sweep ($0.2 V \cdot s^{-1}$) to negative potentials. At pH 8 (in Fig. 6) a rather slow irreversible rise of peak 3 and a concomittant disappearance of peak 2 against time at E_s -1.2 V characterizes the interfacial transformation of the adsorbed ds structure into the ss structure.

The increase of peak 3 can be simply formulated by a first order reaction. The first order rate constant dependence on temperature for the appearance of ss structure is shown in an Arrhenius plot representation at E_s - 1.2 V (Fig. 7). The interfacial process requires an activation energy of 9.5 Kcal mole^{-1}. For comparison it must be remembered that the enthalpy of helix formation for DNA is about -10 Kcal/mole of base pair (ref. 20). The half-life times for adsorbed ds DNA structure at -1.2 V are calculated to be 9 s and 2 s at 4 $^{\circ}C$ and 25 $^{\circ}C$ respectively, i.e. in the same time order of recent values (ref. 21). It follows that high sweep rate of potential (>1 V s^{-1}) (ref. 14,22) or low temperature are experimental conditions at pH 8 which limit the "history" effect at potentials around -1.2 V on the voltammetric responses from more anodic adsorption potentials, E_s. At pH 5.6 an additional interfacial parameter characteristic of faradaic responses must be also considered: the protonation of accessible nucleic bases adenine and cytosine. Under these conditions, high potential sweep rates from 1 V•s^{-1} to 50 V•s^{-1} have no effect on the integrated reduction current

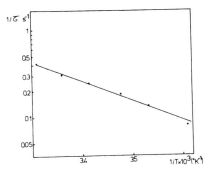

Fig. 7: Arrhenius plot of the dependence of rate constant for CT DNA interfacial transition between ds and ss structures

(charge) (ref. 23). Thus the interfacial protonation, at more cathodic potentials, levels through a pH destabilization the two ss and ds structures into a single signal. Indeed, from non-faradaic response at pH 8, it can be now considered that the faradaic response measuring from E_s -0.4 V, probes the accessibility of nucleic base residues from ds DNA (peak 2). However, the structure of ds DNA at low-charged surface remains a subject under discussion between laboratories. Thus the extent of adsorption of nucleic bases depends partially on the quality of native DNA preparation. However, different points must be briefly reviewed here which can explain possible conformational changes of DNA at a low-charged mercury surface. If we consider a medium with ionic strength 0.1 M, the polyanionic character of DNA is largely shielded by cations (ref. 24). It follows that the adsorption is largely governed by non-electrostatic interactions which are enhanced by the hydrophobic character of mercury around the p.z.c. (ref. 25). It is also clear from the adsorption parameter of nucleic acid components that the DNA interactions with the mercury surface are exothermic (refs. 4,10). For hydrophobic interaction the entropy-driven mechanism results mainly from the disruption of structured water near the two surfaces. Based on thermodynamic data available

for a simple opening of a base pair and for the adsorption of the directly
accessible sugar-phosphate moieties from native DNA, it can be thus calculated
that the probability for opened state for a base pair is at least ten times
higher in contact with a hydrophobic low-charged mercury surface (ref. 15).
Then the possibility of multiple adsorption further spreads a highly exothermic energy to neighbouring base-pairs which contribute to an extended
conformational change of the adsorbed DNA. The smooth surface of mercury in
contrast to a rough carbon surface must also facilitate the flattening of
high molecular native DNA of low flexibility. It can thus be assumed that
this conformational change depends on the relative stability of the AT and
GC base pairs (refs. 5,22,23). A double ladder structure for adsorbed ds
DNA has been proposed by us which could explain the accessibility of nucleic
bases and the double-stranded character (ref. 19).

At high positively charged surfaces ($\geq$ +20 $\mu C\ cm^{-2}$), available with silver
surface, SERS results for native DNA (ref. 3) indicate an electrostatic
adsorption through negatively ionized phosphate groups which supposes
an intact helical structure as also shown by Brabec (ref. 5).

Application

From available adsorption potential range for native DNA at HMDE, it
thus appears that voltammetric responses recorded from E_s around p.z.c.
can be used to characterize the double-stranded nature of DNA. Therefore
any variation of the current intensity of this signal has been used to
probe conformational changes of native DNA in the presence of genotoxic
agents.

From faradaic voltammetric responses pH 5.6 (Fig. 4), the ratio R at
E_s -0.4 V measures the conformational change

$$R = I_{md} - I_n / I_x - I_n \qquad\qquad (5)$$

where I_{md} and I_n correspond respectively to voltammetric responses of modified DNA by a genotoxic agent and to voltammetric response to native DNA.
I_x, a reference signal, depends on the intensity of the I_{md} signal. The first
case, when $I_{md} > I_n$, the increase of accessibility of nucleic bases, related
to a labilization of the helical structure by genotoxic agent, can be measured by R where I_x corresponds to the labilization maximum, $I_x = I_{dn}$ (dn:
denatured DNA). The second case, $I_{md} < I_n$, the decrease of accessibility of

nucleic bases can be related to a steric hindrance in the presence of a genotoxic agent. The steric hindrance is measured by R where I_x corresponds to the hindrance maximum, $I_x = 0$.

R values from (5) can thus express quantitatively conformational changes of DNA in percentage of base pairs, elementary units of ds DNA structure. In case of Calf Thymus (CT) DNA with an average MW of 11×10^6, following results have been obtained. For ionizing γ-irradiations, 3 to 8 percent of base pairs of native CT DNA are labilized per Krad (10 Gy) under different experimental conditions (ref. 26). In case of a chemical methylation of CT DNA, 3 to 4 base pairs of native DNA are also labilized by one methylated base in DNA (ref. 27). However for cis-Pt(NH$_3$)$_2$Cl$_2$, an antitumoral Pt compound, a steric hindrance effect of about 3 base pairs per one platinated nucleic base can be estimated for Pt-modified CT DNA ($r_b = 0.05$). In case of trans-Pt(NH$_3$)$_2$Cl$_2$, a steric hindrance effect of about 9 base pairs has been calculated (ref. 28). It is known that both isomers form interstrand crosslinks, shorten DNA. They prevent thus an interfacial unwinding of the helical structure. However, lower steric hindrance effect of cis-DDP can be related to a concomitant destabilizing effect, while trans-DDP stabilizes the DNA (ref. 29).

Another possibility of rapidly probing the labilization of native CT DNA is to measure at pH 8 the variation of non-faradaic peak 3 intensity in voltammetric response from E_s -0.4 V. Taking into account the maximum intensity of peak 3, obtained for responses from E_s -1.2 V, as an internal reference of ss DNA, rationalized values $I_{-0.4 \text{ V}}/I_{-1.2 \text{ V}}$ can be used to probe the relative extent of labilization of DNA (ref. 13). The effect of temperature on the rationalized values I_r for irradiated DNA is shown in Fig. 8 for very low irradiation doses in*in vitro* conditions (<500 Rad). The quasi-parallel profile of I_r on the irradiation dose dependence at different temperature (5 °C → 20°C) must be connected to the "history" effect at

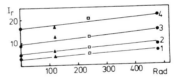

Fig. 8: Dependence of values of non-faradaic peak 3 of CT DNA (E_s -0.4 V) rationalized to their maximal height on the γ-irradiation dose at different temperaturs. 1:T 4.6 °C, 2:T 10 °C; 3:T 15.3 °C; 4:T 19.8 °C. Other voltammetric conditions see Fig. 2. Irradiations conditions c 940 µg ml^{-1}; 0.1 M NaCl, buffer phosphate pH 8; γ-rays from ^{60}Co-source (dose rate 120 Rad/min of 0 °C)

HMDE (see above) observed with native CT DNA for slow potential sweep rates $(0.2 \ V \cdot s^{-1})$. However the slope of I_r dependence on γ-irradition doses can be used as an indicator of labilization.

The previous voltammetric results only concern DNA, but it must be added that the presence of genotoxic agents can also be detected by voltammetry. Thus the complexation of nucleic acids by heavy metals and detection of conformational changes can be simultaneously performed by voltammetric methods (ref. 30).

REFERENCES

1. E. Paleček, Prog. Nucleic Acid Res. Mol. Biol. 9 (1969) 31-73
2. W. Saenger, Principles of Nucleic Acid Structure, Springer-Verlag, New York, 1984, p. 8
3. E. Koglin and J.-M. Séquaris, Topics in Current Chemistry, 134 (1986) 1-57
4. H.W. Nürnberg in G. Milazzo and M. Blank (Eds.), Bioelectrochemistry I, Plenum Press, London 1983, pp. 183-225
5. V. Brabec, Bioelectrochem. Bioenerg., 11 (1983) 245-255
6. P. Valenta and P. Grahmann, J. Electroanal. Chem., 49 (1974) 41-53
7. B. Malfoy, J.-M. Séquaris, P. Valenta and H.W. Nürnberg, J. Electroanal. Chem., 75 (1977) 455-469
8. B. Malfoy and J.-M. Séquaris, no published results
9. J.W. Webb, B. Janik and P.J. Elving, J. Amer. Chem. Soc., 95 (1973) 8495-8505
10. Y.M. Temerk, P. Valenta and H.W. Nürnberg, Bioelectrochem. Bioenerg. 7 (1980) 705-722
11. Y.M. Temerk, P. Valenta, J. Electroanal. Chem., to be published
12. B. Malfoy and J.A. Reynaud, J. Electroanal. Chem., 67 (1976) 359-381
13. J.-M. Séquaris, P. Valenta, H.W. Nürnberg and B. Malfoy, Bioelectrochem. Bioenerg. 5 (1978) 483-503
14. V. Brabec, V. Glezers and V. Kadysh, Collection Czechoslovak Chem. Commun. 48 (1983) 1257-1271
15. J.-M. Séquaris, Doctor Thesis, Orléans 1982
16. J.-M. Séquaris and P. Valenta, J. Electroanal. Chem., to be published
17. P. Valenta and H.W. Nürnberg, Biophys. Struct. Mechanism 1 (1974) 17-26
18. V. Brabec and E. Paleček, Biophys. Chem. 4 (1976) 79-92
19. B. Malfoy, J.M. Séquaris, P. Valenta and H.W. Nürnberg, Bioelectrochem. Bioenerg. 3 (1976) 440 - 460
20. V.A. Bloomfield, D.M. Crothers and I. Tinoco, Physical Chemistry of Nucleic Acids, Harper and Row, New York 1974, p. 313
21. F. Jelen and E. Paleček, Gen. Physiol. Biophys. 4 (1985) 219-237
22. J.-M. Séquaris, P. Valenta, H.W. Nürnberg and B. Malfoy, in: D.H. Everett and B. Vincent (Eds.), Ions in Macromolecular and Biological Systems, Scientechnica, Bristol 1977, pp. 230-234
23. J.-M. Séquaris, B. Malfoy, P. Valenta and H.W. Nürnberg, Bioelectrochem. Bioenerg., 3 (1976) 461-473
24. V.A. Bloomfield, D.M. Crothers and I. Tinoco, Physical Chemistry of Nucleic Acids, Harpen and Row, New York 1974, pp. 379-406
25. A. Frumkin, B. Damaskin, N. Grigoryev and I. Bagotskaya, Electrochim. Acta, 19 (1974) 69-74

26. J.-M. Séquaris, P. Valenta and H.W. Nürnberg, Int. J. Radiat. Biol. 42 (1982) 407-415
27. J.-M. Séquaris, H.W. Nürnberg and P. Valenta, Toxicol. Environm. Chem. 10 (1985) 83-101
28. J.-M. Séquaris, no published results
29. J.-L. Butour and J.-P. Macquet, Biochim. Biophys. Acta, 653 (1981) 305-315
30. J.-M. Séquaris, M.L. Kaba and P. Valenta, Bioelectrochem. Bioenerg., 13 (1984) 225-227

M.R. Smyth and J.G. Vos
Electrochemistry, Sensors and Analysis
© Elsevier Science Publishers B.V., Amsterdam — Printed in The Netherlands

AN ELECTROCHEMICAL METHOD TO DETERMINE BINDING CONSTANTS OF SMALL LIGANDS TO NUCLEIC ACIDS.

J.M.KELLY [1] ,M.E.G.LYONS [1] ,W.J.M. VAN DER PUTTEN [2]

[1] Department of Chemistry, University of Dublin, Trinity College, Dublin.

[2] School of Physical Sciences, National Institute for Higher Education, Dublin.

SUMMARY
    Cyclic voltammetric measurements of methylene blue/DNA complexes show an increase in the redox potential and a decrease in peak-currents in comparison to those of free dye. These changes are explained by firstly deaggregation of dye-dimers and secondly the different hydrodynamic properties of the dye when polymer-bound.This is substantiated using rotating disk experiments. A method has been derived to determine the association constant of the dye/polymer complex and the value so obtained agrees well with literature values. Association constants for the dyes neutral red and cresyl violet to DNA have been determined in a similar way.

INTRODUCTION
    Recently an interdisciplinary research effort has started in the departments of chemistry, physics and genetics of Trinity College Dublin to investigate the photophysical and photochemical properties of a series of small organic and inorganic ligands bound to nucleic acids. The final aim is to utilise base-specific photoproperties to achieve cleavage of the nucleic acid backbone at the different bases and thus achieve a novel base sequencing method. Ligands currently under investigation are porphyrins [1], ruthenium polypyridyl compounds [2] and organic dyes such as methylene blue [3]. Methylene blue (MB) is a well known histological stain and photosensitiser and Friedmann and Brown have shown [4] that irradiation with visible light of a MB/DNA solution yields specific modification at guanine sites. Recently we have shown that irradiation of MB/DNA complexes leads to direct strand

cleavage [5]. The mechanism of this cleavage appears to be both via an indirect singlet oxygen pathway and partly via a direct -radical- reaction. A possible hypothesis for the latter reaction is a redox reaction of the excited singlet state of the dye with the nucleotides [6]. Calculations of the redox potential of the excited singlet state of the dye using the Rehm-Weller formalism suggest oxidation of mainly ground state guanine and quenching of the dye excited state eventually leading to strand cleavage [7]. Previously we have already shown that dye fluorescence quenching is G-C base-pair specific [3]. In order to obtain more evidence for this theory and to investigate the effect of polymer-binding on the electrochemistry of the dye , a cyclic voltammetry study was performed on MB and MB/DNA complexes.

EXPERIMENTAL METHODS

Mb (Fluka "puriss") was purified via gel chromatography on Sephadex LH-20 using methanol as eluent. Purity was checked via TLC (silica, MeOH:acetic acid 9:1). DNA (calf thymus, Sigma Corp.) was purified via repeated phenol extraction as described elsewhere [1]. Neutral red (Fluka "purum") and cresyl violet (Lambda physik, laser grade) were used as received. All solutions were made up in a 25 mM KH_2PO_4, 25 mM K_2HPO_4 buffer.

The electrochemical apparatus consisted of a Metrohm E612 sweep generator, a Thompson "Ministat" potentiostat and a homebuilt voltage follower. C.V. curves were recorded on a Linseis X-Y recorder. The working electrode was glassy carbon, which was polished between each measurement. The counter electrode was a Pt wire and the reference electrode was Ag¦AgCl. Sweep rates in all experiments were 20 mVs^{-1} and solutions were deaerated by bubbling nitrogen. Solutions were thermostatted at 25 C.

RESULTS AND DISCUSSION

Figure 1 shows a typical cyclic voltammogram obtained for a 2×10^{-5} M solution of MB with and without added DNA (P/D = concentration DNA phophate groups / concentration dye) in the 50 mM phosphate buffer at 25 C. Addition of DNA markedly reduces the peak currents both of the anodic oxidation and of the

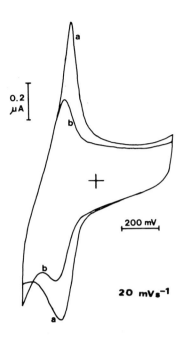

Fig. 1. Cyclic voltammogram of
2x10-5 M MB with and without
added DNA (P/D=42.5) in 50 mM
phosphate buffer.

0.2
µA

200 mV

20 mVs⁻¹

a : **MB⁺**   2.0 10⁻⁵ **M**

b : $\frac{phos}{dye}$ = **43.7**

Fig. 2. Effect of addition of
DNA on E0 of MB (dye
concentration constant 2x10-5 M,
buffer 50 mM phosphate). Also
shown (dotted line) E0 as a
percentage of dye dimer
aggregates (no DNA present).

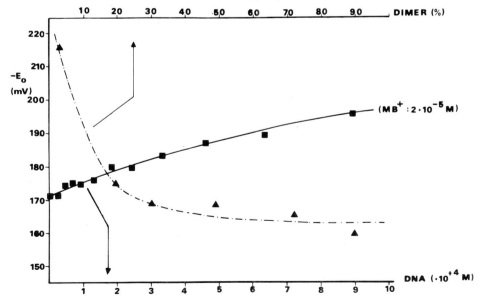

cathodic reduction wave and increases E0 (= .5x(Epa + Epc)). This increase in E0 on addition of DNA is shown in figure 2.

Mb is well known to form dimeric aggregates and figure 2 also shows E0 as a percentage of dimer in solution in the absence of DNA. At the concentration used 26% of the dye is in solution as aggregate (Keq : 1.2x10+4 M [3]). Consequently a number of different species can be involved in the electrochemical reactions at the electrode : monomer dye, dimer dye and monomer:DNA complexes. We have shown previously that binding of dye to the polymer causes deaggregation [3] and thus dimer:DNA complexes need not be considered. Difference in the electrochemical properties of dimer and monomer of MB have been observed earlier [8]. Thus addition of DNA apparently decreases the amount of dimer in solution. Intercalation of (monomeric) dye between the nucleic acid base pairs has been suggested as a possible mode of binding and this is consistent with our observations here. Other effects noted are an increase in the ratio of anodic peak currents over cathodic peak currents on addition of DNA. An exact explanation has to await further investigation but might be due to a difference in binding to the nucleic acids of MB and its reduced form : leuco-MB. The latter is produced in the cathodic wave via a two electron, one proton reduction and is thus not charged. Therefore it is not expected to bind as strongly to the polyanion DNA as the cationic dye. Thus relatively more LMB is free in solution and available at the electrode for oxidation to MB.

The main effect of addition of DNA on the observed electrochemistry is a decrease in the peak curents and this is illustrated in figure 3. This decrease in current is either due to a decrease in diffusion coefficients of the electroactive species or due to a decrease in its apparent concentration. Both explanations are equally valid as bound dye will have the diffusion coefficient of the very large molecular weight polymer whereas intercalated dye will be shielded from electrode reactions by the polymer environment and this appears as an apparent decrease in bulk dye concentration.

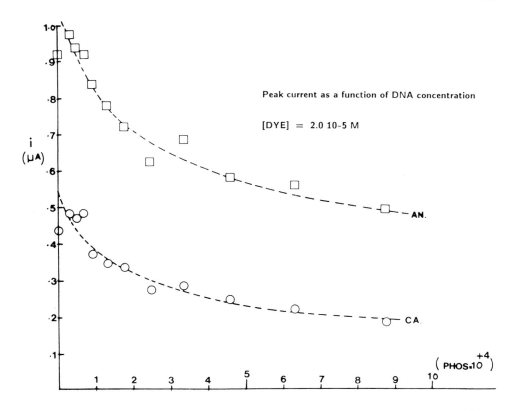

Fig. 3. Anodic (□) and cathodic (O) peak currents of a 2×10^{-5}M solution of MB on addition of calf thymus DNA. Buffer 50 mM phosphate.

Rotating disk experiments essentially confirm the results from the CV experiments. Levich plots are linear, with slopes that decrease on increased DNA concentration consistent with either a lower diffusion coefficient or a lower apparent concentration (not shown). Tafel plots for bound and free dye are identical. Thus a preliminary conclusion that can be drawn at this point is that the observed effects are solely due to binding of the dye to DNA. No new electrochemistry is detected.

The plot of cathodic and anodic currents versus DNA yields a curve not dissimilar from the fluorescence quenching curve (not shown) and suggested the use of electrochemistry in the determination of the association constant of the dye with the polymer. Binding constants can be derived from fluorescence quenching and the most popular method for this is using the Scatchard equation. A good discussion on the properties of the Scatchard equation is given in [9]. Suffice to note that succesfull application of the Scatchard formalism depends crucially on an accurate determination of the amount of free (ie. not bound) dye in solution. This is difficult at high polymer/dye ratios and/or strong binding constants and in these cases the use of the Scatchard equation is questionable.

For the analysis of the electrochemical quenching we have taken a different approach after Berg and Gollmick et al. [7]. First consider the peak currents in the absence and presence of DNA :

$$i0 = A * Ct \tag{1a}$$

$$i = A * (Cf + q * Cb) \tag{1b}$$

where Ct = total dye concentration, Cf and Cb respectively free and bound dye concentration, A a proportionality constant and q a constant denoting the effect of polymer binding on the current ($0 < q < 1$). The apparent association constant of the dye with the polymer is given in equation 2:

$$K * n = [complex]/(Cf * Pf) \tag{2}$$

with Pf the concentration of free sites, K is the binding constant (M^{-1}) and n, the binding number, the maximum number of dye molecules per binding site ($0<n<1$). This is under the assumption of a single binding mode. After some rearrangements and under the assumption that Pt >> Ct equation 3 follows (see appendix)

$$1/Pt = K * n * \{(1-q)/(1-X) - 1\} \tag{3}$$

with X = i/i0, ie. a plot of 1/Pt vs. 1/(1-X) is a straight line with intercept -K * n and slope K * n * (1-q).

Figure 4 shows binding curves so obtained for the dyes

MB, neutral red and cresyl violet with values of K * n of respectively $4.7 \times 10^{+3}$, $1.2 \times 10^{+3}$ and $7.3 \times 10^{+3}$ M^{-1}. In the case of MB it was possible to derive the binding number n from a Scatchard analysis as 0.05 (from fluorescence quenching, not shown) which gives a K of $1 \times 10^{+5}$ M^{-1}, in good agreement with values reported elsewhere [10,11].

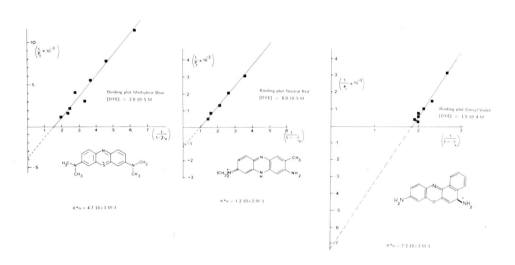

Fig. 4. Binding plots obtained using equation 3 for MB, neutral red and cresyl violet. Buffer 50 mM phosphate (with cresyl violet 5% ethanol was added to assist solubility).

It has to be noted that in contrast to the Scatchard analysis the method described above does not yield the binding number n and as such is somewhat limited. It is however possible to determine n seperately (for example using Hill plots, see [9]) and a combination of these methods then yields K even in the limit of very strong binding.

Concluding, the electrochemistry of the dye MB does not change to a large extent, except for the changes purely "hydrodynamic" in character and these changes can be utilised to determine association constants of the dye to the polymer. This

method of binding determination is applicable to a wide range
of electroactive compounds under the assumptions of a single
binding mode, non-cooperative binding and Pt > Ct. It is
applicable to other forms of analysis and future experiments are
planned using more sensitive electrochemical methods such as
differential pulse polarography.

ACKNOWLEDGEMENTS

The authors acknowledge British Petroleum Venture research, A.
Guinness & Co Ltd. and Trinity Trust for support.

REFERENCES

1 J.M.Kelly, M.J.Murphy, D.J.McConnell, C.OhUigin, Nucleic Acid
   Res. (1985) 13 167-183
2 J.M.Kelly, A.B.Tossi, D.J.McConnell, C.OhUigin, Nucleic Acid
   Res. (1985) 13 6017-6034
3 J.M.Kelly, W.J.M.van der Putten, D.J.McConnell, submitted
   for publication
4 Th.Friedmann, D.M.Brown, Nucleic Acid Res. (1978) 5 615-622
5 C.OhUigin, D.J.McConnell, J.M.Kelly, W.J.M.van der Putten, to
   be published.
6 G.Loeber, L.Kittler, Studia Biophysica (1978)73 25-30
7 H.Berg, F.A.Gollmick, H.E.Jacob, H.Triebel, Photochem.
   Photobiol. (1972)16 125-138
8 E.Rabinowitsch, L.Epstein, J.Am.Chem.Soc. (1941) 63 69-78
9 K.E.van Holde, Physical Biochemistry, (1985) Prentice-Hall
   Inc.,Englewood Cliffs NJ.
10 M.Hogan, J.Wang, R.H.Austin, C.L.Morito, S.Hershkowitz,
   Proc.Natl.Acad.Sci.USA (1982) 79 3518-3522
11 S.Tanaka, Y.Baba, A.Kagemoto, Makromol.Chem. (1982) 182
   1475-148

APPENDIX

With $Ct = Cf + Cb$ equations 1a and 1b can be rearranged
to give Cb as a function of q and Ct:

$$Cb = \{1/(1-q)\} * (1-i/i0) * Ct \hspace{3cm} (A1)$$

With [complex] = Cb and Pf = Pt - Cb equation 2 can be rewritten as:

$$K * n = Cb/\{(Pt - Cb) * (Ct - Cb)\} \hspace{3cm} (A2)$$

Substituting (A1) in (A2) then leads to :

$$Pt = Q * (1-X)/[K * n * \{1 - Q * (1-X)\}] + Q * (1-X) * Ct \hspace{1cm} (A3)$$

with Q = 1/(1-q) and X = i/i0. In general q and X will not be too small and are of the order 0.5. Then when Pt >> Ct equation A3 reduces to :

$$Pt = Q * (1-X)/[K * n * \{1-Q*(1-X)\}] \hspace{3cm} (A4)$$

This can be rewritten as :

$$1/Pt = K * n * \{(1-q)/(1-X) - 1\} \hspace{3cm} (A5)$$

THE EFFECTS OF INTRACEREBRAL AND SYSTEMIC ADMINISTRATION OF DRUGS ON
TRANSMITTER RELEASE IN THE UNRESTRAINED RAT MEASURED USING
MICROCOMPUTER-CONTROLLED VOLTAMMETRY

ROBERT D. O'NEILL

Department of Chemistry, University College, Belfield, Dublin 4, (Ireland)

SUMMARY

The ascorbic acid and homovanillic acid signals, recorded using linear sweep voltammetry with carbon-paste electrodes, were used as an index of excitatory amino acid release and dopamine release, respectively. The effects of a wide variety of drugs administered either systemically or directly into discrete brain regions were measured to investigate factors regulating the release of these two neurotransmitters.

INTRODUCTION

Since the first reports of the use of oxidative voltammetry to monitor changes in the extracellular concentration of electroactive compounds in the central nervous system (ref. 1), various combinations of electrode material and shape of applied potential have been used to improve the sensitivity, resolution and stability of the nanoampere currents recorded in brain tissue (refs. 2-4). One such combination is linear sweep voltammetry with a carbon-paste working electrode (LSVCPE). In the range 0-900 mV (Ag wire as reference) at a scan rate of 5 mV/s, four well-resolved peaks are obtained in brain regions with a significant dopaminergic innervation; peak three is absent from recordings in dopamine-deficient areas (ref.4). In addition to the good resolution, the main advantage of LSVCPE is the stability of the voltammograms over several weeks of continuous recording.

The identification of the compounds responsible for the voltammetric signals recorded in vivo is based on oxidation potentials in vitro (refs. 5,6), micro-infusion of candidate compounds and enzymes beside implanted electrodes (refs. 5, 7-9) biochemical assays (refs. 10,11) and on the effects of pharmacological agents on the recorded current. For LSVCPE, peak 1 (approx. 250 mV) is due to the oxidation of ascorbic acid, peak 2 (approx. 400 mV) to uric acid and peak 3 (approx. 550 mV) to the dopamine metabolite, homovanillic acid (HVA). Peak 4 (approx. 780 mV) is associated with the oxidation of tyrosine- and tryptophan-containing compounds, indoles and presumably other electroactive species (ref. 4). Analysis of the shapes of the peaks, and the relationship between peak height and scan rate indicate that implanted electrodes sample a restricted compartment (ref. 12). The absolute size of each signal will, therefore, depend on scan frequency, compartment volume and on the concentration of the substrate. Since the volume sampled remains relatively constant, changes in the height of a peak are proportional to changes in the extracellular concentration of the substrate provided the wait period between scans is constant.

The neurochemical significance of changes in the extracellular concentration of ascorbic acid, uric acid and HVA has been investigated. In the absence of anaesthesia, the ascorbate signal has been used as an index of excitatory amino acid (EAA) release; in particular, changes in this signal recorded in the striatum reflect variations in the release of EAAs predominantly from cortico-striatal terminals (refs. 3,13). The HVA signal has been used to monitor both spontaneous and drug-induced changes in the release of dopamine (refs. 14–17). The significance of variations in the urate signal is unclear but there is preliminary evidence that they may reflect changes in the release/metabolism of adenosine (refs. 8,9,18).

Since the signals recorded are only indirectly related to transmitter release, caution is needed in interpreting the results. For example, Commiss-iong (ref. 26) has pointed out that the levels of monoamine metabolites do not necessarily reflect changes in the release of the parent transmitter. This is especially true when drugs which interfere with the metabolic pathways or with metabolite clearance from the extracellular fluid are present at the site of detection. In the absence of such agents, however, there is evidence that extracellular metabolite concentration is a faithful index of release. Even in the presence of these drugs, metabolite levels can be used as an index of release up to one hour after drug administration (ref. 15).

METHODS

The technical aspects of voltammetry in vivo have been described in detail (ref. 19); the methodology of microcomputer-controlled LSVCPE has also been published (refs. 5). Male Sprague-Dawley rats were stereotaxically implanted, under chloral hydrate anaesthesia, with a carbon-paste electrode (300 um o.d.) in the appropriate brain region; silver wires placed in the cortex served as auxiliary and reference electrodes. After a 2-day recovery period, the animals were placed in recording cages and linear sweep voltammograms recorded at 12-min intervals at a sweep rate of 5 mV/s between 0–650 mV. The background current for each electrode was measured in situ by scanning continuously, and subtraction from voltammograms recorded at 12-min intervals gave well-resolved ascorbate, urate and HVA peaks (ref. 4).

The equipment consisted of a 380Z microcomputer (Research Machines, Oxford) linked to a general-purpose interface (Chemistry Department, Imperial College, London). The interface included: two DACs to generate the output potential ramps; an 8-channel differential multiplexer so that up to four rats each with two electrodes could be monitored simultaneously; input/output ports to control the lighting in the recording room; and graphics DACs and EPROMs linked to an X-Y plotter.

Recordings started at 12 noon and all injections took place at 16.00 h to control for the normal circadian changes in transmitter release. The effects of drugs on the signals were calculated by comparing the average height of the peak 1 h before injection, with the average recorded 1 h after drug administration. The significance of the results was calculated using paired t-tests on the pre- and post-injection averages for each electrode. All the drug effects reported here are significant with $P < 0.05$; $n \geqslant 4$. In control experiments where saline was injected, the effects were not significant.

RESULTS AND DISCUSSION

A knowledge of the effects of systemically administered drugs on transmitter release in the central nervous system is important to an understanding of their therapeutic action. However, because of the numerous sites of action, both peripheral and central, this type of investigation throws little light on the mechanism of drug action. The effects of infusions of the drug into discrete brain regions is a much more direct approach to this question. Voltammetry _in vivo_ has been used to investigate these two aspects of pharmacology.

Systemic administration of drugs

Receptors on the nerve terminal and on the cell body are two sites of action for drugs which are injected intraperitoneally. The effects of the dopamine-receptor agonist, apomorphine (2 mg/kg i.p.), and antagonist, haloperidol (0.5 mg/kg i.p.), on the HVA signals recorded simultaneously in frontal cortex, nucleus accumbens and striatum are given in Table 1. The finding that both drugs lead to large changes in accumbens and striatum with only small changes in frontal cortex supports the view that there is much less feed-back inhibition of dopamine release in the frontal cortex compared with the other two regions (ref. 20). These results, in conjunction with the effects of the two enantiomers of the dopamine-receptor ligand 3PPP, also indicate that auto-inhibition of dopamine release is more pronounced in striatum than in accumbens (ref. 16).

The amino acids, glutamic and aspartic acids, are two principal excitatory neurotransmitters within the nervous system (ref. 21). Thus the finding that there is a correlation between motor activity and the release of these acids in brain regions which regulate movement is not surprising (ref. 22). The ascorbate signal has also been used to investigate the effects of drugs on EAA release in the striatum (ref. 3). Table 2 shows these results.

TABLE 1
Percentage changes in the HVA signal recorded in three brain regions after injection of apomorphine (2 mg/kg i.p.) and haloperidol (0.5 mg/kg i.p.)

Brain region	Apomorphine	Haloperidol
Frontal cortex	−24 +/− 6	50 +/− 17
Nucleus accumbens	−63 +/− 7	306 +/− 55
Striatum	−74 +/− 8	365 +/− 81

TABLE 2
The effects of i.p. administration of drugs on the ascorbate signal recorded in the striatum

Drug	Dose	Percentage change
Amphetamine	5 mg/kg	97 +/− 19
Caffeine	50 mg/kg	102 +/− 18
Haloperidol	0.5 mg/kg	−49 +/− 9
Clonidine	0.1 mg/kg	−40 +/− 10
Pentobarbitone	20 mg/kg	−43 +/− 6

Stimulants, whether socially acceptable (caffeine) or not (amphetamine), increase the release of glutamate from cortico-striatal terminals. Consistent with the ascorbate/EAA hypothesis, the sedatives haloperidol, clonidine and pertobarbitone depress this signal.

Intracerebral administration of drugs

More detailed information about the site of action of drugs and the interrelationship between different transmitter systems can be obtained from injections into discrete brain areas. The effects on striatal dopamine release of microinfusion of the inhibitory transmitter, taurine, and the putative taurine-receptor antagonist, 6-aminomethyl-3-methyl-4H-1,2,4-benzothiadiazine-1,1-dioxide (TAG), into the dendritic field of the dopamine cells which project to the striatum are shown in Table 3.

TABLE 3
The effects on the striatal HVA signal of 2 ul microinfusions into
the ipsilateral substantia nigra

Infusion	Dose	Percentage change
Saline	-	21 +/- 15
Taurine	100 ug	77 +/- 15
TAG (2 rats)	20 ug	-10 +/- 1
TAG (4 rats)	20 ug	60 +/- 19

TABLE 4
Percentage changes for infusions of 2ul directly into the striatum

Infusion	Dose	Ascorbate signal	HVA signal
Saline	-	-6 +/- 5	-23 +/- 19
2-Chloroadenosine	15 ug	-67 +/- 5	-92 +/- 5
Adenosine deaminase	1 unit	51 +/- 8	100 +/- 30

The results are consistent with the suggestion that the infusion of taurine leads to inhibition of inhibitory interneurones within the pars reticulata (ref. 23). The effect of TAG, although varying between animals, indicates that taurine may be an endogenous neuromodulator in the substantia nigra (ref. 24).

Voltammetry can also be used to investigate the role of pre- synaptic receptors in regulating transmitter release. In these experiments both the working electrode and the injection cannula are placed in the terminal region. The effects on striatal EAA and dopamine release of infusing the adenosine-receptor agonist, 2-chloroadenosine, and an enzyme which metabolises endogenously released adenosine, adenosine deaminase, is given in Table 4. These results provide evidence that released endogenous adenosine may inhibit the release of transmitter from both glutamtergic cortico-striatal and dopaminergic nigro-striatal terminals (ref. 25).

ACKNOWLEDGEMENTS
This paper was written on the basis of research carried out in the University Laboratory of Physiology, Oxford, U.K. Professor W. John Albery (Imperial College, London), Dr. Marianne Fillenz and Dr. Richard A. Grunewald have contributed much to the development of this technique.

REFERENCES

1 P.T. Kissinger, J.B. Hart and R.N. Adams, Brain Res., 55 (1973) 209-213.
2 J.A. Stamford, Brain Res. Rev., 10 (1985) 119-135.
3 M.H. Joseph, M. Fillenz, I.A. MacDonald and C.A. Marsden (Eds.),
   Monitoring Neurotransmitter Release During Behaviour, Ellis Horwood Health
   Science Series, Chichester, 1986.
4 R.D. O'Neill, J. Instit. Chem. Ireland, 1983, 5-10.
5 R.D. O'Neill, M. Fillenz, W.J. Albery and N.J. Goddard, Neurosci., 9 (1983)
   87-93.
6 M.P. Brazell and C.A. Marsden, Brit. J. Pharmacol., 75 (1982) 539-547.
7 M.P. Brazell and C.A. Marsden, Brain Res., 249 (1982) 167-172.
8 R.D. O'Neill, M. Fillenz, R.A. Grunewald, M.R. Bloomfield, W.J. Albery,
   C.M. Jamieson and J.A. Gray, Neurosci. Lett., 45 (1984) 39-46.
9 F. Crespi, T. Sharp, N. Maidment and C.A. Marsden, Neurosci. Lett., 43
   (1983) 203-207.
10 F. Gonon, M. Buda, R. Cespuglio, M. Jouvet and J.-F. Pujol, Brain Res., 223
   (1981) 69-80.
11 R. Cespuglio, J. Histochem. Cytochem., 30 (1982) 821-823.
12 W.J. Albery, T.W. Beck, M. Fillenz and R.D. O'Neill, J. Electroanal. Chem.,
   161 (1984) 221-233.
13 R.D. O'Neill, M. Fillenz, L. Sundstrom and J.N.P. Rawlins, Neurosci. Lett.,
   52 (1984) 227-233.
14 J.A. Clemens and L.A. Phebus, Brain Res., 267 (1983) 183-186.
15 R.D. O'Neill and M. Fillenz, Neurosci., 14 (1985) 753-763.
16 R.D. O'Neill and M. Fillenz, Neurosci., 16 (1985) 49-55.
17 F. Crespi, J. Paret, P.E. Keane and M. Morre, Neurosci. Lett., 52 (1984)
   159-164.
18 K. Mueller, R. Palmour, C.D. Andrews and P.J. Knott, Brain Res., 335 (1985)
   231-235.
19 C.A. Marsden (Ed.) Measurement of Neurotransmitter Release In Vivo, J.
   Wiley, Chichester, 1984.
20 M.J. Bannon and R.H. Roth, Pharmac. Rev., 35 (1983) 53-68.
21 P.J. Roberts, J. Storm-Mathisen and G.A.R. Johnston (Eds.) Glutamate:
   Transmitter in the Central Nervous System, J. Wiley, Chichester, 1981.
22 R.D. O'Neill and M. Fillenz, Neurosci. Lett., 60 (1985) 331-336.
23 R.D. O'Neill, Brain Res., in press.
24 G.E. Martin, R.J. Bendesky and M. Williams, Brain Res., 299 (1981)
   530-535.
25 R.D. O'Neill, Neurosci. Lett., 63 (1986) 11-16.
26 J. Commissiong, Biochem. Pharmacol., 34 (1985) 1127-1132.

M.R. Smyth and J.G. Vos
Electrochemistry, Sensors and Analysis
© Elsevier Science Publishers B.V., Amsterdam — Printed in The Netherlands

ELECTROANALYSIS OF TOYOCAMYCIN AND SANGIVAMYCIN

E. Bojarska[1], K. Pawlicki[2], M.R. Smyth[3] and B.Czochralska[4]
[1]Institute of Biochemistry and Biophysics, Academy of Sciences, 02-532 Warszawa, Poland.
[2]Institute of Biophysics, Silesian Academy of Medicine, 40-952 Katowice, Poland.
[3]School of Chemical Sciences, National Institute for Higher Education, Dublin 9, Ireland.
[4]Department of Biophysics, Institute of Experimental Physics, University of Warsaw, 02-089 Warszawa, Poland.

SUMMARY

A study has been made of the direct current (D.C.) polarographic and cyclic voltammetric behaviour of the structurally-related pyrrolopyrimidine nucleosides toyocamycin and sangivamycin. Both compounds exhibit one 3-electron polarographic wave in the pH range 1-6. It was found that reduction of both compounds occurs in the pyrimidine ring, leading to two reduction products, one of which (λ =306 nm) is photochemically reversible to the parent compound. The analysis of these compounds has been investigated using differential pulse polarography.

INTRODUCTION

The structurally-related pyrrolopyrimidine nucleoside antibiotics toyocamycin (I) and sangivamycin (II) have been isolated from 13 Streptomyces cultures (ref.1). The pyrrolopyrimidine ribosides represent a new group of nucleoside antibiotics that are highly cytotoxic to mammalian cell lines in culture and inhibitory to the growth of bacteria and fungi, as well as to RNA and DNA viruses. The close structural relationship of toyocamycin and sangivamycin to adenosine has made these nucleosides extremely valuable biochemical tools for studying many cellular and enzyme reactions. Toyocamycin has been shown to be significant antitumour compound (ref.2).

R = Ribose

MATERIALS AND METHODS

Toyocamycin and sangivamycin were synthetised by Z. Kazimierczuk (University of Warsaw) according to the procedure of Tolman et al. (ref.3); their ultraviolet spectra agreed with the literature data.

Britton-Robinson buffers were used in the pH range between 2 and 8. A Precision Digital pH Meter type OP - 208 (Radelkis, Hungary) was employed for pH measurements.

D.C. polarography was carried out using a Radiometer Polariter PO4 (Copenhagen, Denmark). The characteristics of the dropping mercury electrode were m = 2.5 mg/s, t = 3.7 s at 60 mm Hg. Cyclic voltammetry (C.V.) and differential pulse polarography (D.P.P.) were carried out using a PAR Model 174 Polarographic Analyser. Polarographic curves were recorded at room temperature, and all potentials are reported relative to the saturated calomel electrode. Solutions were rendered oxygen-free by flushing with argon.

Preparative electrolysis was performed on a mercury electrode (surface area 12 cm^2) in a three compartment cell, at a constant potential controlled by a Potentiostat type OH - 404/A (Radelkis, Hungary). The concentration of the depolarizer was 10^{-3} M. Magnetic stirring with a Teflon - covered bar was used. Argon was continuously passed through the cell during electrolysis. An integrator type OH - 404/C (Radelkis, Hungary) was used to determine the number of electrons involved in the reduction process.

UV absorption spectra were recorded with a Specord UV - VIS spectrometer using 1 and 10 mm pathlength quartz cuvettes. Irradiation at 254 nm of electrolysed solutions in 10mm pathlength quartz cuvettes was achieved using a Philip TUV 6W lamp; radiation below 240 nm was eliminated with the aid of a 5mm filter of 33% acetic acid. A Philip Osram 125W HQW Wood's lamp was employed for irradiation at wavelengths higher than 320 nm. Cuvettes were maintained at room temperature with the aid of a stream of cooled air.

Paper chromatography was carried out using ethanol:ammonium acetate pH 7.5 (7:3) as solvent system.

RESULTS AND DISCUSSION

D.C. Polarography

D.C. polarographic investigations of toyocamycin and sangivamycin were performed in Britton-Robinson buffers at a concentration of $4X10^{-4}M$. Both compounds exhibited a single, diffusion-controlled cathodic wave (or C.V. peak) in the pH range 1-6. Above pH 6 the reduction wave (or cathodic peak) disappeared, as for adenosine (ref.4) and 4-aminopyrimidine (ref.5). Coulometric measurements at pH 4 gave a value for n = 3.4 for both compounds. On increasing pH, half-wave potentials shifted towards more negative potentials, whereas the diffusion current decreased (Table 1).

The cyclic voltammetric behaviour of toyocamycin and sangivamycin showed consistent results compared to D.C. polarography. No anodic peak complimentary to the reduction peak appeared on the return sweep at pH 0.4 to 6 at v = 0.5 V/s, indicating an irreversible electrode process.

TABLE 1

D.C. polarographic behaviour of $4x10^{-4}M$ toyocamycin and sangivamycin in Britton-Robinson buffers.

pH	$E_{1/2}$ (V)	Wave slope (mV) $(E_{1/4} - E_{3/4})$	i_d (A)	$I_d{}^a$
		toyocamycin		
1.00[b]	-1.250	35	9.0	9.78
2.12	-1.265	35	5.0	5.43
2.50	-1.285	40	3.9	4.24
2.98	-1.295	45	3.1	3.37
3.59	-1.330	48	3.0	3.32
4.05	-1.350	50	2.9	3.26
4.52	-1.385	55	2.8	3.04
4.97	-1.410	60	2.4	2.61
		sangivamycin		
3.05	-1.370	40	3.7	4.02
3.59	-1.390	45	2.9	3.10
4.05	-1.415	45	2.4	2.61
4.97	-1.450	50	2.3	2.55
6.05	-1.495	55	1.3	1.36

$^aI_d = i_d/Cm^{2/3}t^{1/6}$ ^{b}0.1 M HCl

Electrolysis at controlled potential

Preparative electrolysis of toyocamycin and sangivamycin, using
a large mercury electrode, was carried out at pH 4.0. Following
electrolysis at a potential of -1.4 V, the characteristic
absorption maximum at 279 nm (E_{max} = 12200) for sangivamycin,
disappeared and a new absorption peak appeared at 306 nm for both
compounds. After electrolysis,the polarographic wave due to the
reduction of toyocamycin and sangivamycin disappeared and no other
wave was seen.

Paper chromatography of the electrolysed solutions showed two
reduction products, one of which absorbed at 306 nm (40%), the
other one absorbed at 250 nm. Photochemical transformation of the
reduction products was attempted in order to see if the reduction
products of toyocamycin and sangivamycin are photochemically
oxidized to the corresponding parent compound, as found for the
reduction products of 4-aminopyrimidine (ref. 5) and
2-oxopyrimidine (ref. 6). Irradiation of the reduction products
of toyocamycin and sangivamycin in acidic solution of pH 0.4 at
254 or 320 nm in the presence of metallic mercury resulted in a
decrease of the 306 nm peak and simultaneous growth of an
absorption peak at 279 nm with all intermediate absorption curves
passing through an isosbestic point (Fig. 1).

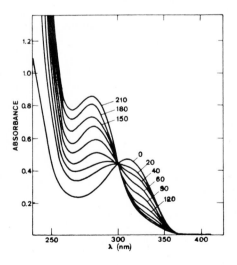

Fig. 1. Photochemical transformation at pH 0.4 on irradiation of
the reduction product of sangivamycin electrolysed at pH 4.0, in
the presence of mercury. (Number beside each curve corresponds to
time of irradiation in minutes).

Photochemical exposure in the presence of metallic mercury was examined because the presence of the mercury significantly increased the efficiency of the photochemical transformation as found for 4-aminopyrimidine (ref. 5). UV spectrophotometry, polarographic analysis and paper chromatography all demonstrated that for toyocamycin and sangivamycin, the photochemically transformed product is the oxidized form of the parent compound.

The photooxidation of the reduction products of toyocamycin and sangivamycin occurs only at pH < 1. At pH 1-6, rearrangement and hydrolysis of the products occurs as for 4-aminopyrimidine.

After exposure to air, the reduction products turned pink. However, electrolysed solutions kept in a solution of pH 0.4 and stored at 0°C under nitrogen atmosphere were stable over a period of one week.

Controlled potential electrolysis of toyocamycin and sangivamycin at -1.3 V at pH 1.0 revealed that the primary reduction product absorbing at λ_{max} = 306 nm is a transient, which following 2 hrs. electrolysis, was transformed into the product absorbing at λ_{max} = 250 nm, as was found for the 4e reduction of adenosine (ref.4). At the same time, the coulometric measurements gave n = 4.2 for the reduction process.

A positive Nessler test for NH_3 (ref.7) was found for solutions of toyocamycin and sangivamycin electrolysed at pH 1 for 2-3 hrs. The amount of the dimeric reduction product estimated polarographically by its photooxidation to the parent compounds was found to be 40%. This is in accord with results of Kwee and Lund (ref.8) pointing to deamination of the reduced adenosine in 2M HCl.

On the basis of the above results it is obvious that the reduction of toyocamycin and sangivamycin occurs in the pyrimidine ring. This is further supported by the polarographic non-reducibility of pyrrole in aqueous media (ref.9).

Although identification of the primary electrolysis products of toyocamycin and sangivamycin is difficult because of the instability of both primary electrolysis products, comparison of the electrochemical behaviour of toyocamycin and sangivamycin with those of related compounds (4-aminopyrimidine, adenine and purine) suggests the reduction mechanism shown in Figure 2. The initial 1e nucleophilic attack on the pyrimidine ring results in the reduction of the 1,6 N=C bond and formation of

6,6'-hydrodimers, by the free radical produced. Since reduced pyrimidines are more basic than the non-reduced species (ref.10), and since the hydrodimers produced in the present study are generally protonated, protonation may be responsible for the shift to longer wavelengths on ultraviolet absorption. A similar shift is seen for cationic forms of 2,2'-bipyridyl (ref.11) and reduced 4-aminopyrimidine (ref.5).

If the proton activity is sufficiently large (pH<3), the carboanion path is favoured (Figure 2) since the carboanion is rapidly protonated to the dihydropyrimidine. At pH < 1 the reduction involving the fourth electron addition is accompanied by catalytic hydrogen discharge. The reduction products can undergo hydrolytic and rearrangement reactions.

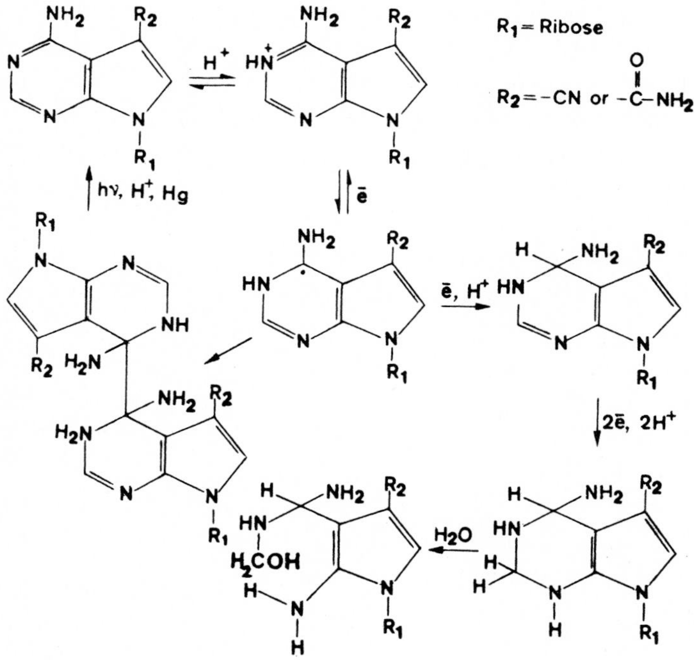

Fig. 2   Proposed electrode reaction mechanism for toyocamycin and sangivamycin.

Analytical Applications

The differential pulse polarographic (D.P.P.) behaviour of
these compounds was investigated in Britton-Robinson buffers of
pH 2.0-8.0. Although the peak currents were found to be larger
in solutions of pH 2.0 and 4.0, the peaks were masked to a
considerable extent by the decay of the supporting electrolyte.
A Britton-Robinson buffer of pH6.0 was therefore chosen for
analytical purposes. The D.P.P. curves for sangivamycin in the
range $1x10^{-7}M$ to $5x10^{-6}M$ is shown in Figure 3.

The calibration curves for sangivamycin and toycamycin in
this medium were both found to be linear in the range $1x10^{-5}$ -
$1x10^{-6}M$ and the method has a limit of detection of $5x10^{-7}M$
for sangivamycin and $8x10^{-7}M$ for toyocamycin.

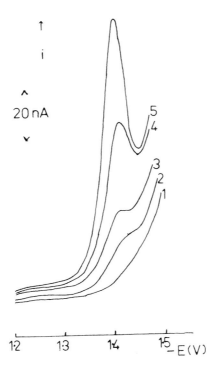

Fig. 3. Differential pulse polarograms of sangivamycin in
Britton-Robinson buffer pH6.0 (1.blank 2. $5x10^{-7}M$ 3.
$1x10^{-6}M$ 4. $2.5x10^{-6}M$ 6.$5x10^{-6}M$).Conditions : scan rate
2mV/s, pulse amplitude 50mV, droptime 1s.

ACKNOWLEDGEMENTS

This investigation was supported by the Polish Ministry of Science, Technology and Higher Education (CPBP 01.06) and the National Institute for Higher Education, Dublin. The authors wish to thank Dr. Z. Kazimierczuk for the synthesis of toyocamycin and sangivamycin and Pawel Przybora and Michelle Connor for technical assistance.

REFERENCES
1. R.J. Suhadolnik, Nucleoside Antibiotics, Wiley Interscience, New York, 1970, pp. 298-353.
2. M. Saneyoshi, R. Tokuzen and F. Fukuoka, Gann, 56 (1965) 219.
3. R.L. Tolman, R.K. Robins and L.B. Townsend, J.Amer.Chem. Soc., 90 (1968) 524; J.Amer.Chem.Soc., 91 (1969) 2102.
4. B. Janik and P.J. Elving, J.Amer.Chem.Soc., 92 (1970) 235.
5. B. Czochralska and P.J. Elving, Electrochimica Acta, 26 (1981) 1755.
6. B. Czochralska and D. Shugar, Biochim. Biophys. Acta, 281 (1972) 1.
7. M.J. Taras, in Colorimetric Determination of Nonmetals, (D.F. Boltz, Ed.), Interscience, New York, 1985, pp. 75-160.
8. S. Kwee and H. Lund, Acta Chemica Scandinavica, 26 (1972) 1195.
9. G. Dryhurst, Electrochemistry of Biological Molecules, Academic Press, New York, 1977, p.327.
10. D.J. Brown, The Pyrimidines: Supplement I, John Wiley, New York, 1970, pp. 355-6 and 370.
11. DMS UV Atlas of Organic Compounds, Vol. III, Butterworth and Verlag Chemia, London and Weinheim, 1967.

M.R. Smyth and J.G. Vos
Electrochemistry, Sensors and Analysis
© Elsevier Science Publishers B.V., Amsterdam — Printed in The Netherlands

DETERMINATION OF SOME ADENINE COMPOUNDS BY DIFFERENTIAL PULSE POLAROGRAPHY

Y.M. TEMERK, Z.A. AHMED, M.E. AHMED and M.M. KAMAL

Chemistry Department, Faculty of Science, Assiut University, Assiut , Egypt.

SUMMARY
    The applicability of differential pulse polarography for the trace determination of 1-methyladenine, deoxyadenosine, deoxy-adenosine monophosphate, adenylyl-(3'—5')-adenosine and adenylyl-(3'—5')-guanosine was studied at various pH values. Detection limits are: 1-Methyladenine 0.9 μM, Deoxy-A 0.93 μM, Deoxy-5'-AMP 1.8 μM, APA 0.93 μM and APG 0.91 μM. The validity of this method is supported by the constancy of the i_{DPP}/C values and a statis-tical analysis is included on calibration curve parameters and observed concentrations.

INTRODUCTION

    Modern methods of electrochemical analysis and differential pulse polarography (dpp) have been applied to the determination of some biological compounds at trace level concentrations (refs.1-6). Recent studies of interactions of mercury with nucleic acid com-ponents have shown that the purine series nucleosides and nucleo-tides can be determined by means of C.S.V., while pyrimidine nucleo-sides and nucleotides are inactive (ref. 6).

    In continuation of the applicability of differential pulse polarography for the quantitative trace determination of some rare nucleic acid components, the present investigation is focussed on the analytical determination of 1-methyladenine, deoxyadenosine (deoxy-A), deoxyadenosine-5'-monophosphate (deoxy-5'-AMP), adenylyl-(3'—5')-adenosine (APA) and adenylyl-(3'—5')-guanosine (APG) under the optimum conditions.

EXPERIMENTAL

    1-Methyladenine, Deoxy-A, Deoxy-5'-AMP, APA and APG were pro-ducts of Serva (Heidelberg F.R.G.). The solutions containing dif-ferent concentrations of the compounds investigated were prepared by dissolving a known amount of the chemically pure product into a specific volume of McIlvaine buffer. pH was measured with a digital Radiometer pH-meter Model pH M 64, accurate to±0.005 unit.

Differential pulse polarograms were carried out with a Model 174A polarographic analyser equipped with a PAR Model 172 A drop timer and electrode assembly. Polarograms were recorded on an advanced X-Y recorder (Hewlett-Packard 7045 A). A thermostated Metrohm cell with a three electrode systems was used. All measurements were performed at $22 \pm 0.5°C$.

RESULTS AND DISCUSSION

Differential pulse polarographic current voltage curves of the investigated compounds in acid and moderate buffer solutions (pH<7) show a well-developed faradaic peak corresponding to the subsequent reduction of the $N(I) = C(6)$ and $N(3) = C(2)$ double bonds for adenine moiety according to the mechanism established by Janik and Elving (ref. 1).

The differential pulse peak is pH-dependent in that it shifts to more negative potentials with rise of pH (Table 1). The resolution of the reduction peaks from the background discharge was improved with decreasing pH, whereas at pH< 2.2 the reduction peak current is inconstant as the result of the effect of strong hydrogen evolution, according to the general mechanism established for N-heterocyclic compounds. Therefore a well developed peak for 1-methyladenine, deoxy-A and APG was found at pH 3.2, and at pH 4.2 for deoxy-5'-AMP and APA.

TABLE 1

Maximum peak current, $(i_p, \mu A)$ and peak potential (E_p, V) at different pH values for the investigated compounds.

pH	1-Methyladenine		Deoxy-A		Deoxy-5'AMP		APA		APG	
	i_p μA	$-E_p$ V	i_p μA	$-E_p$ V	i_p μA	$-E_p$ V	i_p μA	$-E_p$ V	i_p μA	$-E_p$ V
2.2	6.4	1.10	7.8	1.16	7.2	1.19	16.0	1.09	9.1	1.08
3.2	6.0	1.13	7.7	1.18	6.8	1.21	18.2	1.19	10.9	1.15
3.8	3.8	1.21	6.8	1.20	-	-	18.2	1.20	8.8	1.19
4.2	3.7	1.21	5.3	1.27	5.7	1.30	19.1	1.21	7.2	1.22
4.6	-	-	4.9	1.30	5.4	1.37	19.3	1.25	5.1	1.27
5.2	2.5	1.26	-	-	3.0	1.44	-	-	-	-

Although an increased scan rate, ν, is analytically advantageous because of reduced analysis time, ν = 5 mV/s distorts the res-

ponse of reduction, decreases resolution and reduces sensitivity.
Because response for $\nu = 2$ mV/s was very similar to $\nu = 1$ mV/s, the
former was used for the optimum resolution of the investigated biological compounds.

Differential pulse polarograms of the investigated compounds
indicate that the peak height increases linearly as pulse amplitude is increased. The resolution of 1-methyl adenine, deoxy-A and
APA reduction peaks from the background discharge is sufficient at
a pulse amplitude 25-100 mV. The improved resolution of the reduction peak of deoxy-5-AMP and APG is found at a ΔE of 50-100 mV.
However, at a pulse amplitude of 100 mV the differential pulse
peaks of the investigated biological compounds gave the highest
sensitivity and well-defined peaks; hence a 100 mV pulse was used in
subsequent studies.

The optimum conditions for the analytical determination of
1-methyladenine, deoxy-A, deoxy-5'-AMP, APA and APG by dpp were
found to be the following settings: 2 mV s^{-1} scan rate, 100 mV
pulse amplitude and 2 s drop time, whereas the dpp peak of deoxy-
5'-AMP showed better resolution from the background discharge at
a drop time of 1 s. The investigated biological compounds display
a maximum value of differential pulse current at the optimum pH
value. Response of differential pulse peak height to concentration
of each compound is a straight line passing through the origin.
The validity of the method is supported by the constancy of the
i_{dpp}/C values.

TABLE 2

Straight line constants, values of i_p/conc. and detection limits
for the differential pulse determination of the investigated
compounds at the optimum pH.

Compound	pH	$a(\pm$ s.d.$)^{(1)}$	$b(\pm$ s.d.$)^{(2)}$	$r^{(3)}$	$i_{DPP}/C.10^5$	Detection limit μM
1-Methyladenine	3.2	1.35(0.01)	0.09(0.003)	0.999	1.32±0.04	0.9
Deoxy-A	3.2	1.80(0.12)	0.08(0.003)	0.997	1.76±0.33	0.93
Deoxy-5'-AMP	4.2	0.77(0.18)	0.15(0.02)	0.997	0.94±0.13	1.8
APA	3.2	3.90(0.1)	0.10(0.003)	0.998	4.12±0.11	0.93
APG	4.2	1.92(0.06)	0.18(0.02)	0.998	1.86±0.21	0.91

(1) Slope   (2) Intercept   (3) Regression coefficient

The variation of i_p with the concentration of each compound is
represented by the straight line equation $\underline{i_p} = \underline{ac} + \underline{b}$. The data

for three to five replicated measurements were subject to a least squares refinement and the values of the regression coefficient are computed and assembled together with the straight line constant in table 2.

REFERENCES

1   B. Janik, P.J. Elving; J. Am. Chem. Soc., 95 (1973) 99.
2   E. Palecek; Anal. Lett., 13 (1980) 331.
3   T.E. Cummings, J.R. Fraser, P.J. Elving; Anal. Chem., 52 (1980) 558.
4   Y.M. Temerk, M.M. Kamal; Z. Anal. Chem., 305 (1981) 200
5   Y.M. Temerk, M.M. Kamal, M.E. Ahmed and M. Ibrahim; Bioelectrochem. Bioenerg., 11 (1983) 449.
6   E. Palecek, J. Osteryoung, R.A. Osteryoung; Anal. Chem., 54 (1982) 1389.

MODIFIED ELECTRODES AND SENSORS

M.R. Smyth and J.G. Vos
Electrochemistry, Sensors and Analysis
© Elsevier Science Publishers B.V., Amsterdam — Printed in The Netherlands

SOME THEORETICAL ASPECTS OF MODIFIED ELECTRODES

C.P. ANDRIEUX

Laboratoire d'Electrochimie de l'Université de Paris 7, Unité Associée au CNRS N°438, 2 place Jussieu, 75251 Paris Cedex 05 (France)

ABSTRACT

Redox polymer coated electrodes appear as particularly attractive modified electrodes for the catalysis of electrochemical reactions. The catalytic properties depend on three main kinetic factors :
- propagation of charge through the coating;
- diffusion of the substrate;
- rate of the chemical reaction between the substrate and the mediator.

A theory and a procedure for analyzing the kinetic behaviour of such a system is presented. As an experimental example, the practical applicability of this procedure is demonstrated with electrodes coated with poly(4-vinylpyridine) bound ruthenium complexes.

INTRODUCTION

Redox polymer coated electrodes have been the object of active investigation during the past ten years. Besides applications to analysis or corrosion protection, the main motivation of these studies is the catalysis of electrochemical reactions (refs. 1,2). The reason of the interest they aroused in this respect is the expectation that they may combine the advantages of monolayer derivatized electrodes with those of homogeneous catalytic systems. With redox polymer coatings as with monolayer derivatized surfaces, high local concentrations of catalytic sites can be achieved even though the total amount of catalyst remains small. The two systems also share the advantage of an easy **separation** of the reaction products from the catalyst. On the other hand, redox polymer coatings as homogeneous catalytic systems, offer a three-dimensional dispersion of the reacting centers as opposed to the two-dimensional arrangement prevailing at bare electrodes and at monolayer derivatized surfaces. Thus catalysis can be obtained with redox polymer coatings and homogeneous systems, but not with monolayer derivatized surfaces, even in the cases where the catalyst simply exchanges electron in an outer-sphere manner with the substrate (ref. 3). This "redox catalysis" thus results from physical rather than chemical reasons. In the case of "chemical catalysis" (ref. 4) where an essential step is the transient formation of an adduct between the catalyst and the substrate, catalysis occurs at monolayer derivatized surfaces as a consequense of their particular chemical properties. The potential advantage of redox polymer coatings is then the possible multiplication of the effect as the number of equivalent monolayers of catalyst increases. It follows that catalytic efficiencies of redox polymer coatings are expected to increase with the amount of redox polymer deposited on the electrode surface (ref. 3). However, this potentiality may be counteracted by limitations imposed by the rates of substrate diffusion and charge propagation accross the coatings.

DIFFUSIONAL CHARGE TRANSPORT THROUGH FIXED SITE ELECTROACTIVE POLYMER

In the current practice, three kinds of redox polymer modified electrodes can be used according to the nature of the bonding of the active centers (chemical coordination or ionic bonding).

In all cases the charge transfer follows the laws of diffusion and can be characterized by a charge transfer diffusion coefficient D_E (often referred to as D_{ct} (ref. 1)).

Although other phenomena such as flow of counter ions or segmental motion of the polymer chain can occur, the today accepted mechanism of charge transfer is that sites become oxidized or reduced by successive electron transfer between neighbour redox sites (ref. 5).

Yet the litterature contains only few examples where the dependence of charge propagation as a function of experimental parameters such as the concentration of electroactive group are studied (ref. 6).

The most comprehensive (ref. 6) was devoted to a copolymer of osmium and ruthenium complexes where the osmium redox sites were diluted without changing the polymer structure. The comparison with the Dahms-Ruff equation (ref. 7) obtained in the case of solutions, was analyzed. D_E was composed of two parts, D_S : physical displacement and D_{ET} : electron exchange;

$$D = D_S + D_{ET} = D_S + \frac{k_{ex} \delta^2 C}{4}$$

δ being the distance between the centers of reacting redox sites and C is their concentration.

An agreement with the Dahms-Ruff equation was only obtained at high osmium concentration (ref. 6c). Recently we re-examined this problem in the case of poly(4-vinyl-pyridine) film containing coordinated ruthenium complexes first described by Haas and Vos (ref. 8). In our study the number of ruthenium per pyridines varies from 1/5 to 1/100 and we can safely assume an identical structure for each polymer (Fig. 1).

The diffusion coefficients are measured by chronoamperometry when a Cottrell equation is obtained :

$$i = \frac{nF}{\sqrt{\pi}} \sqrt{D_E} C t^{-1/2}$$

(i : current density, C : concentration)

For long times a saturation decay occurs.

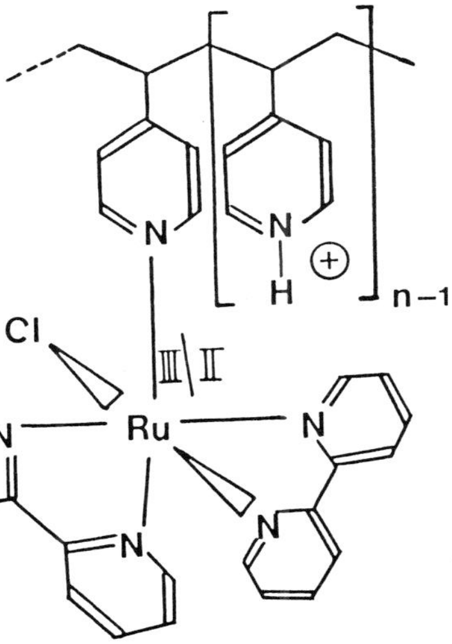

Fig. 1. Chemical structure of the ruthenium polymer.

Experimental values of $\sqrt{D_E}C$ in $HClO_4$ medium are measured between 5°C and 45°C for five concentrations of ruthenium complexes. These values are transformed taking in account the concentration factor as given in fig. 2 in term of $\sqrt{D_E}C°$, $C°$: concentration of ruthenium in 1/5 polymer, this concentration is not accurately known because the exact swelling of the polymer in the solution is difficult to determine.

Some remarks can be made :

- There is no statistically significant difference of $\sqrt{D_E}C$ for $Ru^{II} \longrightarrow Ru^{III}$ or for the reverse process and no influence of electrolyte concentrations.

- The temperature strongly influences the observed pattern but surprisingly $\sqrt{D_E}C°$ increases for two values of C.

- Untill now the literature contains only examples which correspond to the pattern observed at 5°C.

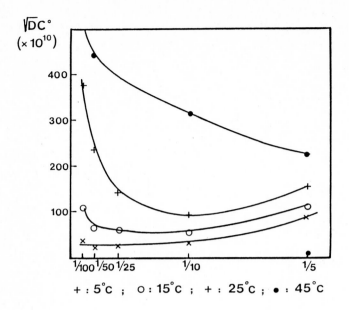

Fig. 2. Dependence of $\sqrt{D_E C^0}$ on concentration C.

These results are represented according to an Arrhenius plot in fig. 3.

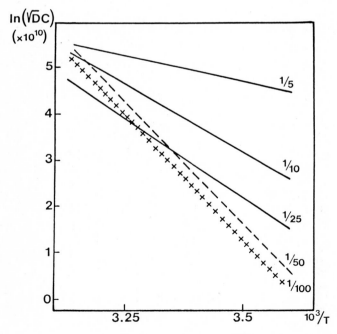

Fig. 3. Arrhenius plots $\ln \sqrt{D_E C}$ versus $1/T$.

These plots are approximately linear and the slopes lead to the determination of activation energies E_a for the diffusion coefficient D_E.

Polymer	1/5	1/10	1/25	1/50	1/100
E_a (Kcal/mole)	8.8	21.5	25.9	39.8	39.8

Values such as 8.8 kcal/mole are typical of the movement of ions in a polymer (ref. 9) or of rotation of side chains in a polymer (ref. 10) while activation energies of 40 kcal/mole are typical for motions of large segments of the polymer (ref. 10) . Thus our results can be rationalized as large chain motions required to bring the electroactive center in contact in the 1/100 polymer and as small distance changes (counter ions or active center) for the 1/5 polymer.

KINETICS OF ELECTROCHEMICAL REACTION MEDIATED BY REDOX POLYMER FILM

We shall now describe the kinetics of the mediated electrochemical reaction for the case of an irreversible cross-exchange reaction and in the context of rotating disc electrode voltammetry.

P/Q is the mediator couple incorporated in the film, A the substrate and B the initial product after reduction (or oxidation). The film thickness ϕ is considered as constant.

Thus we can sketch the system by :

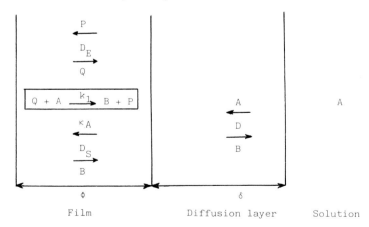

Film	Diffusion layer	Solution

With the following definitions of the various symbols : C_A^o : bulk concentration of the substrate, C_P^o : total concentration of the redox mediator in the film, Γ^o : total surface concentration of the redox mediator in the film ($\Gamma^o = C_P^o \phi$), D : diffusion coefficient of substrate in the solution, D_E : "diffusion" coefficient for the diffusion-like propagation of electrons in the film, D_S : diffusion of substrate in the film, κ : partition coefficient of the substrate between the film and the solution, k_1 : second order rate constant of the reaction between the substrate and the active form of the mediator in the film. F : Faraday's constant, S : electrode surface area, δ : diffusion layer thickness, a dimensionless formulation of the whole problem is obtained through the following changes of variables. The distance to the electrode is normalized according to the film thickness, $y = x/\Phi$, the concentration of substrate according to its bulk value and to the partition coefficient, $a = C_A / \kappa C_A^o$ and the catalyst concentration according to its maximum value $q = C_Q/C_P^o$.

We are now looking for the expression of the plateau current i_1 for the mediated process, thus the kinetic problem (ref. 11) is expressed as :

$$\frac{d^2 q}{dy^2} - (k_1 \ \Phi^2 \ K \ C_A^o \ / \ D_E) \ a \ q = 0$$

$$\frac{d^2 a}{dy} - (k_1 \ \Phi^2 \ C_P^o \ / \ D_S) \ a \ q = 0$$

$y = 0 : q_0 = 1 \quad (da/dy)_0 = 0$

$y = 1 : (dq/dy)_1 = 0 \quad 1-a_1 = (D_S \delta K/D\Phi)(da/dy)_1$

$$i_1 = - \frac{F \ S \ D_E \ C_P^o}{\Phi} \ (\frac{dq}{dy})_0$$

We introduce :

- the four characteristic currents that represent particular values of i_1 (ref. 12) :

(1) $i_A = FSC_A^o D/\delta$: substrate solution diffusion current which is equal to the plateau current at a bare electrode of same surface area;

(2) $i_S = FSC_A^o \kappa D_S/\Phi$: substrate film diffusion current (S case);

(3) $i_E = FSC_P^o D_E/\Phi = FS\Gamma^o D_E/\Phi^2$: film electron diffusion-like transport current (E case);

(4) $i_k = FSC_A^o \kappa k_1 C_P^o \Phi = FSC_A^o \kappa k_1 \Gamma^o$: cross-exchange reaction current (R case).

- normalized value of a towards the concentration of A at the film solution interface $a^* = a(1-i_1/i_A)$ (ref. 13) and the corresponding normalized characteristic currents :

$$i_k^* = i_k(1-i_1/i_A) \quad , \quad i_S^* = i_S(1-i_1/i_A)$$

Thus, the overall kinetics can be expressed in term of q (or of a^*) (ref.14)

$$\frac{d^2 q}{dy^2} - \frac{i_k^*}{i_E} \ q \ \{1 + \frac{i_E}{i_S^*} \ [q-q_1+(1-y)(\frac{dq}{dy})_0]\}$$

with $i_1 = -i_E(dq/dy)_0$

This general expression leads to limiting values of i_1 for extreme values of parameters i_S^*/i_k^* and i_E/i_k^* and correspondingly the concentration profiles of A and Q present characteristic patterns (Fig. 4).

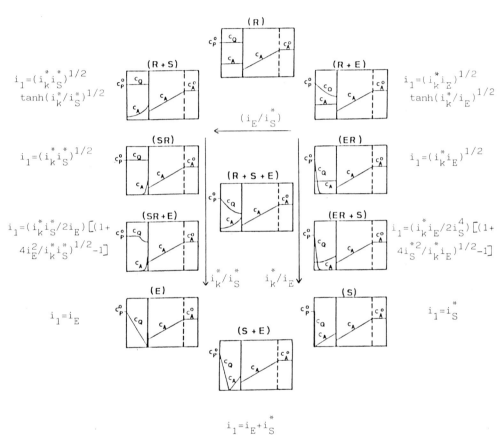

Fig. 4. Limiting expressions of current plateau for mediated process and corresponding concentrations profiles.

For example, in the left part where $i_E \gg i_S^*$ according to the increasing values of i_k^*/i_S, the kinetic control is first by the chemical reaction (R) then by the mutual compensation between catalytic reaction and substrate diffusion (SR) and at last by electron transfer (E) when the i_S^* is so low that the substrate cannot penetrate in the film.

From a practical point of view, experimental conditions lead, with a given accuracy, to limiting behaviours in almost all the cases. Only a small range of possible values of the two parameters i_S^*/i_k^* and i_E^*/i_k^* leads to the general case (Fig. 5).

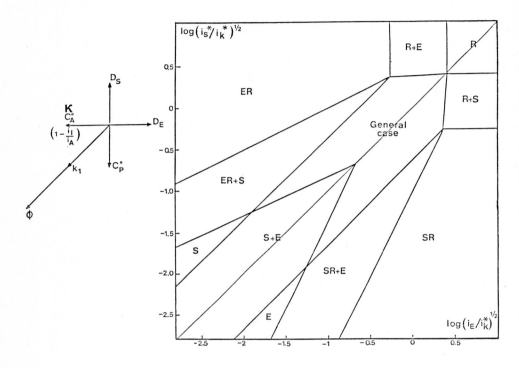

Fig. 5. Kinetic zone diagram.

On the same figure are represented the predicted variations with the experimental parameters (Φ, C_A^o, C_P^o for example). The rotation speed (ω) is often used as an experimental parameter, its effect appears in the quantity $(1-i_1/i_A)$.

In previous papers, we have described (ref. 15) the evolution of half wave potential for the same mechanism and for a reversible exchange reaction according to the values of the same parameters.

MEDIATION OF THE Fe(III)/Fe(II) OXIDO-REDUCTION BY POLY(4-VINYL-PYRIDINE) BOUND RUTHENIUM COMPLEXES.

The applicability of the above kinetic analysis is illustrated in the experimental example of the 1/5 polymer described in Fig. 1. This polymer is used as a mediator for the oxidation of Fe(II) in Fe(III) in HCl medium.

First we can determine the kinetic control of this process, Koutecky-

Levich criteria ($1/i_1$ versus $\omega^{-1/2}$) are used for this purpose.

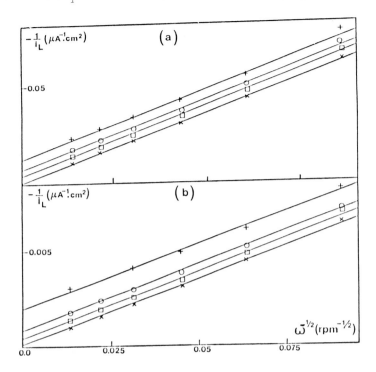

Fig. 6. Koutecky-Levich plots for mediated reduction of Fe(II). a) $C^o_{Fe(II)}$ = 0.1 mM. b) $C^o_{Fe(II)}$ = 1 mM. Γ^o (M.cm^{-2}) : 1.4 10^{-10} (+), 2.5 10^{-10} (o), 6.7 10^{-10} (□), 2.9 10^{-9} (x).

In fig. 6, for two concentrations of Fe(II) and different surface cove-rages Γ^o, linear Koutecky-Levich plots are obtained. At important coverage this plot is identical to a plot for a bare electrode (but with an important gain of potential).

$\frac{1}{i_1} = \frac{1}{i_A}$ = cst x $\omega^{-1/2}$, but for lower values of Γ^o the linear plot inter-cept with the Y axis is inversely proportional to $C^o_{Fe(II)}$ and to Γ^o. This is in agreement with a kinetic control by the chemical electron exchange reac-tion between Ru(III) and Fe(II), case R.

Through additional experiments described in more details in (ref. 16) it is possible to determine all the characteristic currents for this system as functions of $C^o_{Fe(II)}$ and Γ^o.

$i_A = 1.42 \ 10^{-2} \ C^o_{Fe(II)} \omega^{-1/2}$

$i_E = 2.9 \ 10^{-10}/\Gamma^o_P$

$$i_S = 1.35 \ 10^{-8} \ c^{\circ}_{Fe(II)}$$

$$i_k = 4.5 \ 10^9 \ c^{\circ}_{Fe(II)} \Gamma^{\circ}$$

$i(A.cm^{-2})$, $c^{\circ}_{Fe(II)}$ $(M.1^{-1})$, Γ° $(M.cm^{-2})$, ω (rpm).

Thus this polymer is able to catalyze the oxidation of Fe(II) into Fe(III) in 1 M HCl quite efficiently. Total catalysis giving rise to plateau currents, solely controlled by solution mass transport, is reached for surface coverage of the order of 10^{-8} $M.cm^{-2}$. Catalysis then occurs at a half-wave potential which is only 75 mV positive to the Fe(II)/Fe(III) standard potential. Up to this value of the surface coverage, which appears as an optimum both in terms of current and potential catalytic efficiencies, quite substantial portions of the coating are catalytically active. In this respect, the investigated system is a good example of the interest of redox polymer coatings in catalysis of electrochemical reactions deriving from the three-dimensional dispersion of the catalytic centers.

ACKNOWLEDGMENT

The author wishes to thank Drs. J.M. Dumas-Bouchiat, H. Gampp, O. Haas, J.M. Savéant and J.G. Vos for their active participation to this work.

REFERENCES

1 R.W. Murray, "Chemically Modified Electrodes", Electroanalytical Chemistry, A.J. Bard, Ed., Dekker, New York, 1984, vol. 13, pp. 191-368.
2 M. Majda and L.R. Faulkner, J. Electroanal. Chem., 169 (1984) 77-95.
3 (a) : C.P. Andrieux and J.M. Savéant, J. Electroanal. Chem., 93 (1978) 163-168. (b) : C.P. Andrieux, J.M. Dumas-Bouchiat and J.M. Savéant, J. Electroanal. Chem., 123 (1981) 171-187.
4 (a) : D. Lexa, J.M. Savéant and J.P. Soufflet, J. Electroanal. Chem., 100 (1979) 159-172. (b) : F.C. Anson, C.L. Ni and J.M. Savéant, J. Am. Chem. Soc., 107 (1985) 3442-3450.
5 (a) : C.P. Andrieux and J.M. Savéant, J. Electroanal. Chem., 111 (1980) 377-381. (b) : E. Laviron, J. Electroanal. Chem., 112 (1980) 1-9.
6 (a) : K. Shigehara, N. Oyama and F.C. Anson, J. Am. Chem. Soc., 103 (1981) 2552-2558. (b) : S. Nakahama and R.W. Murray, J. Electroanal. Chem., 158 (1983) 303-322. (c) : J.S. Facci, R.H. Schmehl and R.W. Murray, J. Am. Chem. Soc., 104 (1982) 4959-4960.
7 (a) : H. Dahms, J. Phys. Chem., 72 (1968) 362-3 . (b) : I. Ruff and V.J. Friedrich, J. Phys. Chem., 105 (1971) 3297-3302.
8 O. Haas and J.G. Vos, J. Electroanal. Chem., 113 (1980) 139.
9 W.J. Albery, M.G. Boutelle, P.J. Colby and A.R. Hillman, J. Electroanal. Chem., 133 (1982) 135-145.
10 D.A. Seanor in "Polymer Science", A.D. Jenkins (Ed.), North Holland Publishing Comp., Amsterdam, 1972, chap. 17, pp. 1188-1280.
11 C.P. Andrieux, J.M. Dumas-Bouchiat and J.M. Savéant, J. Electroanal. Chem., 131 (1982) 1-35.
12 C.P. Andrieux and J.M. Savéant, J. Electroanal. Chem., 134 (1982) 163-166.
13 W.J. Albery and A.R. Hillman, Annual Report C., (1981), The Royal Chemistry

Society, London, (1983) pp. 317-437.
14 C.P. Andrieux, J.M. Dumas-Bouchiat and J.M. Savéant, J. Electroanal. Chem.,
    169 (1984) 9-21.
15 C.P. Andrieux, J.M. Dumas-Bouchiat and J.M. Savéant, J. Electroanal. Chem.,
    142 (1982) 1-30.
16 C.P. Andrieux, O. Haas and J.M. Savéant, J. Am. Chem. Soc., submitted.

M.R. Smyth and J.G. Vos
Electrochemistry, Sensors and Analysis
© Elsevier Science Publishers B.V., Amsterdam — Printed in The Netherlands

THE CHEMORECEPTOR - TRANSDUCER INTERFACE IN THE DEVELOPMENT OF BIOSENSORS

MICHAEL THOMPSON[1] and U.J. KRULL[2]

[1]Department of Chemistry, University of Toronto,
80 St. George Street, Toronto, Ontario, M5S 1A1 (Canada)

[2]Department of Chemistry, Erindale College,
University of Toronto, Mississauga, Ontario, L5L 1C6 (Canada)

SUMMARY
    Ideal properties of biosensors and design of major device structures are reviewed. Their common dependance on steady-state or equilibrium measurements is evaluated. An alternative route to a sensing strategy is the perturbation of a sensor energy barrier involving transition between initial and final states, possibly in a transient manner. Such technology is apparently employed in natural sensing mechanisms such as olfaction (smell) and neurotransmission. In the present paper we consider our work in the area of chemoreception coupled with alteration of the Arrhenius-Eyring energy barrier to ion conduction of the lipid membrane and acoustic transmission through interfaces as examples of out-of-equilibrium "energy barrier" perturbation.

INTRODUCTION

Research and development in the field of chemical and biosensor technology is progressing at a rapidly accelerating rate as evidenced by the recent appearance of several texts and escalating frequency of conferences dealing with the subject (ref. 1). Possible applications include on-line monitoring and control of industrial chemical processes, real-time environmental analysis, automation of clinical biochemical analyses, feedback control of artificial organs, *in vivo* probe analysis of natural species and drugs, control of robots, and strategic and tactical monitoring of chemical warfare. These applications lead us to a set of attributes that biosensors might possess (ref. 2); selectivity towards a particular chemical species, high sensitivity, capability for fast response, reversibility of selective binding, small size, ruggedness, low cost (especially for the "discard" mentality of the medical world), capability for facile calibration, biocompatible and straight forward telemetry. It goes without saying that these criteria constitute significant challenges for the sensor specialist. Particularly severe problems, although not exclusive, are the reversibility and calibration issues. With regard to the former it is often remarked that reversibility is not required if the device will be used in the "one-shot" mode. Such systems can hardly be categorized as true sensing devices and will not be employable in a number of the exciting applications mentioned above. Reliable calibration without continual removal of the device from the analyte matrix presents obvious constraints.

EXAMPLES OF BIOSENSOR ARCHITECTURE

We can divide the types of the sensor that have been studied very arbitrarily into four classes:

1. The first area involves what might be termed basic electrochemistry such as potentiometry and amperometry. An earlier example is the enzyme electrode which is based upon well-established ion selective electrode technology in a building-block philosophy. These devices have inherent limitations associated with control of ionic strength, non-specific adsorption, enzyme activity etc. These factors can lead to slow non-Nernstian response.

The integrating action of an amperometric sensor designed to consume the products of an enzyme-substrate reaction can significantly improve the detection limit of a simple enzyme-based selective device. New developments in the use of enzymes are based on coupled reactions where two or more immobilized enzymes and/or substrates are employed concurrently. These act sequentially on a substrate or enzyme-substrate product to satisfy reversibility criteria and drive reactions forward to provide maximum analytical product (ref. 3,4).

Antibodies are another major class of selective protein binding agents which have been modified and immobilized for use in potentiometric and amperometric systems. Though some work is directed at developing Nernstian potentials directly from antibody-antigen binding interactions, it has been experimentally and theoretically shown (ref. 5) that these systems are unlikely to provide reliable results. Coupling of the immunochemical interaction with an ionophore (ref. 6) or enzyme (ref. 7,8) provides a chemical amplification of the selective binding event, and improves the analytical signal-to-noise relationship by overcoming limitations such as those imposed by mixed potential effects (ref. 5).

A further interesting concept which has emerged in the area

of electroanalytical sensors is the use of natural chemoreceptive
mechanisms. This involves either modelling natural cell membrane
function (ref. 9) or employing intact animal and plant tissues to
act as biocatalytic materials for the synthesis of products,
which can be further analyzed by potentiometric or amperometric
means (ref. 10, 11).

2. Optical methods represent the oldest and best established
forms of analysis for biochemicals and organics. The inherent
limitations of the Beer-Lambert law with respect to values of the
extinction coefficient has largely negated the development of
miniaturized absorption-based devices for low concentrations of
organics. However, the high sensitivity obtainable by
fluorescence methods combined with the miniaturization of fibre-
optic technology has provided a viable biosensing strategy. The
optical design offers advantages such as small size, elimination
of the reference signal as required in potentiometric electrode
methods, simultaneous multi-wavelength and multichannel operation
providing for distributed sensing and self-calibration, and
elimination of electrical noise so that remote sensing is greatly
simplified (ref. 12,13). Of course these systems have some
disadvantages, but the most significant unique problems such as
stray radiation and limited linearity of response can be
controlled and accepted in view of substantially improved
detection limits. Extrinsic devices with the indicator chemistry
attached to the end of an optical fibre represent the most
common, and perhaps the most primitive system being constructed.
These systems have been tested for pH, inorganic gas and ion
detection (ref. 12,13), but little work has been directed towards
development of biosensors. A substantial body of scientific
information is available for fluorescent immunoassay (ref. 14,15)
and biochemical chemiluminescence techniques (ref. 16), and this
technology will soon be employed for dedicated optical
biosensors. A more sophisticated form of fibre optic sensor
known as an intrinsic sensor has also been described. In this
system, the length of the optical fibre is coated with reagent,
and this selective surface can be optically excited by evanescent
radiation, which is a salient feature of total internal
reflection (ref. 17). This technique is being investigated as
the basis for construction of optical biosensors using

immunochemical reactions (ref. 18) and specialized lipid membrane matrices (ref. 19).

3.   Applications of semiconductor devices dominate the field of chemical sensors (ref. 1,20), but rarely are they described as experimental systems designed for biochemical analysis.   The devices have much to offer in the way of physical attributes, including small size, ruggedness, electrical pathlength miniaturization and potential for on-chip signal processing (ref. 21).   While a wide variety of semiconductor sensors have been described (ref. 1), very few structures have been modified to act as biosensors.   The best known device is the field effect transistor, (FET) which physically interfaces to an aqueous environment by means of a ultrathin non-conductive membrane capable of supporting ionic charge.   The potential developed at this point can control an analytically measurable underlying electronic conduction process located within the semiconductor. The variable potential is derived from identical ion selective-, enzyme- or antibody-electrode interactions previously described above and therefore offers no distinct chemical advantages with respect to selectivity or detection limit.   Examples of interfacing selective biochemical processes, such as enzyme based reactions (ref. 22-24), onto the FET continue to be reported. The real advance in the use of this device will appear when the on-chip processing abilities for background drift and interference control finally mature.

4.   Piezoelectric devices consist of specially cut quartz crystals which can mechanically oscillate under the influence of an applied alternating electrical potential.   Two types of crystal exist and are classified by their respective modes of oscillation.   Bulk acoustic wave devices have been used extensively as gas sensors (ref. 25), and more recently as liquid phase devices with potential for biosensing (ref. 26,27). Surface acoustic wave (SAW) devices have also been described as gas (ref. 28) and liquid phase sensors (ref. 29), though some significant doubt remains as to the mode of operation of SAW's in aqueous media.   The theoretical descriptions for liquid phase operation accepted-to-date (ref. 30) imply that the frequency of crystal oscillation can be modulated by mass changes on a crystal

surface, and therefore mass changes such as those associated with antibody-antigen binding should cause dramatic frequency alterations. Experimental evidence does indicate however that the chemical properties of the interface between the coating on a crystal and an aqueous medium may be very significant in the determination of energy loss, and therefore frequency alteration of a piezoelectric device (ref. 27).

CHEMORECEPTIVE TRANSIENT PERTURBATION OF A POTENTIAL ENERGY BARRIER TO ION CONDUCTION IN THE LIPID MEMBRANE

Almost all of the chemical selectivity reported for these four major sensing devices has been based on the evolution of steady state or equilibrium concentration of analyte resulting from the biochemical action of a recognition compound. Of all the deficiencies which can be attributed to the different devices, this is probably the most significant since it is universal and immediately limits the analytical parameters of response time, reversibility, sensitivity and detection limit. This fact indicates a major flaw in the philosophy of biosensor development and will always limit the development of the field if it remains unresolved.

There has been much interest in the biochemical and physiological attributes of natural chemical sensing of the neural and olfactory systems. The results of extensive studies lead us to draw one important conclusion about nature's method of sensitive and selective chemical sensing. The process employed in natural chemoreception rarely makes use of equilibrium or steady state conditions. Rather, signals are generated by transient phenomena which exist for brief time periods, but have effects much greater than those previously mentioned, while concurrently regenerating the system at a recovery rate which could never be achieved by the previously described sensing structures. This implies that the analytical signal is derived from the perturbed physical state which exists after an initial steady state is altered by receptor-stimulant interaction, but before the system has reached a new steady state which can accomodate the new energetics induced by the selective interaction. Of paramount importance is the physical alteration which occurs in moving between states rather than the energy difference between a defined initial and final state.

In order to examine the hypothesis that "out-of-equilibrium" alteration of the physical chemical state of the bilayer lipid membrane constitutes a potentially valuable electrochemical signal, we have studied the effect of a number of different probes on membrane ion conduction. The latter is apparently governed by membrane molecular packing/fluidity and by the magnitude of the surface dipole potential. Both of these parameters can be thought of, albeit arbitrarily, as components of an Arrhenius-Eyring energy barrier. Our calculations have shown that the dipole potential is perturbable through an angular variation in the dipoles which constitute the lipid headgroup zone (ref. 31). Additionally, perturbation of packing can be seen as an alteration of a "steric" factor which can lead to modulation of intramembrane interstitial cavities. The importance of both of these parameters has been highlighted by the use of a number of oxidized species of the membrane component, cholesterol (ref. 32). In our experiments, such derivatives were mixed with phosphatidyl choline to produce bilayers (for ion conduction measurement) and Langmuir-Blodgett monolayers (for packing and surface potential determination). Correlation of the results of these two experiments show essentially that the steric factor is predominant, but that the dipole potential can be important in the "trapping" of cations at the membrane or monolayer surface. In an overall sense a barrier of 100 meV correlates with a packing area charge of $0.01 nm^2$ for the lipid molecules.

For practical transient signal development it is required to perturb the membrane intrinsic structure in a temporary fashion with full relaxation of lipid molecular orientation. Such an effect is manifested in our work on the alteration of membrane ion current with the hydrophobic carrier, valinomycin (ref. 33), and the dipolar probe, phloretin (ref. 34). The ion current increases caused by these species can be of much greater magnitude, and developed much more rapidly, than the final steady state signal. The work described above was performed with the delicate free bilayer, but recently similar research has been achieved with a bilayer supported on hydrogel as a first attempt at a crude sensor configuration (ref. 35).

Finally, we have shown that the transient perturbation of

the energy barrier to lipid membrane ion conduction can be effected by selective sensing reactions. Incorporation of lectins such as concanavalin A or antibodies into a free bilayer for selective reaction with polysaccharide and antigen, respectively lead to a "state" change in current. The signal is obtained from $\sim 10^{-9}$ M concentrations of analyte and is highly selective. In recent work, similar experiments with Langmuir-Blodgett monolayers which involve infusion of the reactants into an in situ subphase have demonstrated the origin of the transient perturbation (ref. 36).

PERTURBATION OF THE ACOUSTIC INTERFACE

As a second example of non-steady state alteration of a parameter for the development of a sensor signal, we consider the propagation of low energy acoustic waves through a solid-liquid interface. Here, we have devised a system that allows the operation of bulk-wave piezoelectric crystals in liquid media in a flow injection configuration (ref. 27). To achieve this we have used a wideband marginal oscillator to function with an automatic gain control to obviate the effects of mechanical damping. The device is stable to $\pm$ 1Hz in a volume of approximately 50μl of water.

In order to examine the influence of the crystal/electrode interface we have placed different chemistries on the crystal surface, with particular emphasis on transient hydrophilic or hydrophobic character (including the use of Langmuir-Blodgett films.) For example, a silanized surface (with $(CH_3)_2SiCl_2$) which begins with hydrophobic character, becomes hydrophilic after a period of oscillation in water. The interfacial change, which does not correspond to a mass effect, is reflected in the appearance of a transient peak in the frequency response of the crystal. In order to explain this result we have to consider the physics of propagation of the acoustic shear wave through the interface and then out into bulk liquid (ref. 37). The latter can be obtained via the Navier-Stokes theory for a Newtonian isotropic fluid:

$$\rho[\partial v/\partial t + (v.\nabla)v] = -\nabla p + \rho g + \eta \nabla^2 v \qquad (1)$$

where ρ is density, p is pressure, η is bulk viscosity and v is the velocity vector. Solution of this equation results in the damped wave expression:

$$\xi = \xi^0 \exp(-\alpha z)\exp[i(\omega t-(\omega z/vx))] \tag{2}$$

where α is a damping coefficient which includes a viscosity term.
This equation shows that the wave will propagate into water to a
level of about a micron. A second important factor is the nature
of the interface at a given time during oscillation. The stress
will be (x axis is crystal plane, z is axis of propagation):

$$\sigma_{xz} = d\gamma/dx + (n_s + \eta'_s)d^2vx/dx^2 \tag{3}$$

where γ is the interfacial free energy, and n_s and η'_s are two
coefficients of <u>interfacial</u> viscosity. The boundary condition
for acoustic propagation states that the outside solid particles
must move in phase uniformly with the first liquid layer.
Inspection of the last equation shows that change of γ, n_s or η'_s
with time or x results in a lag between crystal oscillation and
the boundary liquid layer. Such an effect can be viewed as an
impedance which leads to loss of acoustic transmission across the
interface.

Accordingly, the crystal frequency rises. However, on
reaching equilibrium the frequency drops resulting in an apparent
transient effect. This process has been shown to occur on
antibody-antigen reaction at the interface and has been suggested
as a possible sensing mechanism for immunochemistry (ref. 37).

FUTURE PERSPECTIVES

1. It is becoming increasingly apparent that the engineering
   aspects of sensor fabrication have outpaced our ability to
   interface devices to appropriate chemistry. This is not
   simply a case of instigating new advances in materials
   science, but is associated with the problems involved in
   chemical transducer mechanisms which require further work.

2. Biotechnologists are increasingly being attracted to the
   visualization of bimolecular interactions, in the case of
   enzyme-substrate chemistry, by molecular graphics. Clearly,
   chemical sensor specialists will have significant interest in
   this field in view of the potential for simulating selective
   interactions for sensor technology.

3. Although we have been almost entirely restricted to enzyme and
   immunochemistry for our selective sensor receiving sites, the
   future may well take the route provided to us by molecular
   receptors. In order for us to be able to employ the requisite

chemistry afforded by receptors, we will have to learn a lot more about their tertiary structure and conformation, mode of action and energetics.

4. The chemometric approach, it is said, can lead to the use of multisensor assays where each sensing element responds selectively, but not specifically, to a particular determinant. However, caution must be invoked here since it has been demonstrated that appropriate mathematical treatments require knowledge of the complete chemical assay. In other words, the chemometric assay will have difficulty dealing with a completely unknown set of compounds.

5. Research into new selective sites for chemoreception and transduction of the chemical binding event will no doubt grow in momentum. This will be fueled by the recognition that "optimization" of existing technology will not suffice.

ACKNOWLEDGEMENTS

We are indebted to the Natural Science Engineering Research Council of Canada, Ministry of the Environment of the Province of Ontario and DARPA for support of our work.

REFERENCES

1   D. Schuetzle, and R. Hammerle, eds., Fundamentals and Applications of Chemical Sensors, American Chemical Society Symposium Series, Vol. 309, (1986) Ch. 21, 351.

2   M. Thompson and U.J. Krull, Trends Anal. Chem., 3 (1984) 173.

3   C.P. Pau, and G.A. Rechnitz, Anal. Chim. Acta, 160 (1984) 141.

4   F. Schubert, D. Kirstein, K.L. Schroeder and F.W. Scheller, Anal. Chim. Acta, 169 (1985) 391.

5   J. Janata, G.F. Blackburn, Immunochemical Potentiometric Sensors, in Annals N.Y. Acad. Sci., 425 (1984) 286.

6   M.Y. Keating and G.A. Rechnitz, Anal. Chem. 56 (1984) 801.

7   T. Fonong and G.A. Rechnitz, Anal. Chem., 56 (1984) 2586.

8   S.B. Brontman and M.E. Meyerhoff, Anal. Chim. Acta, 162 (1984) 363.

9   U.J. Krull and M. Thompson, Trends Anal. Chem. 4 (1985) 90.

10 N. Smit and G.A. Rechnitz, Biotech. Lett., 6 (1984) 209.

11 B.J. Vincke, M.J. Devleeschouwer and G.J. Patriarche, Anal. Chem., 18 (1985) 1593.

12 O.S. Wolfbeis, Trends Anal. Chem., 4 (1985) 184.

13 W.R. Seitz, Anal. Chem., 56 (1984) 16A.

14 J.N. Miller, Analyst, 109 (1984) 191.

256

15   H.T. Karnes, J.S. O'Neal and S.G. Schulman, in Molecular Luminescence Spectroscopy. Methods and Applications: Part 1, Wiley, New York (1985) 717.

16   W. Butt, Practical Immunoassay: The State of the Art, Marcel Dekker, New York, (1984).

17   D. Axelrod, N.L. Thompson and T.P. Burghardt, J. Microscopy, 129 (1983) 19.

18   J.D. Andrade, R.A. Van Wagenen, D.E. Gregonis, K. Newby and J.-N. Lin, IEEE-Electron Devices, 32 (1985) 1175.

19   U.J. Krull, C. Bloore and G. Gumbs, Analyst, 111 (1986) 259.

20   T. Seiyama, K. Fueki, J. Shiokawa and S. Suzuki, Proc. Int'l. Meeting on Chemical Sensors, Fukuoka, Japan, 1983, Elsevier-Kodanska, New York, Tokyo.

21   J.N. Zemel, J. Van der Spiegel, T. Faire and J.C. Young, in Fundamentals and Applications of Chemical Sensors, D. Schuetzle and R. Hammerle, (eds.) American Chemical Society Symposium Series, Vol. 309, 1986, 2.

22   S.D. Caras, J. Janata, D. Saupe and K. Schmitt, Anal. Chem., 57 (1985) 1917.

23   S.D. Caras and J. Janata, Anal. Chem., 57, (1985) 1924.

24   S.D. Caras, D. Petelenz and J. Janata, Anal. Chem., 57 (1985) 1920.

25   G.G. Guilbault in Methods and Phenomena, Their Applications in Science and Technology, C.S. Lu and A.W. Czanderna, (eds.) Elsevier, New York, 7 (1984) 251.

26   P.L. Konash and G.J. Bastiaans, Anal. Chem., 52 (1980) 1929.

27   M. Thompson, C.L. Arthur and G.K. Dhaliwal, Anal. Chem., 58 (1986) 1206.

28   H. Wohltjen, Sens. Actuators, 5 (1984) 307.

29   J.E. Roederer and G.J. Bastiaans, Anal. Chem. 55 (1983) 2333.

30   K.K. Kanazawa and J.G. Gordon, Anal. Chim. Acta, 175 (1985) 99.

31   W.H. Dorn, "The Dipole Potential in Monolayers, Bilayers and Biological Membranes", M.Sc. Thesis, University of Toronto, 1985.

32   U.J. Krull, M. Thompson, E.T. Vandenberg and H.E. Wong, Anal. Chim. Acta, 174 (1985) 83.

33   M. Thompson and U.J. Krull, Anal. Chim. Acta, 141 (1982) 33..

34   M. Thompson, U.J. Krull, I. Lundström and C. Nylander, Anal. Chim. Acta, 173 (1985) 129.

35   A. Arya, U.J. Krull, M. Thompson and H.E. Wong, Anal. Chim. Acta, 173 (1985) 331.

36   M. Thompson and H.E. Wong, submitted for publication.

37   M. Thompson, G.K. Dhaliwal, C.L. Arthur and G.S. Calabrese, IEEE Trans. Ultrasonics, Ferroelectrics and Frequency Control, in press.

M.R. Smyth and J.G. Vos
Electrochemistry, Sensors and Analysis
© Elsevier Science Publishers B.V., Amsterdam — Printed in The Netherlands

REDOX BEHAVIOUR OF METAL AND METAL OXIDE ELECTRODES IN AQUEOUS
SOLUTION WITH PARTICULAR REFERENCE TO ELECTROCATALYTIC DETECTION
OF DISSOLVED ORGANIC AND INORGANIC SPECIES.

LAURENCE D. BURKE and VINCENT J. CUNNANE
Chemistry Department, University College Cork, Cork, Ireland.

SUMMARY

Platinum and gold electrodes are frequently used in
voltammetry work involving electroorganic oxidation reactions in
aqueous media, and the use of on-line electrochemical detectors in
HPLC systems, based on the electrocatalytic response at platinum
substrates, has already been demonstrated elsewhere. In the
present work a new interpretation of the electrocatalytic
behaviour of noble metal anodes is outlined; it is based on the
concept of an interfacial redox cycle involving participation of
in situ-generated, incipient, hydrous oxide species. The
provision of a more logical interpretation of the important
features of such processes should enhance the application of
electrocatalysis in analytical and other fields.

INTRODUCTION

As pointed out by Austin et al. (ref. 1) there is
significant interest, from the electroanalytical viewpoint, in
the anodic oxidation of organic compounds at noble metal
substrates. Since the oxidation response in such circumstances
is apparently determined mainly by reaction of the organics with
active surface oxides generated in situ during the anodic sweep,
it is usually impossible to quantitatively identify components in
a mixture by the direct approach. However, this can be achieved
by combining on-line electrocatalytic detection with prior
separation using a HPLC column. The application of this type of
amperometric detection to a range of inorganics, e.g. Cl^- and Br^-
ions, and organics, e.g. carbohydrates, amino acids and sulphur
compounds, has been outlined (ref. 1) and detection limits well

below the ppm level were achieved in some instances.

The substrate most commonly employed for electrocatalytic amperometric detection is platinum metal. However, the approach is not simple as a reproducible active surface state for the detector electrode is neither readily attained or maintained. There are various factors contributing to this irreproducibility, e.g. variations in the degree of surface roughness, nature of the oxide species present, and both the character and extent of strongly adsorbed organic deactivating species. In practice this problem is largely alleviated by repetitive multipulsing procedures (refs. 1 and 2); basically the adsorbed carbonaceous material is removed by applying an anodic pulse sufficient to generate at least a monolayer of oxide, followed by a negative step to reduce the latter, and then a small anodic step to initiate electrooxidation of the species of interest. The main variables in such repetitive triple-step waveforms in potential-pulse amperometry, as applied for instance in conjunction with HPLC separation (ref. 1), are the potential values, arrest durations, and overall cycling frequency.

The scope and sensitivity of the voltammetric detector system employed in this area is governed to a very large extent by the electrocatalytic response of the indicator electrode. As pointed out in a recent review of electrocatalysis by Pletcher (ref. 3), there is a significant gulf between theory and practice in this field. The degree of uncertainty extends even to the mechanism of electrooxidation of simple organic compounds, e.g. methanol, formaldehyde and formic acid, which are of considerable interest in fuel cell development (one of the most active areas of electrocatalysis research over the past two decades). While some authors (ref. 3) emphasize the role of activated chemisorption, viz.

$$
\begin{array}{ccc}
\overset{\displaystyle O}{\overset{\|}{H - C - OH}} & \overset{\displaystyle O}{\overset{\|}{H - - - C - OH}} & \quad H \quad\ \ COOH \\
| \quad | \quad | & | \quad\ | & \quad\ |\quad\ \ | \\
- M - M - M - & - M - M - M - & - M - M - M -
\end{array}
\qquad (1)
$$

in such reactions, others (ref. 1) stress the importance of surface oxides. While oxide species are clearly not involved in certain anodic reactions, e.g. hydrogen gas oxidation, they apparently are in others, e.g.

$$CH_3OH + H_2O = CO_2 + 6H^+ + 6e^- \qquad (2)$$

The oxygen involved in CO_2 formation in the latter reaction is assumed to be inserted via some form of a Pt-O or Pt-OH complex.

To understand the nature of these electrocatalytic processes, and increase both the sensitivity and versatility of the amperometric approach in analytical chemistry, it is important to develop a better understanding of these electrocatalytic processes. A basic problem here is that although organic compounds oxidize on platinum (ref. 4) at potentials as low as ca. 0.2 V (RHE) in acid media, surface oxidation on this metal, i.e. formation of $Pt.OH_{ads}$.-type species (ref. 5), is normally regarded as commencing, under anodic sweep conditions, only at potentials in the region of ca. 0.8 V (RHE). How then can surface oxides be involved in a reaction occurring at a much lower potential? The solution to this problem is to be found in recent work in this laboratory devoted largely to hydrous, hyperextended - as opposed to regular, compact - oxides. It is worth noting first of all that the term _oxide_ is frequently used in electrochemistry in a rather broad sense to denote not just pure oxides, M_mO_n, but also hydroxides, $M(OH)_x$, oxyhydroxides, $MO_y.(OH)_z$ and mixtures of variable composition in terms of not only O and OH but also OH_2 content. In the case of nickel oxide for example the anhydrous materials, in the ideal case, consist of an extended array of brucite-type layers of $Ni(OH)_2$ or NiOOH (ref. 6); in the hydrous materials neighbouring OH or O layers are separated from one another by water molecules (plus, in the case of the oxidized state, some electrolyte species). In the case of platinum the hydrous material involved is now assumed to be a layer of hexahydroxyplatinic acid, $H_2Pt(OH)_6$, or one of its salts. Such materials, which can be produced on platinum by d.c. polarization or potential cycling techniques (ref. 7), reduce in acid at ca. 0.2 V (RHE) - precisely the region where the oxidation of organic materials commences on this metal in acid.

It is quite widely assumed (ref. 8) that in the case of platinum in aqueous media a clear separation exists between the hydrogen and oxygen adsorption regions. However, this view may have to be modified as on the cathodic sweep the hydrous oxide reduction peak at ca. 0.2 V (RHE) strongly overlaps with the hydrogen adsorption region (ca. 0.0 to 0.35 V (RHE)). A further complication here is that although the hydrous oxide reduction is quasi-reversible (the peak current is proportional to sweep-rate), the peak potential drifts anodically with decreasing sweep-rate and the reverse process (hydrous oxide formation from the metal) occurs only to a negligible extent at the lower region of the anodic sweep (overall in the latter case, compact monolayer oxide formation above ca. 0.8 V (RHE) is the main reaction). In addition, hydrous platinum oxide reduction cannot readily be achieved, even at low sweep-rates (10 mV s^{-1}), in base; vigorous hydrogen gas evolution occurs prior to the oxide removal process at higher pH (ref. 7). Such results may be rationalized by the folowing reaction scheme, viz.

$$Pt(OH)_6{}^{2-} + 6H^+ + 4e^- = Pt^* + 6H_2O \quad (reversible) \tag{3}$$
$$Pt^* = Pt_L \quad (irreversible) \tag{4}$$

The difficulty in reducing the film in base can be explained in terms of equation (3); since the H^+/e^- ratio involved is 3/2, the potential/pH shift according to the Nernst equation is 3/2 (2.303 RT/F) V/pH unit (SHE scale) or 1/2(2.303 RT/F) V/pH unit (RHE scale), i.e. if the peak potential is 0.2 V (RHE) in acid it must be at ca. -0.2 V (RHE) in base. Furthermore, since the initial reduction product, the adatom-type species Pt*, is followed by incorporation of the latter into regular lattice sites (Pt$_L$ is a much less reactive species), the lack of overall reversibility is understandable. At a polycrystalline surface there will always be a low level of metal atoms present with a low degree of bulk metal lattice coordination. However, it is virtually impossible to observe the current response due to oxidation of the latter on the anodic sweep - the small, probably broad, peak is submerged in the currents associated with adsorbed hydrogen oxidation. However, this process, formation of submonolayer traces of active hydrous oxide material prior to the normal monolayer oxide formation peak, is vital from an electrocatalytic viewpoint. For this reason attention in the present work is concentrated on gold where the reactions of hydrous and anhydrous oxide films are easily distinguished (refs. 9 and 10), especially in base, and there is no overlap with hydrogen adsorption/desorption processes - gold in fact has only a quite low affinity for hydrogen (ref. 8).

EXPERIMENTAL

The working and counter electrodes consisted of gold wire (Goodfellow Metal, 99.95% pure, 1.0 mm diam., ca. 0.65 cm^2 exposed area) sealed directly into glass. The surface was cleaned by brief immersion in aqua regia, following by washing with triply distilled water; abrasion with fine emery paper followed by further washing completed the pretreatment. Solutions were generally made up with triply distilled water and Analar grade, or redistilled, reagents.

A conventional three compartment cell was used with a glass frit separating the working and counter compartments. A hydrogen reference electrode in the same solution (RHE) was used, together with a Luggin capillary to minimize iR effects. Solutions were deoxygenated and stirred as required with a flow of purified nitrogen gas. Potential control was achieved with a potentiostat (Wenking, Model PGS 81) combined with a Metrohm VA-Scanner, type E612. Cyclic voltammograms were plotted with the aid of a Rikadenki, Model BW-133, X-Y recorder.

RESULTS AND DISCUSSION

Typical examples of cyclic voltammograms for gold in acid and base are shown in Fig. 1. Although, in agreement with

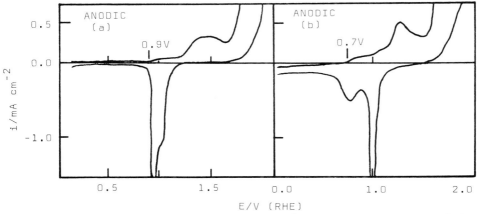

Fig. 1. Cyclic voltammograms (40 mV s^{-1}, 0 - 2.2 V, 5th cycle) for Au in (a) 1.0 M H$_2$SO$_4$ and (b) 1.0 M NaOH.

conventional thermodynamic data (ref. 12), most authors assume
that the metal does not commence to oxidize below about 1.35 V
(RHE) in acid, or at a slightly lower value (ca. 1.15 V) in base
(refs. 8 and 13), the onset of small faradaic currents can be
observed on the anodic sweeps in Fig. 1 as low as ca. 0.9 V (RHE)
in acid and ca. 0.70 V (RHE) in base. These processes, which
occur prior to the initiation of regular monolayer oxide
formation, and in some instances – especially in base (refs. 14
and 15) – have given rise to one or more discrete anodic/cathodic
pairs of peaks, have occasionally been attributed to impurities
(ref . 8). However, since thick hydrous oxide layers, produced
on gold by vigorous anodization, reduce at the potentials in
question (ref. 9), the reactions involved are assumed to be of
the following type

$$Au_2(OH)_9^{3-} + 9H^+ + 6e^- = 2Au^* + 9H_2O \text{ (acid)} \qquad (5)$$
$$Au_2(OH)_9^{3-} + 6e^- = 2Au^* + 9OH^- \text{ (base)} \qquad (6)$$

On the anodic sweep there are only a limited number of adatoms
(Au*), i.e. gold atoms virtually protruding from the surface
about which six hydroxy groups can coordinate; hence the
majority of surface gold atoms react in the conventionally
accepted manner as follows

$$Au + H_2O = Au(OH)_{ads} + H^+ + e^- \qquad (7)$$

The resulting hydroxy species are assumed to undergo
place-exchange type reactions, eventually forming the
conventional, largely anhydrous, compact monolayer, usually
considered to be Au_2O_3 or $Au(OH)_3$.

The assignment of the formula $Au_2(OH)_9^{3-}$ to the active
species formed on oxidation of adatoms at the gold/solution
interface must be regarded as tentative. The argument for such
species to date (refs. 10 and 11) is mainly electrochemical; in
situ structural data for species present at a submonolayer level
at the metal/solution interface is exceedingly difficult to
obtain. However, it is interesting that the unusual,
super-Nernstian, E/pH shifts described here for platinum and gold
have also been reported for organic coatings in redox modified
electrode systems (ref. 16). Such super-Nernstian shifts are
interpretable on a thermodynamic basis; however, it is necessary

to assume interaction between the acid-base and redox properties
of the electroactive centres in such films (ref. 11).

Two examples are given here (Fig. 2) to illustrate the
involvement of incipient, anionic, hydrous oxide species in the
electrocatalytic behaviour of organic materials at gold anodes in

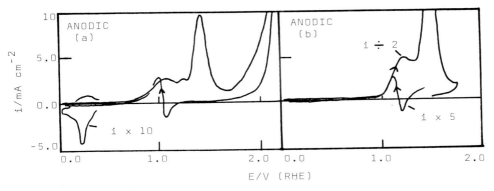

Fig. 2. Cyclic voltammograms for Au in (a) 1.0 M NaOH + 0.1 M
pyrrolidine and (b) 1.0 M H$_2$SO$_4$ + 0.1 M oxalic acid (40 mV s^{-1},
0 - 2.2 V).

acid and base. It is clear that oxidation commences (anodic
sweep) and terminates (cathodic sweep) at the potential
corresponding to the adatom/hydrous oxide transition. It is
assumed here that an interfacial surface redox transition is
involved in which the incipient hydrous oxide species mediate the
organic oxidation process, viz.

$$R = O + ze^- \tag{8}$$
$$O + X = R + \text{oxidation products} \tag{9}$$

R is the reduced state, i.e. Au*, and O is the oxidized form
Au$_2$(OH)$_9{}^{3-}$ (the H$_2$O or OH$^-$ species involved in these reactions
are not shown here); this initial redox transition of the
surface - bonded species is assumed to be rapid, i.e. it
generally occurs in a thermodynamically reversible manner. The
second stage, reaction of surface bound oxidant with the
dissolved organic molecules, (X) is the rate determining stage.
The detailed mechanism of the latter reaction is uncertain; it
may well be that the organic molecule is activated to some degree

by formation of an $[Au(1)X_n]^+$ cationic complex - some evidence for formation of the latter (see the lower cathodic peak in Fig. 2(a)) will be discussed later. During the conversion of this complex to the anionic hydrous oxide, $Au_2(OH)_9{}^{3-}$, state hydroxy species and the organic ligand are held in close proximity at adjacent coordination sites at the same central metal ion; consequently, the probability of reaction is quite large.

Simple kinetic treatment of this type of interfacial cyclic redox process (ref. 17) suggests that the i/E response under potential sweep conditions should resemble that observed in polarography, with the onset/termination potential coinciding (as observed here) with the reversible potential for the surface couple (in this case the $Au^*/Au_2(OH)_9{}^{3-}$ transition - modification of the latter due to Au^* interaction with X does not appear to be marked). However, the limiting current (i_{max}) in the present case is determined not by diffusion but the rate of the second reaction, equation (9), in the overall interfacial process (ref. 17)

$$i_{max} = zFkKC_X[A] \tag{10}$$

[A] is the maximum coverage of the oxidized state (0) at the interface, C_X is the bulk concentration of the organic reactant X, and both k and K are constants. From an analytical viewpoint the absence of diffusion control is probably a disadvantage. Also the limiting current value, i_{max}, will be influenced by the active site, i.e. $Au_2(OH)_9{}^{3-}$, density at the interface - a factor which is not easily controlled.

It is clear that the electrooxidation behaviour outlined here is complex; for example in the case of pyrollidine in base, Fig. 2(a), three peaks, rather than a plateau, are observed on the anodic sweep in the region above the onset of organic oxidation. There may be contributions due to adsorption and transport phenomena involved here but the main feature, especially in the case of the main peak at ca. 1.4 V, is probably changes in both the character and concentration of the interfacial oxyspecies. As demonstrated earlier (ref. 13) the species produced in the initial stages of conventional monolayer oxide formation of ca. 1.1 V (RHE) are hydrated to a significant degree - this is reflected here in the sharp increase in anodic

current in this region. However, as the oxide coverage increases
at higher potential the activity for electrooxidation decreases.
Evidently as the conventional oxide monolayer coverage increases
there is either a loss of hydrous species (due to dissolution or
conversion to anhydrous oxide) or the gold adatoms necessary for
activation of the organic species are lost due to conversion to
monolayer oxide. In either event the activity of the surface
decreases and remains low during the subsequent anodic sweep
until significant reduction of the monolayer oxide material takes
place. Once this occurs, gold adatoms are again present at the
interface and organic oxidation recommences - note the sharp
switch from cathodic to anodic current on the cathodic sweep. As
the sweep continues in the negative direction the oxidation
current decreases, eventually dropping to zero at ca. 0.70 V
(RHE) at which point the interfacial cyclic redox mechanism can
no longer operate as the oxidized state (O) of the surface
couple, equation (8), is no longer produced at the interface.

Conventional descriptions of the voltammetric behaviour of
gold in aqueous media (ref. 8) are usually confined to one anodic
and one cathodic peak, corresponding to compact monolayer oxide
formation and reduction, respectively, in the region between
hydrogen gas evolution at one end of the range and oxygen gas
evolution at the other. However, as outlined here in Fig. 3 more

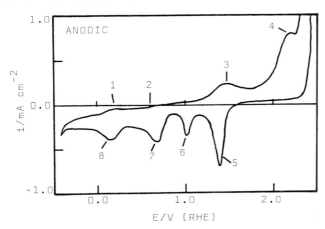

Fig. 3. Cyclic voltammogram (40 mV s^{-1}, 0 to 2.5 V) for Au in
0.2 M Na$_2$SO$_4$ + NaOH, pH = 11.65, + 0.1 M pyridine: sweep stopped
at 2.5 V for 30 s.

complex behaviour, involving a total of eight peaks (some rather ill-defined, e.g. 1 and 2), may be observed under certain conditions. Peaks 3 and 6 correspond to the generation and removal, respectively, of the compact oxide monolayer. Peak 5 is somewhat of an artifact; it is most clearly observed when the electrode is held for a significant period at the upper limit, i.e. above 2.2 V (RHE). In the latter region vigorous oxygen gas evolution occurs according to the reaction

$$H_2O = 1/2O_2 + 2H^+ + 2e^- \qquad\qquad (11)$$

and the production of protons lowers the pH of the solution surrounding the working electrode. Consequently, on resuming the sweep in the cathodic direction some of the compact oxide material is reduced in an acidic environment – elementary Nerstian analysis shows that with respect to the pH invariant reference electrode, in this case Pt/H_2(1 atm), H^+ (pH = 11.65), the Au_2O_3 material in acid will reduce at a higher potential than the Au_2O_3 material in base. Thus peak 5 corresponds to the same type of reaction as peak 6 – the splitting of the compact oxide reduction peak being due to a local pH change.

The pH change in this case is obviously quite dramatic as most of the compact material is reduced in an acidic environment (peak 5 is clearly much larger than peak 6 and the charge ratio of 5 to 6 increased with increased holding time above 2.2 V). This is scarcely surprising as holding the potential above 2.2 V also results in significant conversion of the outer region of the compact film to the hydrous oxide (the presence of the latter inhibits pH readjustment); the reduction of the later is assumed to give rise to peak 7.

Peak 4 is due to a phenomenon that is observed with gold in solutions of high pH (the effect is also seen in earlier work (ref. 13) where the rise in anodic current in the oxygen gas evolution region occurs at much lower potential in base as compared with acid). A clear peak can be seen in this region in some cases and release of oxygen gas was observed at potentials about the maximum. However, the effect here is somewhat transitory in character and may be due to the presence of an active intermediate for oxygen gas evolution which either dissolves or decays to a less active state; the currents in this region decay with time if the anodic sweep is stopped at, or before, the maximum and under the conditions used here no indication of reaction was recorded in this region during the subsequent cathodic sweep.

The final cathodic peak, i.e. peak 8, is observed only when organic bases, e.g. pyridine, pyrrolidine or piperidine are present in solution (see also Fig. 2(a)). Such organic species are capable of stabilizing the Au(1) state; $Au(pyridine)_2^+$ is an established species (ref. 18) and is assumed to be produced here at gold adatom sites at low potential, i.e. after reduction of the compact and hydrated oxide materials. The reaction giving rise to peak 8 is, therefore, attributed to the reaction

$$Au(pyridine)_n^+ + e^- = Au^* + n \text{ pyridine} \qquad (12)$$

The two minor peaks (1 and 2) on the anodic sweep, which in fact generally appear as small current increases rather than discrete peaks, are assumed to be due to the formation of the pyridine complex (peak 1) and hydrous oxide material (peak 2) i.e. the anodic counterparts of the reactions giving rise to peaks 7 and 8, respectively. The much lower response in this region on the anodic, as compared with the cathodic, sweep is assumed to be due to the reduced level of gold adatoms at the interface in the former case. On the cathodic sweep reduction of both the compact and hydrous oxide layers yields a high level of gold adatoms which are quickly stabilized by reaction with pyridine. However, after reduction of the pyridine complex, equation (12), the adatoms become incorporated into regular surface lattice sites and, therefore, only a much lower level is available for reaction on the subsequent anodic sweep.

Further support for this new approach will be found in the work of Lamy et al. (ref. 4). The onset/termination potential for formic acid oxidation on platinum occurs at ca. 0.2 V (RHE), precisely the region where hydrous oxide reduction takes place for this metal. Under comparable conditions methanol reacts more slowly and the onset/termination potential is slightly higher, ca. 0.4 V (RHE). Methanol may be less capable of reacting with, and thereby stabilizing, Pt* species. An additional complication with such species on platinum is the existence of a severe deactivating effect due to carbonaceous residues; this results in significant distortion of the i/E behaviour during the anodic sweep.

CONCLUSIONS

The analytical applications of noble metal based, on-line electrocatalytic detectors for HPLC work has been described by Austin et al (ref. 1); their data for carbohydrates, and other organics, are impressive and the approach has apparently been commercialized. However, it is obvious that since the response is electrocatalytic in origin, a clearer model of electrocatalysis is essential if the sensitivity and scope of this detection procedure is to improve. The present work should provide a solid basis for further endeavour in this area.

REFERENCES

1   D.S. Austin, J.A. Polta, T.Z. Polta, A.P.-C. Tang, T.D. Cabelka and D.C. Johnson, J. Electroanal. Chem., 168 (1984) 227.
2   S.B. Brummer, J.I. Ford and M.J. Turner, J. Phys. Chem., 69 (1965) 3424.
3   D. Pletcher, J. Appl. Electrochem., 14 (1984) 403.
4   C. Lang, J.M. Leger, J. Clavilier and R. Parsons, J. Electroanal. Chem., 150 (1983) 71.
5   H. Angerstein-Kozlowska, B.E. Conway and W.B.A. Sharp, J. Electroanal. Chem., 43 (1973) 9.
6   L.D. Burke and T.A.M. Twomey, in R.G. Gunther and S. Gross (Eds.), Proc. Symp. on the Nickel Electrode, The Electro-chemical Society, Pennington, New Jerseym 1982, pp. 75-96.
7   L.D. Burke and M.B.C. Roche, J. Electroanal. Chem. 164 (1984) 315
8   R. Woods, in A.J. Bard (Ed.), Electroanal. Chem. Vol. 9, Marcel Dekker, New York, 19, pp. 1-162.
9   L.D. Burke and M. McRann, J. Electroanal. Chem., 125 (1981) 387.
10 L.D. Burke, M.E. Lyons and D.P. Whelan, J. Electroanal. Chem., 139 (1982) 131.
11 L.D. Burke and D.P. Whelan, J. Electroanal. Chem., 162 (1984) 121.
12 M. Pourbaix, Atlas of Electrochemical Equilibria in Aqueous Solutions, Pergamon Press, Oxford, 1966 pp. 399-405.
13 L.D. Burke, M.M. McCarthy and M.B.C. Roche, J. Electroanal. Chem., 167 (1984) 291.
14 H. Angerstein-Kozlowska, B.E. Conway, B. Barnet and J. Mozota, J. Electroanal. Chem., 100 (1979) 417.
15 J. Gonzales-Velasco and J. Heitbaum, J. Electroanal. Chem., 54 (1974) 147.
16 G. Inzelt, J.Q. Chambers, J.K. Kinstle and R.W. Day, J. Amer. Chem. Soc., 106 (1984) 3396.
17 F. Beck and H. Schulz, Electrochim. Acta, 29 (1984) 1569.
18 P.J. Saddler, Structure and Bonding, 29 (1976) 176.

M.R. Smyth and J.G. Vos
Electrochemistry, Sensors and Analysis
© Elsevier Science Publishers B.V., Amsterdam — Printed in The Netherlands

AN ELECTROCHEMICAL STUDY OF THE OXIDATION OF $[Fe(CN)_6]^{4-}$ MEDIATED BY ELECTRODES MODIFIED WITH RUTHENIUM-CONTAINING POLYMERS.

JOHN F. CASSIDY, ANTHONY G. ROSS AND JOHANNES G. VOS
School of Chemical Sciences, National Institute of Higher
Education, Dublin 9, Ireland

SUMMARY
   The mediating properties of two ruthenium-containing polymers
have been investigated as a function of polymer backbone, pH,
background electrolyte anion, and substrate concentration. For
the thin coatings and low $[Fe(CN)_6]^{4-}$ concentrations used,
the polymer mediated oxidation is not affected very strongly by
a change of these parameters. The mediating process is
controlled by diffusion of the substrate to the electrode
surface. In rotating disk experiments non-steady state
voltammograms are obtained for higher scan rates (>1 mV/s). The
current-voltage curves thus obtained yield additional
information about the nature of the mediating process. The
specificity of the modified electrodes towards dissolved oxygen
suggests their possible use as oxygen sensors.

INTRODUCTION

   Poly(4-vinylpyridine) has been widely used for the binding of
electrochemical species to electrode surfaces. The electroactive species can be bound to the polymer backbone by either
electrostatic (refs.1-3) or chemical means (refs.4-6).

   The aim of our work is to examine the mediated oxidation of
$[Fe(CN)_6]^{4-}$ by electrodes modified with metal-containing
polymers of the second type as a function of pH, polymer
backbone, background electrolyte anion and substrate concentration. In this paper preliminary results for the polymers
$[Ru(bpy)_2Cl(PVP)_5]Cl$ (polymer I) and $[Ru(bpy)_2-Cl(PVP)(MMA)_4]Cl$ (polymer II) are given where MMA denotes
poly(methylmethacrylate) and PVP is poly(4-vinylpyridine).

EXPERIMENTAL

   The polymers were prepared according to literature procedures
(refs.4-5). Known quantities of solutions of the polymer in
methanol were placed on a glassy carbon disk (r=0.15cm) and the

solvent was allowed to evaporate. The resulting polymer coated electrode was left to 'cure' for a few days. It was found that a longer curing process improved the stability of the modified electrode. The coatings were characterised by cyclic voltammetry in aqueous solution (0.2 M $NaClO_4$ or 0.2 M KCl). For the electrochemical measurements a three electrode system was used with a Pt wire as auxiliary and a saturated sodium chloride calomel electrode as a reference electrode, in connection with a rotating disk assembly (Pine Instruments), a PAR model 174A polarographic analyser and a PAR 175 Universal Programmer. The amount of electroactive material on the electrode surface was obtained from the area under the cyclic voltammograms and by using a PAR 379 Digital Coulometer. Freshly prepared $K_4[Fe(CN)_6]$ solutions in the appropriate background electrolyte were used as substrates. Experiments were carried out at $20° \pm 2°C$ in a thermostatted cell.

RESULTS AND DISCUSSION

Cyclic voltammetry of the polymer modified electrodes in background electrolyte yielded curves which are characteristic of surface bound species. In chloride containing buffers the peak to peak separation decreased from 80 mV at pH 6 to 20 mV at pH 2. This behaviour may be attributed to a decrease in film resistance as the chains are protonated and opened to the solution. In perchlorate medium there was no change in either peak to peak separation (80 mV) or peak position with pH. The polymers were found to be insoluble in perchloric acid and are expected to be less swollen in this electrolyte. This most likely results in a reduced efficiency for the movement of counter ions into the layer and therefore is a slower redox process resulting in a rather large peak to peak separation. A similar behaviour has been reported by Niwa and Doblhofer (ref.7) for crosslinked methyl quarternised poly-(4-vinylpyridine) with electrostatically attached $[Fe(CN)_6]^{3-}$ groups. In chloride containing buffers the behaviour of electrodes coated with polymers I and II is quite different. In a high pH chloride buffer containing equimolar amounts of $[Fe(CN)_6]^{4-}/[Fe(CN)_6]^{3-}$ an anodic current was observed for the mediated oxidation of $[Fe(CN)_6]^{4-}$ by the polymer I. At low pH an additional peak due to the oxidation of $[Fe(CN)_6]^{4-}$ at the bare electrode was observed. Under the

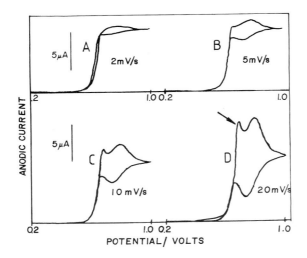

Figure 1. RDE voltammogram of GC electrode coated with polymer I
at various scan rates. $\Gamma_{Ru} = 7 \times 10^{-9}$ moles/cm^2, rotation
rate = 60 RPM, electrolyte; 0.2M NaClO$_4$, pH = 6.5, substrate
$[Fe(CN)_6]^{4-}$ = 0.17 mM.

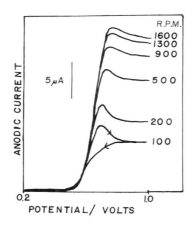

Fig.2. RDE voltammograms of a GC electrode coated with polymer I
at various rotation rates; $\Gamma_{Ru} = 7 \times 10^{-8}$ moles/cm^2 ;
$[Fe(CN)_6]^{4-}$ = 0.86 mM ;0.2 M NaClO$_4$ pH = 2.0; scan rate ;
5 mV/s; rotation rates as indicated in RPM.

conditions used no reduction of $[Fe(CN)_6]^{3-}$ was observed.
This indicates that unlike the results obtained by Niwa et al.
(ref.7) for quarternised PVP, $[Fe(CN)_6]^{3-}$ can not reach the
electrode surface in this particular case. Little or no
oxidation of ferrocyanide at the bare electrode was observed
under similar conditions for an electrode coated with polymer
II. This suggests a less open structure due to the absence of
protonated pyridine groups in this polymer film. Under the
experimental conditions used ($\Gamma'_{Ru} = 10^{-8}-10^{-10}$
moles/cm^2; $[Fe(CN)_6]^{4-}$<2mM) non-stationary current-
voltage curves as in Fig.1 or Fig.2 were obtained. The shape of
these curves can be explained by a combination of the stationary
wave of the bulk $[Fe(CN)_6]^{4-}$ oxidation, coupled with the
oxidation and reduction of the surface bound mediator. As
expected, a faster scan rate gives rise to an increased cyclic
voltammetric peak on top of the stationary wave (Fig 1A - 1D).
Another feature is the appearance at higher scan rates of a
sharp prepeak on top of the wave (see arrow Fig. 1D). This
prepeak can be explained by the oxidation of $[Fe(CN)_6)^{4-}$
already present in the film. The magnitude of this current does
therefore depend on diffusion of the substrate from the bulk
solution to the electrode surface. Once this amount of substrate
has been oxidised the current is controlled by substrate diffusion in the solution. Since there is a rapid increase in
current long before all the ruthenium in the film has been
oxidised, curves such as Fig. 1 indicate that initially the
reaction between the mediator and the substrate occurs well
within the film, close to the electrode surface. The differences between Figs. 1 and 2 may be explained by a more open
structure of the polymer layer in Fig. 2. This will allow a
more efficient diffusion of the substrate into the coating. The
substrate is then able to reduce Ru(III) faster than the electrode. Curves as in Fig. 2 have been reported in the literature
(refs. 7-10).

The position and size of the prepeak in Fig. 1. could yield
information about where the mediation is taking place within the
polymer and the ease with which the $[Fe(CN)_6]^{4-}$ enters into
the film. The presence of this peak does not interfere with the
subsequent limiting current plateau. It was found that the
current voltage curves did not change shape in perchlorate
solution as a function of pH for either polymer I or II.

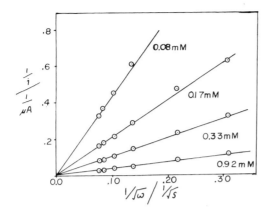

Fig.3. Koutecky-Levich plot for a GC electrode coated with polymer II; $\lceil_{Ru} = 5 \times 10^{-9}$ moles/cm^2; RDE experiment carried out at various $[Fe(CN)_6]^{4-}$ concentrations in mM in 0.2 M NaClO$_4$; pH = 6.5, scan rate = 5 mV/s.

Typical Koutecky-Levich plots are shown in Fig. 3 as a function of the $[Fe(CN)_6]^{4-}$ concentration. Under the experimental conditions used, the slope changes with $[Fe(CN)_6]^{4-}$ concentration, while the intercept remains zero within experimental error. The change in intercept was not significant as a function of substrate concentration, pH or polymer backbone. The slope of the curves was used to estimate the apparent diffusion coefficient, D, of the substrate in bulk solution (ref.11); using the formula;

$$SLOPE = 4.98 \, \nu^{1/6}/(FSC_A^0 \, (D^{2/3}))\tag{1}$$

where ν is the kinematic viscosity, S the electrode area, and C_A^0 the $[Fe(CN)_6]^{4-}$ bulk concentration.

The calculated diffusion coefficient for $[Fe(CN)_6]^{4-}$ depended on the background electrolyte anion. In perchlorate buffer, a value similar to that in the literature was obtained ($6.5 \pm 1.5 \times 10^{-6}$ cm^2/s). In chloride a significantly higher but inconsistent value for D was obtained ($11 \pm 5 \times 10^{-6}$ cm^2/s). There was no difference in the calculated D as a

function of the polymer backbone. Both polymers I and II
yielded consistent results in chloride and perchlorate electrolytes. There was no difference in the calculated value as a
function of pH.

   Under normal experimental conditions, it was found that while
over 90% of the original polymer remained after the series of
experiments in perchlorate, there was sometimes less than 50%
remaining after a similar set of experiments in chloride at low
pH. Loss of polymer from the electrode surface does not seem to
affect the limiting current under these conditions.

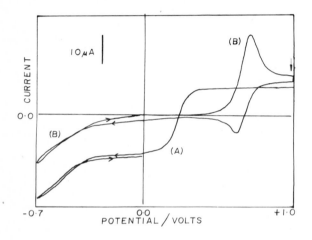

Fig.4. RDE voltammetry experiment carried out in
$[Fe(CN)_6]^{-3/-4}$ (10 mM) ;0.2 M $NaClO_4$; pH = 6.3; scan rate
= 10 mV/s; rotation rate = 200 RPM. using; (A) a bare glassy
carbon electrode, (B) a GC electrode coated with polymer I
(Γ_{Ru} = 1x10^{-8} moles/cm^2)

   Fig. 4 shows the RDE voltammogram for both a bare glassy
carbon electrode and an electrode coated with polymer I in an
equimolar solution of $[Fe(CN)_6]^{4-}$ and $[Fe(CN)_6]^{3-}$ in
perchlorate buffer. When the potential is held at +1.0 V the
current is seen to drop to a steady-state value. At the
modified electrode no reduction current was observed for
$[Fe(CN)_6]^{3-}$. In low pH chloride containing buffers a
similar behaviour is observed with some reduction of
$[Fe(CN)_6]^{3-}$ at the electrode surface. It is interesting to

note that for similar experiments carried out in KCl using quaternised poly(4-vinylpyridine) an efficient reduction of ferricyanide at the electrode surface was observed (ref.7). This suggests that the quarternised polymer has a more open structure so that substrate can diffuse better to the electrode surface. The important observation is that while ferrocyanide is not reduced, oxygen is reduced at about -350 mV vs. SSCE, as at a bare electrode. Although the reduction of oxygen has been under examination by various groups (refs.9,12), this specificity has not been cited. Specificity of this nature is a prerequisite for a useful sensor. Possible applications of this system are under investigation.

CONCLUDING REMARKS

An analysis of the current voltage curves obtained in this study using Koutecky-Levich plots yields constant intercepts as a function of substrate concentration. In earlier experiments with $[Fe(H_2O)_6]^{2+}$ as a substrate, a variation of the intercept with the substrate concentration was found (refs. 13,14). In our studies $[Fe(CN)_6]^{4-}$ was used as a substrate and the results obtained suggest that this ion, because of its negative charge, penetrates more easily into the polymer layer, than the positively charged $[Fe(H_2O)_6]^{2+}$ ion. This is expected to result in a faster mediating process. As a result, the rate determining step for the mediated oxidation of ferrocyanide is diffusion from the bulk solution to the electrode surface. If one therefore wishes to study the influence of the polymer back-bone, pH, etc. on the mediated oxidation of $[Fe(CN)_6]^{4-}$, thicker layer and/or higher substrate concentrations are needed.

ACKNOWLEDGEMENTS

JC would like to acknowledge a Department of Education (Ireland) post-doctoral fellowship. The National Board for Science and Technology is thanked for financial assistance.

REFERENCES

1   N.Oyama and F.Anson, J. Am. Chem. Soc., 101, (1979), 3450.
2   K.Shigehara, N.Oyama and F.C.Anson, J. Am. Chem. Soc., 103, (1981), 2552.
3   P.W.Geno, K.Ravichandran and R.P.Baldwin, J. Electroanal. Chem., 183, (1985), 155.
4   A.G.Ross and J.G.Vos, unpublished results
5   J.M.Clear, J.M.Kelly, C.M.O'Connell and J.G.Vos, J. Chem. Res.M, (1981), 3039.
6   J.M.Calvert and T.J.Meyer, Inorg. Chem., 21, (1982), 3978.
7   K.Niwa, and K.Doblhofer, Electrochim. Acta, 31, (1986), 549.
8   M.Sharp, D.D.Montgomery and F.C.Anson, J. Electroanal. Chem., 94, (1985), 247.
9   J.M.Baudreay and M.D.Archer, Electrochim Acta, 30, (1985), 1355.
10 E.T.Turner-Jones, J. Electrochem. Soc. 132, (1985), 245C.
11 C.P.Andrieux, J.M.Dumas-Bouchiat and J.M.Saveant, J. Electroanal. Chem., 141, (1982), 1.
12 S.E.Morris, J. Electrochem. Soc., 132, (1985), 477C.
13 S.M.Geraty, A.G.Ross and J.G.Vos Proceedings of the International Society of Electrochemistry Meeting, Salamanca, Spain, September 1985, p.02200.
14 C.Andrieux, this volume.

M.R. Smyth and J.G. Vos
Electrochemistry, Sensors and Analysis
© Elsevier Science Publishers B.V., Amsterdam — Printed in The Netherlands

NUMERICAL SIMULATION OF ELECTROCHEMICAL REACTIONS AT REDOX POLYMER ELECTRODES

CLAUDE DAUL[1] AND OTTO HAAS[2]

[1] Institute of Inorganic Chemistry, University of Fribourg, Perolles, CH-1700 Fribourg, Switzerland

[2] Swiss Federal Institute for Reactor Research, CH-5303 Würenlingen, Switzerland

SUMMARY

The diffusion and redox kinetics within a redox polymer film at the surface of an inert electrode are of interest for an understanding of catalytic processes at modified electrodes. It can be described by a system of nonlinear differential equations. Numerical methods are required to obtain a general solution. Using Pereyra's algorithm we have been able to solve the system of differential equations for all experimental cases, and thus to simulate current-potential curves expected for polymer-coated rotating disc electrodes.

INTRODUCTION

Savéant et al.(ref.1) and Albery et al.(ref.2) showed that the mediation of electrochemical reactions by redox polymer films may be described by a system of two 2nd-order differential equations which is subject to given boundary conditions. For special cases it is possible to obtain analytical solutions. Savéant showed that the expected limiting current of rotating disc electrodes coated with a redox polymer film may be discussed in terms of four characteristic currents (Table 2 of ref.1a). A complete general solution of the problem is only possible with numerical methods. Such methods also allow to take into consideration the heterogeneous kinetics at the electrode underneath the coating. The present paper shows that with the numerical method described it is possible to simulate current-potential curves for all possible experimental cases.

SYSTEM AND NOTATIONS

The redox reaction taking place within the film is given by the following equation:

$$Q + A \overset{k}{\underset{k/K}{\rightleftharpoons}} B + P \qquad 1)$$

where Q/P is the polymer-fixed redox couple and A/B is the substrate redox couple dissolved in the solution.

At the electrode surface the heterogeneous reactions:

$$P + e \underset{k^-_{PQ}}{\overset{k^+_{PQ}}{\rightleftharpoons}} Q \qquad\qquad 2)$$

$$A + e \underset{k^-_{AB}}{\overset{k^+_{AB}}{\rightleftharpoons}} B \qquad\qquad 3)$$

may occur.

Using Savéant's dimensionless notation, but employing heterogeneous rate laws to define the boundary conditions at the electrode surface (x=0), we can describe the process within the coating as follows:

$$0 < x < 1 : \qquad d^2a/dx^2 = æaq - (æ/K)(1-a)(1-q) \qquad\qquad 4)$$

$$d^2q/dx^2 = (æ/ʃ)aq - (æ/ʃK)(1-a)(1-q) \qquad\qquad 5)$$

$$x = 0 : \qquad da/dx - r_1 a + r_2(1-a) = 0 \qquad\qquad 6)$$

$$dq/dx + s_1(1-q) - s_2 q = 0 \qquad\qquad 7)$$

$$x = 1 : \qquad \sigma(da/dx) = \curlyvee - a \qquad\qquad 8)$$

$$dq/dx = 0 \qquad\qquad 9)$$

where r_1, r_2, s_1, s_2 are given by a dimensionless formulation of Tafel's law:

$$r_1 = \Phi k^+_{AB}/D_S = \Phi k^o_{AB}/D_S \; \exp[-\alpha_{AB} F(E-E^o_{AB})/RT] \qquad\qquad 10)$$

$$r_2 = \Phi k^-_{AB}/D_S = \Phi k^o_{AB}/D_S \; \exp[(1-\alpha_{AB})F(E-E^o_{AB})/RT] \qquad\qquad 11)$$

$$s_1 = \Phi k^+_{PQ}/D_E = \Phi k^o_{PQ}/D_E \; \exp[-\alpha'_{PQ} F(E-E^o_{PQ})/RT] \qquad\qquad 12)$$

$$s_2 = \Phi k^-_{PQ}/D_E = \Phi k^o_{PQ}/D_E \; \exp[(1-\alpha'_{PQ})F(E-E^o_{PQ})/RT] \qquad\qquad 13)$$

The symbols used are listed at the end of the communication.

NUMERICAL SOLUTION OF THE BOUNDARY-VALUE PROBLEM

To obtain a general solution of the non-linear differential equations we seek a numerical method. A wide variety of solutions for ordinary differential equations with boundary-value problems is now available (ref. 3,4). After several trials with different algorithms we adopted Pereyra's code (ref. 4). A subroutine based on this algorithm is available in most mathematical and scientific program libraries. In our case we use the routine DVCPR from IMSL (International Mathematical and Statistical Libraries).

The practical way to solve the system of two 2nd-order differential equations is given below.

$$y_1(x) := a(x) \qquad y_2(x) := (da/dx)_x$$
$$y_3(x) := q(x) \qquad y_4(x) := (dq/dx)_x$$

Thus the system of four 1st-order differential equations to be solved becomes:

$$y_1' = y_2 \qquad y_2' = æy_1 y_3 - (æ/K)(1-y_1)(1-y_3)$$
$$y_3' = y_4 \qquad y_4' = (æ/ϛ)y_1 y_3 - (æ/Kϛ)(1-y_1)(1-y_3)$$

The Jacobian matrix P of the partial derivatives $P_{ij} = \partial y_i'/\partial y_j$ which is required by DVCPR is given below:

$$
P = \begin{bmatrix}
0 & 1 & 0 & 0 \\
æy_3 + (æ/K)(1-y_3) & 0 & æy_1 + (æ/K)(1-y_1) & 0 \\
0 & 0 & 0 & 1 \\
(æ/ϛ)y_3 + (x/ϛK)(1-y_3) & 0 & (æ/ϛ)y_1 + (æ/ϛK)(1-y_1) & 0
\end{bmatrix}
$$

and the boundary conditions are:

$$y_2(0) - r_1 y_1(0) + r_2[1-y_1(0)] = 0$$
$$y_4(0) + s_1[1-y_3(0)] - s_2 y_3(0) = 0$$
$$\sigma y_2(1) + y_1(1) - ɣ = 0$$
$$y_4(1) = 0$$

To achieve a precision of 10^{-5} we have to use a 50 pts mesh for the concentration profile.

RESULTS

In order to test the program we took the example of current-potential curves obtained at [Ru(bpy)$_2$Cl(PVP)]Cl-coated rotating disc electrodes immersed in a "substrate" solution (10^{-3} M Fe(II) in 1M HCl). This system has been studied extensively (ref. 5). All the characteristic currents (see below) could be measured by using independent electrochemical methods. For this system the electron transfer of the fixed redox couple (Ru II/III) is assumed to be fast whereas Fe(II) exhibits very slow electron exchange with electrode underneath the coating.

The shape, the limiting current and the half-wave potential of the current-potential curve contain a lot of information about the electrocatalytic process going on in the coating, and thus about the characteristic current controlling the process. Especially the half-wave potential and the limiting current are sensitive to the surface-concentration of the redox polymer and the Fe(II) concentration in the electrolyte solution. To test the computer program described some current-potential curves

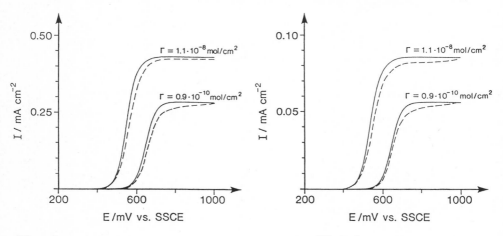

Fig. 1

---Experimental and simulated current-potential curves of two redox polymer coated rotating disc electrodes
[Fe(II)] = 10^{-3} M ; w = 1000 rpm

Fig. 2

Same curves as in fig.1
[Fe(II)] = $2*10^{-4}$ M

involving different surface concentrations and Fe(II) concentrations have therefore been recorded. The curves were then simulated by using the data set in ref. 5c. The results obtained are depicted in figs. 1,2, where the experimental and the simulated curves are depicted. The experimental data and the parameter used for the simulations are given in table 1. They have been deduced from experimental data in ref. 5c.

It can be shown that the characteristic currents (ref. 1) are directly related to the dimensionless parameters used in the simulation. These relations are shown below:

$$\ae = i_K/i_S \qquad \varrho = i_E/i_S \qquad \sigma = i_S/i_A$$

<u>Table 1</u>   Experimental data and the parameter used for the computer simulation in figs. 1,2

Γ [mol/cm^2]	$0.9*10^{-10}$	$1.1*10^{-8}$	$0.9*10^{-10}$	$1.1*10^{-8}$
Fe(II) [M]	10^{-3}	10^{-3}	$2*10^{-4}$	$2*10^{-4}$
i_A [mA/cm^2]	0.450	0.450	0.090	0.090
i_E [mA/cm^2]	3300	26	3300	26
i_S [mA/cm^2]	155	1.23	31	0.246
i_K [mA/cm^2]	0.787	99.5	0.157	19.9
i_K/K [mA/cm^2]	$9.57*10^{-5}$	$1.21*10^{-2}$	$1.91*10^{-5}$	$2.42*10^{-3}$
k^o_{AB} [cm/sec]	10^{-10}	10^{-10}	10^{-10}	10^{-10}
k^o_{PQ} [cm/sec]	1	1	1	1
E^o_{AB} [mV]	439	439	439	439
E^o_{PQ} [mV]	670	670	670	670

Figs. 1+2 show that there is reasonably good agreement between the experimental and simulated curves.

To obtain further insight in the mechanism it is interesting to look at the concentration profiles in the redox polymer coating. Fig. 3 shows the concentration profiles of the species A, B, P and Q calculated for a potential of 1000 mV vs.SSCE for an electrode with low surface concentration of the redox polymer. Fig. 4 shows the concentration profiles at the same potential for an electrode with a much higher surface concentration. It is readily apparent that in the second case the diffusion of Fe(II) becomes much more important, whereas for low surface concentrations this seems not to be a rate-limiting factor.

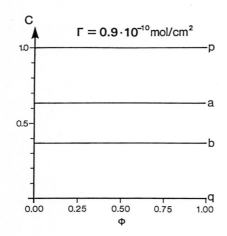

Fig. 3
Calculated normalised concen-
trationprofiles at 1000 mV
vs. SSCE for a rotating disc
eletrode covered with the
redox polymer (P/Q) immersed
in a substrate solution
$[Fe(II)] = 2*10^{-4}$ M

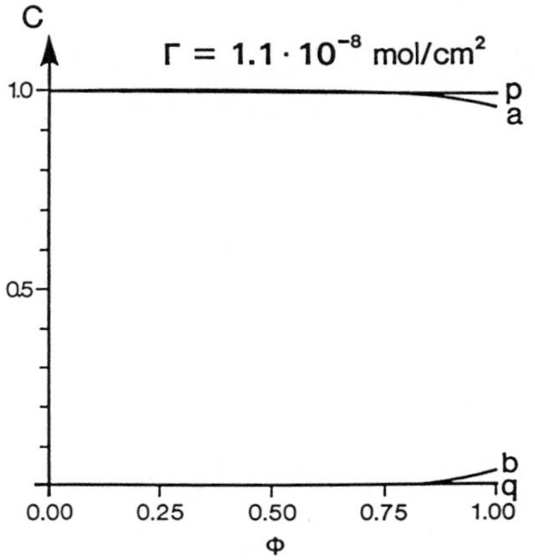

Fig. 4
Same profiles as in fig.3
but for much higher sur-
face concentrations (Γ) of
the redox polymer.

The computer simulation of the current-potential curves is most interesting for parameter studies. A practical application is shown in figs. 5, 6 where the influence of the heterogeneous parameter is illustrated. Fig. 5 shows that the current-potential curves are shifted as soon as the heterogeneous rate constant for the fixed redox couple is slower than 1 cm/sec. Fig. 6 shows the influence of k^o_{AB} (heterogeneous rate constant of the substrate) on the current-potential curve. Comparison with the experimental curves shows that in our example $k_{AB}^o < 10^{-3}$ cm/sec.

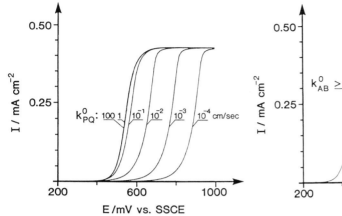

Fig. 5
Influence of the heterogeneous rate constant k_{PQ}^o on the current-potential curve.

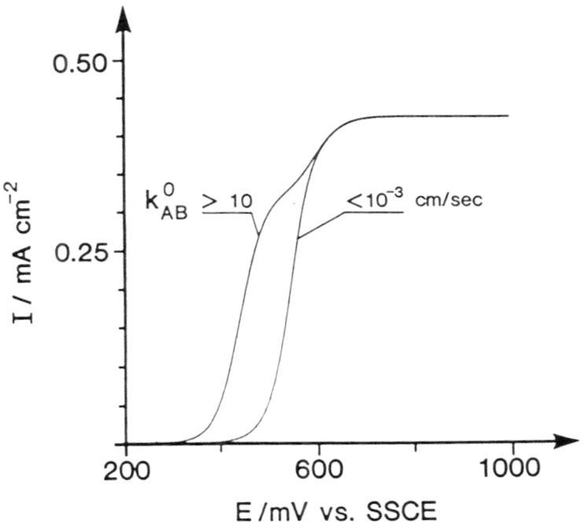

Fig. 6
Influence of the heterogeneous rate constant k_{AB}^o on the current-potential curve.

CONCLUSIONS

The method described is a powerful tool to simulate processes occurring in chemically modified electrodes. It may be used to test the different parameters influencing the processes. Since the simulation is fast we think that it should be possible to use the program in a nonlinear regression.

LIST OF SYMBOLS

a,b   dimensionless concentrations of A and B ($C_A/\varkappa C_A^o$) and ($C_B/\varkappa C_B^o$)

x   distance from the electrode surface (film surface x=1)

k   second-order rate constant for the cross-exchange reaction

K   equilibrium constant

k_{PQ}^+, k_{PQ}^- heterogeneous rate constants for the fixed redox couple

k_{AB}^+, k_{AB}^- heterogeneous rate constants for the redox couple in solution (substrate)

$\varkappa$ partition coefficients of A and B between the film and the solution

284

i_K $(FSC_A \kappa k\Gamma^o)$ characteristic current of the cross-exchange reaction

i_E $(FS\Gamma^o D_E/\Phi^2)$ characteristic current of the diffusion-like electron transport in the film

i_A $(FSC_A^o D/\delta)$ characteristic current for the diffusion of the substrate (FeII) in solution

i_S $(FSC_A^o \kappa D_S/\Phi)$ characteristic current of the substrate Fe(II) in the film

D_E diffusion coefficient of the electrons in the film

D_S diffusion coefficient of the substrate in the film

D diffusion coefficient of the substrate in the solution

Φ film thickness

δ Levich diffusion layer thickness

Γ^o surface concentration of the fixed redox couple P/Q

F Faraday constant

S surface concentration

κ $= A_o/(A_o + B_o)$

ACKNOWLEDGEMENT

This project was supported by the Swiss National Science Foundation grant No 2.915-0.85

REFERENCES

1 a) C.P. Andrieux, J.M. Savéant
   J. Electroanal. Chem., 134 (1982) 163

   b) C.P. Andrieux, J.M. Dumas-Bouchiat, J.M. Savéant
   J. Electroanal. Chem., 131 (1982) 1

2 W.J. Albery, A.R. Hillman
  Annu. Rep. Prog. Chem. Sect. C 78 (1981) 377

3 S.Burlisch and Stoer, Einführung in die numerische Mathematik II, Taschenbücher Band 114, Springer Verlag Berlin 1973 pp.170

4 V. Pereyra, Lect. Notes Comp. Sci. 76 (1978) pp. 67-88
  Springer Verlag Berlin

5. a) O. Haas, J.G. Vos
   J. Electroanal. Chem. 113 (1980) 139

   b) O. Haas, B. Sandmeier
   submitted for publication

   c) C.P. Andrieux, O. Haas, J.M. Savéant
   submitted for publication

© Elsevier Science Publishers B.V., Amsterdam — Printed in The Netherlands

ASPECTS OF THE ELECTROCHEMICAL BEHAVIOUR OF METHYLENE BLUE AND NEUTRAL RED
MODIFIED ELECTRODES IN AQUEOUS SOLUTION

MICHAEL E.G. LYONS*, HUGH G. FAY, CARMEL MITCHELL and DECLAN E. McCORMACK

University of Dublin, Chemistry Department, Trinity College, Dublin 2, Ireland.

*To whom correspondence should be addressed.

SUMMARY

This paper presents preliminary results pertaining to the in-situ electrodeposition of conductive multilayered coatings of methylene blue (MB) and
neutral red (NR) on glassy carbon electrodes via potential cycling of aqueous
dye solutions. A detailed mechanism of polymerization is proposed. Information
pertaining to the nature of the electroactive moieties in the coating is
obtained by examination of the variation of the redox potential of the bound
dye species with solution pH.

INTRODUCTION

In recent years much interest has arisen in the preparation, properties and
application of derivated or chemically modified electrodes (1,2). These systems
can be prepared by several different techniques – most notably by direct
covalent attachment, adsorption or by electrochemical deposition. Fundamental
studies of in-situ surface modification via electrochemical deposition have
received little attention to date (3-5). This paper considers aspects of the
mechanism of formation and the general redox behaviour of methylene blue (I)
and neutral red (II) films deposited on glassy carbon electrodes in aqueous
solution.

MB
(I)

NR
(II)

EXPERIMENTAL

A conventional two compartment cell was used with a Platinum counter
electrode and a Metrohm Ag, AgCl (KCl sat.) reference electrode (E = +0.197V vs
NHE). The working electrode consisted of a Metrohm glassy carbon disc (exposed
geometric area = 0.181 cm^2) encapsulated into a rotating disc assembly using a

PTFE polymeric body and separate electrical contacts. The potential of the
working voltammetry experiments triangular potential sweeps were applied with
the aid of a Metrohm VA E 612 scanner. Current was measured by recording the
potential drop across a standard resistor. Voltammetric profiles were recorded
using a Linseis LY 17100 X - Y recorder. All solutions were prepared using
triply dis-tilled water and Analar grade chemicals. The cell solution was
thermostatted at $25 \pm 0.1^{\circ}C$ and the solution around the working electrode was
purged with purified nitrogen gas prior to each experiment.

RESULTS AND DISCUSSION

   The voltammetric behaviour of solution phase MB and NR ([MB] = 4 x 10^{-5} mol
dm^{-3}, [NR] = 1 x 10^{-4} mol dm^{-3}) in phosphate buffer solution pH 6.72 under
repetitive potential cycling conditions is outlined in Fig. 1 and Fig. 2 below.

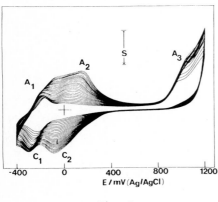 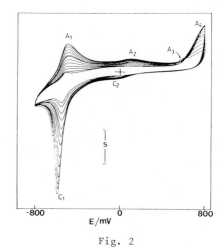

Fig. 1

Fig. 2

Fig. 1. Effect of repetitive potential cycling on the voltammetric response of
an aqueous MB solution ([MB] = 4.2 x 10^{-5} M, phosphate buffer pH 6.7) at a
glassy carbon electrode.
Sweep rate = 20 mVs^{-1}; S = 3μA; T = 298K.

Fig. 2. Effect of repetitive potential cycling on the voltammetric response of
an aqueous NR solution ([NR] = 1 x 10^{-4} M, phosphate buffer, pH 6.7) at a
glassy carbon electrode. Sweep rate 20 mVs^{-1}; s = 4μ; T = 298 K.

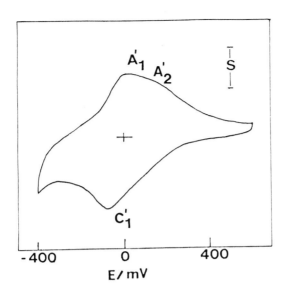

Fig.3. Typical voltammetric response of a MB coated glassy carbon electrode in phosphate buffer solution, pH 6.7.
Sweep rate 20 mVs^{-1}; s = 3µA; T = 298 K

The redox potential of the solution phase dye differs markedly from that for the surface bound species - the latter being observed at more positive potentials that the former. For instance the steady state voltammetric response of a dilute MB solution in phosphate buffer pH 6.7 is characterised by an anodic peak at -160 mV and a cathodic counterpart at -195 MV. However as illustrated in Fig. 3 in the voltammetric profile recorded for the bound dye free solution at the same pH two broad ill defined anodic peaks A_1', A_2' at +30 mV and +200 mV and one cathodic peak C_1' at -75 mV are observed. Similar behaviour is observed for NR modified electrodes except that the peaks observed in dye-free buffer solution are considerably more well defined.

Laviron (6) has shown that for an ideal surface electron transfer reaction (facile electrode kinetics, equivalent fixed site adsorption) the peak potentials for the anodic and cathodic processes coincide i.e. $\Delta Ep = 0$ and the peak shapes are symmetrical with half widths $\delta = 90.6/n$ mV at 25°C. It is clear from Fig. 3 that ΔEp is non zero and all peaks are extremely broad with δ-values greatly exceeding the theoretical value. This deviation is probably caused by redox center repulsive interactions within the dye layer.

288

We assume that film formation occurs via a pathway involving the electropolymerisation of oxidatively formed radical cations. Such a mechanism is outlined in Scheme 1 for MB. We propose that the resonance form (i) predominates (some quantum mechanical calculations are currently being conducted to confirm this hypothesis) and that carbon-carbon coupling occurs between two molecules of (i) to form the dimeric species outlined in Scheme 1. This process can be continued resulting in the eventual formation of an insoluble oligomeric coating on the glassy carbon surface. We note that each monomeric unit retains its electroactive heterocyclic nitrogen atom with the result that the deposited film retains its redox activity. The charge transfer process within the MB film can be represented in the manner illustrated in Scheme 2.

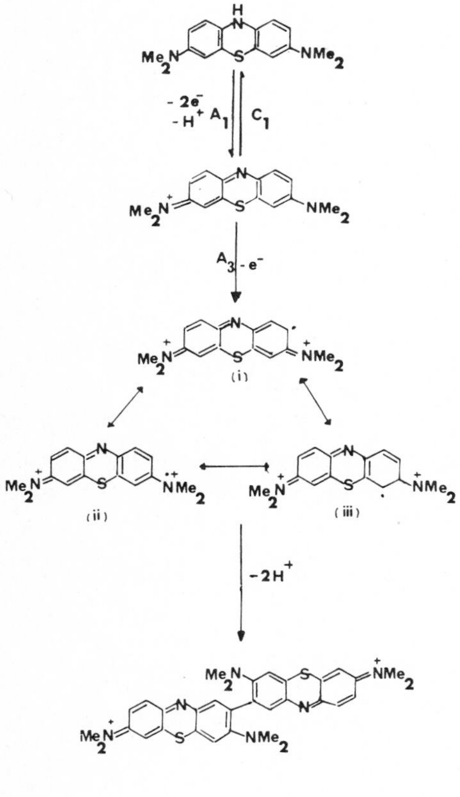

Scheme 1

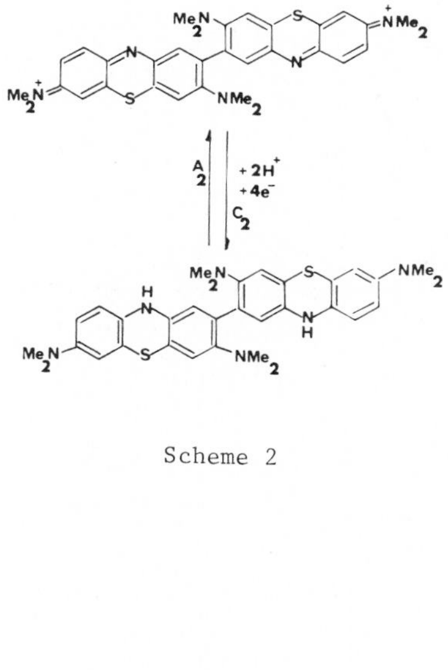

Scheme 2

Scheme 1.   Radical coupling mechanism for MB electrodeposition

Scheme 2.   Redox behaviour of MB modified electrode at pH 6.7

From an examination of the voltammetric response of the dye modified electrode in solutions of varying pH we have recently shown that (7) it is possible to propose a detailed mechanism for film formation and redox behaviour. If the redox process within the film is assumed to be reversible then application of the Nernst equation results in the prediction that the electrode potential (for a given ratio of species in the oxidized and reduced states) should vary in a linear manner with solution pH. The slope of the latter plot yields the ratio of protons to electrons for the particular redox process. Data pertaining to the variation of redox potential with solution pH for NR modified electrodes is outlined in Table 1. A typical plot for the NR modified electrode is illustrated in Fig. 4.

Table 1. Redox potential variation with solution pH

pH range	dEpa/dpH mVdec^{-1}	Reaction Type	dEp.c/dpH mV dec^{-1}	Reaction Type
2.2 – 6.0	-63 ± 4	$2H^+, 2e^-$ $4H^+, 4e^-$	-66 ± 4	$2H^+, 2e^-$ $4H^+, 4e^-$
6.0 –10.0	-44 ± 5	$3H^+, 4e^-$	-84 ± 5	$6H^+, 4e^-$

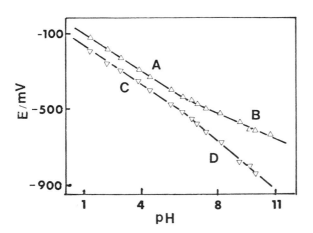

Fig. 4. Variation of redox potential with solution pH for NR coated electrode.
Region A: -63 mV dec^{-1}; Region B: -44 mV dec^{-1};
Region C: -66 mV dec^{-1}; Region D: -84 mV dec^{-1}
Anodic peak (Δ), Cathodic peak (∇)

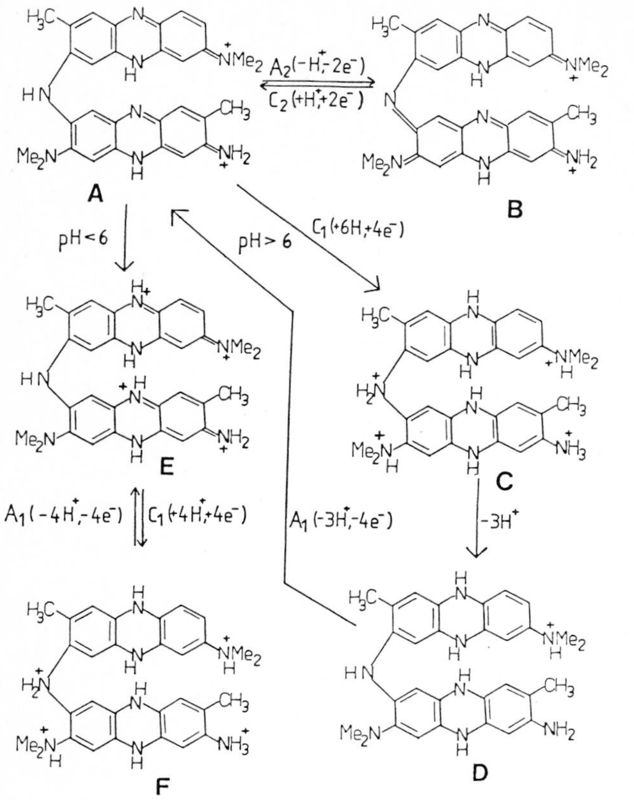

A₂(−H⁺,−2e⁻) — written in scheme as A₂(-H⁺,-2e⁻)

Scheme structures (NR redox behaviour):

$A_2(-H^+,-2e^-)$ / $C_2(+H^+,+2e^-)$ (A ⇌ B)

$C_1(+6H^+,+4e^-)$

pH < 6 ; pH > 6

$A_1(-4H^+,-4e^-)$ ‖ $C_1(+4H^+,+4e^-)$; $A_1(-3H^+,-4e^-)$; $-3H^+$

Labels: A, B, E, C, F, D

Scheme 3

Redox behaviour of NR modified electrode as a function of pH.

These results can be rationalised as follows. Peak A_3 outlined in Fig. 2 can be attributed to the initial production of a radical cation which can undergo coupling via an amino bridging group to form the product (A) illustrated in Scheme 3. Also outlined in this reaction sequence is an analysis of the redox behaviour of the bound dye in dye free solution as a function of solution pH. Consider firstly solutions of high pH. The reduction process (peak C_1) is a $6H^+$, $4e^-$ reaction (theoretical response, −90 mV dec^{-1}). The fact that the experimental slope is slightly less, i.e. −84 mV dec^{-1} demonstrates that the sixth protonation to form species is quite difficult to achieve. This reduced species will exhibit a tendency to lose protons when placed in contact with a medium of high pH and therefore the monoprotonated species D will readily form. Consequently H-oxidation on sweep reversal (peak A_1, Fig. 2) involves a $3H^+$, $4e$ reaction to form species A. The theoretical response of −45 mV dec^{-1} compares well with the experimental value of −44 mV dec^{-1}. Somewhat simpler behaviour is observed for low pH values. It is clear from a perusal of Fig. 4 that a linear potential − pH response of −60 mV dec^{-1} is observed for pH values less than 6. We note from Scheme 3 that the surface bound species A can be readily di-protonated to form species E when the film is in contact with a low pH medium. In this latter situation the redox behaviour is governed by the $4H^+$, $4e^-$ reaction between the species E and F.

Further experiments have shown (7) that good agreement is found between the pH response of bound and solution phase dye. This observation suggests that the dissociation constants of the groups in the bound layer are quite similar to those in the bulk solution. It is also quite probable that the cost is reasonable permeable to H^+ ion and counter ion penetration and that the bound redox centers exhibit a certain degree of hydration.

Further work is currently underway to characterise the electrode kinetics of these dye modified electrode systems as a function of solution pH and will be reported in a subsequent communication.

ACKNOWLEDGEMENTS

MEGL is grateful to the Trinity Trust and the Science Faculty, Trinity College for equipment. We are also grateful to Mrs Corinne Harrison for transformation of the manuscript into camera ready form.

REFERENCES

1   R.W. Murray, Electroanal. Chem. Vol. 13 A.J. Bard (Ed.)
    Marcel Dekker, New York, 1983, 191.
2   W.J. Albery and A.R. Hillman, Ann. Rep. C. Royal Society of Chemistry,
    London, 1981, 377.
3   J.M. Bauldray and M.D. Archer, Electrochim. Acta., 28 (1983) 1515; 30
    (1985) 1355.
4   A.J. McQuillan and M.R. Reid, J. Electroanal. Chem., 194 (1985) 237.
5   O. Haas and M.R. Zumbrunnen, Helv. Chim. Acta., 64 (1981) 854.
6   E. Laviron, Electroanal. Chem., Vol. 12 A.J. Bard (Ed.) Marcel Dekker,
    New York, 1982, 53.
7   M.E.G. Lyons, H.G. Fay, C. Mitchell and D.E. McCormack, JCS Faraday I
    submitted for publication.

M.R. Smyth and J.G. Vos
Electrochemistry, Sensors and Analysis
© Elsevier Science Publishers B.V., Amsterdam — Printed in The Netherlands

METAL ION ANALYSIS USING A POLYPYRROLE-N-CARBODITHIOATE ELECTRODE

Imisides, M.I., O'Riordan, D.M.T. and Wallace, G.G.
Chemistry Department, University of Wollongong, Wollongong, N.S.W. 2500, Australia.

SUMMARY

The use of a polypyrrole-N-carbodithioate electrode for trace metal determinations has been investigated. The electrode shows promise as a sensor for mercury ions in solution and studies into trace metal speciation have been instigated.

INTRODUCTION

The possibility of using chemically modified electrodes (CMEs) for determination of metal ions has attracted the attention of various workers in recent years (1-8). This is due to the limitations imposed using conventional mercury drop or mercury thin film electrodes (MTFEs). Both mercury based electrodes suffer from intermetallic formation when a species which forms an amalgam with the analyte is codeposited on/in the electrode (9,10). Furthermore, both electrodes determine metal ions by monitoring processes at cathodic potentials and therefore dissolved oxygen must be removed from solution. This can often be a time consuming process e.g. up to 30 minutes of nitrogen purging may be required. Another problem with mercury based electrodes is their mechanical instability in flowing solutions.

Proper design of CMEs should help extend the capabilities of voltammetry for trace metal analysis by alleviating the above problems. The development of modified electrodes capable of complexing metals onto the electrode surface is a particularly exciting area. Complexation of metal species has been used extensively in Analytical Chemistry; in combination with liquid-liquid (11,12) or solid phase (13,14) extraction, many useful sample preconcentration procedures have been developed. Complexation is also known to affect redox properties of metal ions (15,16). Using CMEs the complexation can be conveniently carried out where it matters most - at the electrode surface. This is much more efficient than derivatising the total solution prior to voltammetric analysis.

In previous work (2-4) the synthesis of a (poly) pyrrole-N-carbodithioate (PPdtc) electrode was described. The ability of this electrode to uptake copper ions and the subsequent voltammetry of the surface bound species were discussed. The rate of uptake of copper ions and detection limits of the order of 1 ppm encouraged present investigations into the ability of the PPdtc electrode to preconcentrate other metal ions from solution. Hg^{2+}, Co^{2+}, Pb^{2+}, Mn^{2+} and Fe^{2+} were investigated since these metals form dithiocarbamate complexes with a range of stability constants[17]. Also, in this work, the electrode preparation procedure has been optimised further and the use of different voltammetric waveforms to enhance sensitivity has been investigated. Electron Probe Micro Analysis (EPMA) was used to detect the presence of sulfur and metal species on the electrode surface. Scanning Electron Microscopy (SEM) was used to investigate electrode morphology.

EXPERIMENTAL

Reagents and Standard Solutions

All chemicals used were A.R. grade purity unless otherwise stated. L.R. grade pyrrole and carbon disulfide were obtained from BDH. The pyrrole was distilled before use. L.R. grade benzene and 2-propanol were obtained from the Munster Chemical Company. HPLC grade acetonitrile was obtained from Rathburn Chemicals. Metal stock solutions were prepared by dissolving appropriate salts in distilled, deionized water. Platinum wire was obtained from Johnson & Mathey. The buffer employed was a solution of 0.04 $\underline{M}$ acetic acid, 0.04 $\underline{M}$ Phosphoric acid and 0.04 $\underline{M}$ Boric acid. The pH was adjusted adding 0.2 $\underline{M}$ NaOH.

Polypyrrole film formation

Electrodes were pretreated with aqua regia. This resulted in a clean metal surface which gave good film adhesion.

Polypyrrole films were prepared on the electrode surface as described previously (2,3). That is, polypyrrole was plated galvanostatically (I mA) from acetonitrile 0.1 $\underline{M}$ Tetra Ethylammonium Perchlorate (TEAP) containing 1% (v/v) water and 0.1 $\underline{M}$ pyrrole. Polymers of thickness 0.75 μm were prepared with the knowledge that 24 mC/cm^2 yields a thickness of 0.1 μm. (2,3). Alternatively a 0.1 mA current was used for plating (see Results and Discussion).

Preparation of Poly (pyrrole-N-carbodithioate) electrode

Dithiocarbamate groups were grafted onto the polymer surface by placing the electrode in a solution containing 50 mL benzene, 10 mL of 2-propanol, 10 mL of carbon disulfide and 10 mL of 0.3 $\underline{M}$ sodium hydroxide in methanol for 48 hours.

Electrodes prepared in this fashion were stable for at least seven days when stored in air or isopropanol.

Instrumentation

All electrochemical experiments were performed with a Princeton Applied Research Model 174A polarographic analyzer. This instrument was used in conjunction with a Metrohm Model E612 VA Scanner for cyclic voltammetry.

A platinum wire working electrode was employed. A platinum wire auxiliary and a Ag/Ag$^+$(0.1 $\underline{M}$ AgNO$_3$) reference for acetonitrile solution work or a Ag/AgCl (3 $\underline{M}$ KCl) reference for aqueous solution work were employed.

Atomic absorption spectrometric (AAS) measurements were obtained using a Pye Unicam SP191 instrument.

Electron probe microanalysis data was collected using a JEOL JXA58 electron probe microanalyzer. An energy dispersive detector also from JEOL was employed. (Accelerating voltage = 20 kV, beam density = 2×10^{-11} A). Coulometry measurements were performed using a PAR 179 Coulometric Analyzer.

Results and Discussion

In previous work (2-4) optimum analytical responses were obtained using the PPdtc electrode to preconcentrate metal ions from solution and looking at the voltammetry of the metal (copper) species trapped on the electrode surface.

In order to establish if the PPdtc electrode could be employed for voltammetric analysis of other metals the electrode was soaked in relatively concentrated (10 ppm) metal ion solutions for 24 hours and then subjected to cyclic voltammetry. Of the metals investigated (Hg^{2+}, Co^{2+}, Pb^{2+}, Fe^{2+} and Mn^{2+}) only Hg^{2+} uptake experiments resulted in well defined voltammetric responses (Fig. 1). EPMA analysis confirmed the presence of mercury on the electrode suface (Fig. 2).

Using EPMA subsequent to uptake experiments all of the other metal ions investigated were detected on the electrode sufaces (Fig. 3). Presumably the voltammetric responses of these metals are masked by the polypyrrole backbone responses (18) or are trapped in some electroinactive form on the polymer matrix. Of all the metal ions tested to date   only well defined voltammetric responses are obtained for Cu^{2+} and Hg^{2+}. The ability to trap and determine mercury ions is particularly important for environmental monitoring programs since mercury is known to be extremely toxic (19,20).

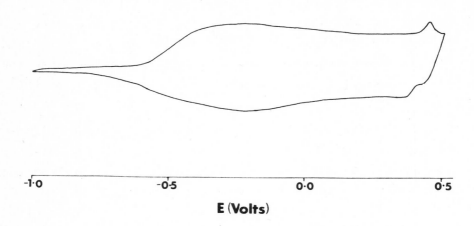

Fig. 1 Cyclic voltammogram of PPdtc electrode after soaking in 10 ppm Hg^{2+} solution for 24 hours. Scan rate = 500 mV/sec.

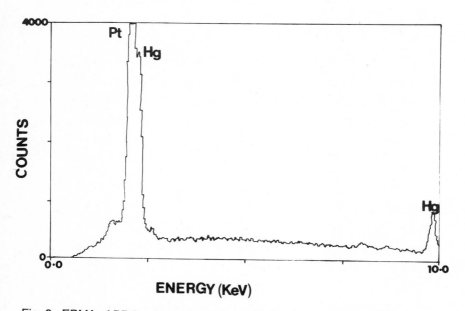

Fig. 2 EPMA of PPdtc electrode after soaking in 10 ppm Hg^{2+} for 24 hours.

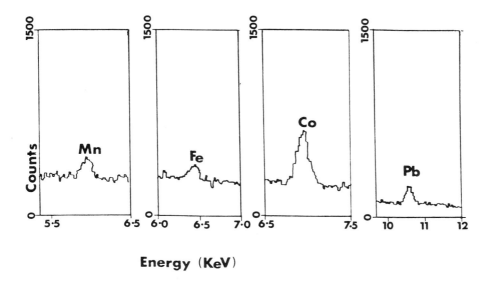

Energy (KeV)

Fig. 3   EPMA of PPdtc electrode after soaking in (a) 10 ppm Mn^{2+}, (b) 10 ppm
Fe^{2+}, (c) 10 ppm Co^{2+}, (d) 10 ppm Pb^{2+}.

Uptake of Mercury Ions

Since only Hg^{2+} uptake resulted in a well defined voltammetric response, which should be suitable for analysis, the uptake of this metal from 1 $\underline{M}$ $NaNO_3$ was investigated in more detail (Table 1). Using cold vapour AAS 1 $\underline{M}$ $NaNO_3$ solutions containing 10 ppm Hg^{2+} and a derivatised PPdtc electrode were monitored with time. With no potential applied to the working (PPdtc) electrode metal uptake did occur. However the initial rate of uptake, 2 x 10^{-8} µg Hg^{2+}/minute for a 1cm^2 electrode, is low and only about 20% of the species is removed from solution even after 60 hours. Previous work (2-4) indicated that uptake under potentiostatic conditions greatly increases the efficiency of the deposition process. A range of potentials in the working range of the polypyrrole substate (-1.00 V to +0.80 V vs Ag/AgCl) were investigated.

Optimum uptake was obtained at -0.10 V (Table 1). The rate of uptake was increased to 3 x 10^{-7} µg Hg^{2+}/minute for a 1 cm^2 electrode and after 5 hours almost 50% of the mercury ions were removed from solution. Presumably application of the potential at -0.10 V results in increased electron density on the ligands which would increase their effectiveness during complexation. Furthermore, application of this potential ensures that the polymer is maintained in the charged conducting state which in turn would ensure a more open, porous, polymer structure(21,22) allowing better access to complexing sites.

The rate and extent of mercury uptake on this PPdtc is greater than those previously observed for copper ions (2-4).

The Voltammetric Response(s)

Following uptake of mercury ions on to the electrode the electrochemical behaviour of the surface bound species was investigated in 1 $\underline{M}$ NaNO$_3$. Using cyclic voltammetry a well-defined oxidation process was observed (Fig. 1).

TABLE 1

Mercury uptake on Pt-polypyrrole-N-carbodithioate.

Stirred solution of 10 ppm Hg^{2+} (aq) containing PPdtc electrode.

Time (hr)	No potential applied. Solution 10 ppm Hg^{2+}(1 $\underline{M}$ NaNO$_3$) Hg^{2+} concentration in solution containing electrode (ppm)	Time (hr)	Hold potential at -0.1 V vs Ag/AgCl in a solution of 10 ppm Hg^{2+} (1 $\underline{M}$ NaNO$_3$ Hg^{2+} concentration in solution containing electrode.
0	10.0	0.0	10.0
5	9.4	0.5	9.1
15	9.0	1.0	8.2
24	8.5	1.5	7.0
40	8.2	2.0	6.3
50	8.2	2.5	5.4
60	8.2	5.0	5.4

Note
1) Mercury concentrations determined by cold vapour hydride generation technique using Atomic Absorption Spectroscopy.

2) Initial mercury concentration in solution = 10.0 ppm.

Scanning to more positive potentials results in an additional response at approximately +0.65 V vs Ag/AgCl. It is envisaged that the mercury(II) ions undergo a redox reaction during complexation and are trapped on the electrode surface as a mercury(I) species or mercury metal. Oxidation of ligand sites during this redox process may explain the limited lifetime of these modified electrodes. Consequently, a regeneration procedure involving chemical reduction is currently being investigated.

The first (less anodic) mercury oxidation response is a well defined, symmetrical, surface response with W 1/2 = 40 mV. If the scan is switched immediately following the first response then it grows with increasing number of scans, becomes constant and eventually starts to decrease again. Both the number of scans required before a constant response is observed and the number of scans for which it is constant are concentration dependent. Both increase with increased concentration. This phenomena has been observed by other workers (21,22) and was attributed to a "break in" process during which the conduction chain which enables charge hopping through the polymer is established. The constant response varies linearly with scan rate up to 50 mV/sec. At faster scan rates (up to 1 V/sec) the response varies linearly with the square root of the scan rate indicating that even with relatively slow scan rates the response becomes limited by the rate of charge hopping through the polymer.

Scanning to more anodic potentials where the second oxidation process was encountered complicated matters considerably. Upon scanning past the second oxidation response both responses diminish rapidly on subsequent scans and the reduction process is shifted in a cathodic direction. This result indicates that the second process removes the first oxidation product from the electrode surface or at least renders it electroinacltive possibly due to polymer restructuring. The small peak width (W 1/2=20 mV) may indicate that the latter is true and that the second response is not in fact due to a faradaic process. For analytical purposes it is recommended that the potential sweep is terminated after the first response.

Differential pulse voltammetry could also be used to detect the mercury species on the electrode surface. A major problem is the background response observed for oxidation of the polypyrrole backbone (Fig. 4). However, by commencing scans at potentials more anodic than $Ep_{(ox)}$ for the polypyrrole backbone this problem was overcome.

It was also found that all voltammetric responses could be improved by modifying the electrode preparation procedure. Plating the polymer at a lower current density had some beneficial affects. For the same thickness polymer but using a 0.1

300

mA rather than a 1mA plating current much more well defined polypyrrole oxidation/reduction responses were observed indicating better conductivity presumably due to the more open, porous structure of the polymer (21,22). This improved conductivity results in much sharper voltammetric responses for the surface bound species and sensitivity is markedly enhanced. The increase in sensitivity may also be due to improved uptake due to the more open structure of the polymer. Another advantage of the lower plating currernt polymers was that

carbon disulfide derivatisation could be achieved in 24 hours rather than the 48 hours previously employed (2,3).

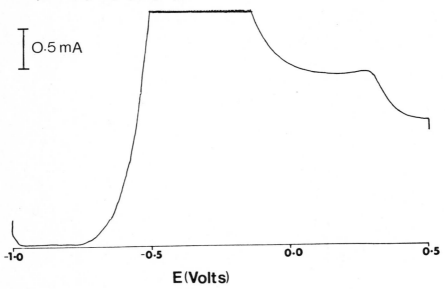

Fig. 4 Differential pulse voltammogram of a PPdtc electrode containing Hg^{2+}.
Uptake Conditions1 ppm Hg^{2+} solution (1 $\underline{M}$ NaNO$_3$)

Applied potential	=	-0.1 V vs Ag/AgCl
Uptake time	=	30 minutes.
Scan rate	=	5 mV/sec. Duration between pulses = 0.5 sec.
Pulse width	=	60 m Sec.

Response (1) Polypyrrole background oxidation.

The voltammetric response (in 1 $\underline{M}$ NaNO$_3$) was monitored after uptake from solutions of varying pH in the range 1 to 8. No affect on the voltammetric response was observed. However, the concentration of supporting electrolyte can affect both steps (uptake and voltammetry) of the analytical process.

The voltammetric response is dramatically improved by uptaking from higher concentration (≥ 1 $\underline{M}$) $NaNO_3$ solutions. (Uptaking from lower concentration salt solutions results in poor responses). This is either due to a salting out affect on the electrode surface or due to the increased polymer porosity in the higher ionic strength media. If voltammetry is performed in media < 1 $\underline{M}$ $NaNO_3$, IR drop affects in terms of slower breakin, broader peaks and Ep being shifted more anodic are observed. Uptaking from a stirred 1 $\underline{M}$ $NaNO_3$ solution for twenty minutes levels of 0.5 ppm Hg^{2+} could easily be detected. It is envisaged that with further optimisation lower detection levels will be attained.

Interferents

Since copper ions are also readily trapped on the PPdtc electrode experiments were carried out to determine if the presence of copper in solution interfered with the mercury response. With a solution containing 10 ppm of each metal and using uptake conditions previously described the mercury response was unaffected. However, the Cu^{2+} response did suffer from the presence of Hg^{2+}. It was smaller in magnitude and much more difficult to detect. Possible interferents from other species are now being considered.

Speciation

In environmental monitoring programs the ability to speciate mercury ions is extremely important (19,20). Consequently the ability of the PPdtc electrode to speciate Hg^{2+} ions and an organomercury species (phenylmercury) was investigated. The response obtained after uptake from a 10 ppm phenylmercury solution occurs at a different potential (less anodic) than that observed after uptake from a mercury(II) solution. The results in this work are preliminary but they highlight an exciting possible application of CMEs in trace metal speciation work.

CONCLUSIONS

The ability of the PPdtc electrode to uptake and determine mercury ions has been demonstrated and slight modification to the electrode preparation method have improved the performance.

Current work in these laboratories involves investigation into electrode regeneration methods, the possibility of using a labile metal complex electrode and more detailed studies on the ability of this electrode to speciate metal ions in some real samples.

302

References
1. G.G. Wallace, Dissolved Oxygen: the electroanalytical chemists dilemma, TrAc. 6 1985, 6, 1425-148.
2. D.M.T. O'Riordan and G.G. Wallace, Poly(pyrrole-N-carbodithioate) electrode for electroanalysis, Anal. Chem. 1986, 58,128-131.
3. D.M.T. O'Riordan and G.G. Wallace, Trace Metal Analysis with no Mercury and no Deoxygenation, Anal. Proc. 22 (1985) 199-200.
4. D.M.T. O'Riordan and G.G. Wallace, Chemically Modified Electrodes Containing Complexing Groups for the Determination of Trace Metals, Anal. Proc. 23 (1986) 14-15.
5. G.T. Cheek, and R.F. Nelson, Applications of chemically modified electrodes to analysis of metal ions, Anal. Lett. A11 (1978) 393-402.
6. J.A. Cox and P.J. Kulesza, Preconcentration and Voltammetric behavior of chromium (VI) at Pt electrodes modified with Poly(4-vinyl pyridine), J. Electro Anal. Chem.159 (1983) 337-346.
7. A.R. Guadalupe and H.D. Abruna, Electroanalysis with chemically modified electrodes, Anal. Chem. 57 (1985) 142-149.
8. R.P. Baldwin; Chemical Preconcentration and voltammetric determination of Ni(II) at a dimethyl glyoxime-containing chemically modified electrode, Pittsburgh Conference and Exposition 1986, Abstract No. 181.
9. A.M. Bond, Modern Polarographic Methods in Analytical Chemistry, Marcel Dekker; New York, 1980.
10. J. Wang, Stripping Analysis, UCH; Deerfield Beach, 1985.
11. G.H. Morrison and H.Freise, Solvent Extraction in Analytical Chemistry, John Wiley & Sons, New York.
12. T. Sekine and Y. Hasegawa, Solvent Extraction Chemistry, Marcel Dekker, New York, 1977.
13. A. Miyazaki and R.M. Barnes, Differential determination of chromium(VI) and chromium(III) with Poly(dithiocarbamate) chelating resin and Inductively coupled plasma atomic emission spectrometry, Anal. Chem.53, (1981) 364-366.
14. J.N. King and J.S. Fritz, Concentration of metal ions by complexation with sodium bis (2-hydroxyethyl) dithiocarbamate and sorption on XAD-4 resin. Anal. Chem. 57, (1985) 1016-1020.
15. J.C. Bailar The Chemistry of the Coordination Compounds. Reinhold, New York, 1956.
16. D. Concouvanis, The Chemistry of the dithioacid and 1,1 dithiolate complexes, Progr. Inorg. Chem. 26, (1979) 301-469.
17. A. Hulanicki, Complexation reactions of dithiocarbamates, Talanta 14 (1967) 1371-1392.
18. R.A. Bull, F.R. Fan, and A.J. Bond, Electrochemical behaviour of polypyrrole coated platinum and tantalum electrodes, J. Electrochem. Soc. 129, (1982) 1009-1015.
19. L. Hodges, Environmental Pollution, Holt, Reinhart and Wilson, New York, 1973.
20. S.E. Manahan, Environmental Chemistry, Willard Grant, Boston, 1979.
21. A.H. Schroeder and F.B. Kaufman, The influence of polymer morphology on polymer film electrochemistry, J. Electroanal. Chem. 113 (1980), 209-224.
22. A.H. Schroeder, F.B. Kaufman, V. Patel, and E.M. Engler, Comparativebehaviour of electrodes coated with thin films of structurally related electroactive polymers, J. Electroanal. Chem. 113, (1980), 193-208.

M.R. Smyth and J.G. Vos   303
Electrochemistry, Sensors and Analysis
© Elsevier Science Publishers B.V., Amsterdam — Printed in The Netherlands

THE USE OF ELECTRODES COATED WITH RUTHENIUM-CONTAINING
POLYMERS AS SENSORS FOR IRON(II)

SUZANNE M. GERATY, DAMIEN W.M. ARRIGAN AND JOHANNES G. VOS
School of Chemical Sciences, National Institute for Higher
Education, Dublin 9, Ireland

SUMMARY
The analytical response towards $[Fe(CN)_6]^{4-}$ and
$[Fe(H_2O)_6]^{2+}$ of rotating disk electrodes coated with
ruthenium containing polymers has been studied. In a pH = 2.0
buffer the linear range for the determination of these
substrates extends to 5×10^{-6} M. In a pH = 6.0 buffer the
linear range obtained for ferrocyanide is again 5×10^{-6} M but
for $[Fe(H_2O)_6]^{2+}$ the lower limit is 5×10^{-5} M. The
mediated oxidation of $[Fe(H_2O)_6]^{2+}$ is controlled by the
polymer layer. The oxidation of ferrocyanide is controlled by
diffusion of the substrate to the electrode for thin coatings
($<10^{-8}$ moles/cm^2), while for thicker layers the mediating
process is controlled by the polymer film.

INTRODUCTION

   Recently, the synthesis, characterisation and electrochemical
properties of electrode surfaces modified with ruthenium
containing polymers have been reported (refs.1-5). These studies
showed that thin layers of metallopolymers such as
$[Ru(bipy)_2(PVP)Cl]Cl$, where bipy = 2,2'-bipyridyl and PVP =
poly(4-vinylpyridine), act as redox catalysts for a number of
substrates. At potentials where the surface bound Ru(II) is
oxidised to Ru(III) the oxidation of Fe(II) is mediated (see
reaction 1). Ce(IV) can be reduced at these electrodes (ref.2).

$$Ru(III) + Fe(II) \xrightarrow{k} Ru(II) + Fe(III) \qquad (1)$$

   Abruna et al. recently reported the use of polymer modified
electrodes as sensors (refs. 6,7). The selective binding of metal
ions into the ruthenium-containing polymer matrix was used as a
preconcentration technique and the amount adsorbed measured

voltammetrically. Similar approaches have been followed by
O'Riordan et al. (refs.8.9) using polymer bound dithiocarbamates
and by Martin using Nafion (ref.10).

The aim of this contribution is to investigate the possible
analytical applications of polymer modified electrodes using
rotating disk electrodes. In a series of rotating disk
experiments the limiting current observed for the polymer mediated
oxidation of Fe(II) was studied as a function of substrate,
substrate concentration and background electrolyte.
The metallopolymers used as mediators were;

$[Ru(bipy)_2(PVI)_5Cl]Cl$ (polymer I) and

$[Ru(bipy)_2(PVI)_{15}Cl]Cl$ (polymer II)

where PVI = poly(N-vinylimidazole). The substrates chosen in this
study were $(NH_4)_2Fe(SO_4)_2 \cdot 6\ H_2O$ and $K_4Fe(CN)_6$.

EXPERIMENTAL

Ruthenium-containing polymers were prepared by reaction of
$Ru(bipy)_2Cl_2 \cdot 2\ H_2O$ and PVI as described in the literature
(refs.11,12). The Ru:PVI ratios used for the syntheses were 1:5
for polymer I and 1:15 for polymer II. Glassy carbon electrodes
were coated as described elsewhere (ref.2). Electrochemical
measurements were carried out using a PAR research Model 174
potentiostat equipped with a PAR Model 175 voltage programmer and
a Pine Instruments Model ASRE rotating disk electrode. The amount
of electroactive ruthenium on the electrode surface was obtained
either by measuring the area under the cyclic voltammograms at low
scan rates, or by using a PAR Model 379 Digital Coulometer.
Electrolytes and substrate solutions were made up using Analar
Grade materials and doubly distilled water.

RESULTS AND DISCUSSION

The polymeric materials used were first characterised using
cyclic voltammetry. In Figure 1 a typical voltammogram is shown.
In contrast with earlier experiments (ref.11a) only one ruthenium
species was obtained as is evidenced by the presence of only one
redox couple in Figure 1. Because of the shorter reflux times
used (6 instead of 24 hours) and the addition of some LiCl, only
reaction 2 is observed;

$$Ru(bipy)_2Cl_2 + xPVI \longrightarrow [Ru(bipy)_2(PVI)_xCl]Cl \quad (2)$$

No further substitution of the second chloride ligand by either solvent or polymer is observed. By changing the Ru:polymer ratio, x can be varied. In these experiments x was five for polymer I and fifteen for polymer II. Of these x imidazole groups, one is bound to the central ruthenium ion and (x-1) are free.

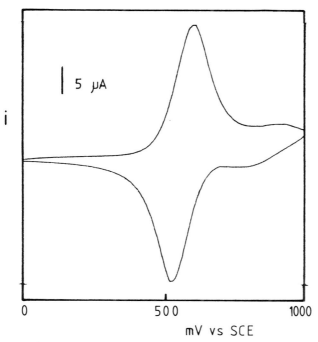

Figure 1. Cyclic voltammogram of $[Ru(bipy)_2(PVI)_5Cl]Cl$.
        Electrolyte 0.1 M NaCl, pH = 2.0. Scan rate 50 mV/sec.

Using glassy carbon rotating disk electrodes the limiting current of the mediated oxidation of both $K_4Fe(CN)_6$ and $(NH_4)_2Fe(SO_4)_2$ was measured as a function of substrate concentration and background electrolyte. For the experiments with the first substrate, 0.1 M solutions of NaCl with a pH of 2.0 and 6.0 were used and a rotation speed of 600 rpm. The experiments with $(NH_4)_2Fe(SO_4)_2$ were carried out at the same pH but in 0.2 M $NaClO_4$ and with a rotation speed of 500 rpm. Substrate concentrations were varied between 10^{-6} and 10^{-2} M. A graphic representation of the results obtained in the form of log-log plots has been given in Figure 2.

From the present data it can be concluded that the linear range for the determination of Fe(II) as $[Fe(CN)_6]^{4-}$ extends down to 5×10^{-6} M irrespective of the pH of the background electrolyte. When Fe(II) is determined as $[Fe(H_2O)_6]^{2+}$ the linear range extends to 5×10^{-6} M at pH = 2.0 and to 5×10^{-5} M at pH = 6.0. The upper limit of the linear range was not determined but log-log plots were in all cases linear to at least 10^{-2} M.

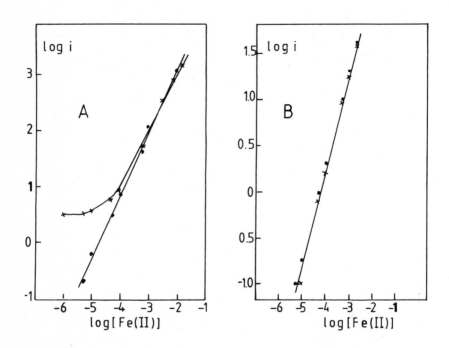

Figure 2. Log-log plots of the limiting current vs substrate concentration.   A) mediator, Polymer II; substrate, $[Fe(H_2O)_6]^{2+}$; Electrolyte ; ●) pH = 2.0, 0.2 M $NaClO_4$; x) pH = 6.0, 0.2 M $NaClO_4$. B) mediator Polymer II; substrate, $[Fe(CN)_6]^{4-}$; Electrolyte ; ●) pH = 2.0, 0.1 M NaCl; x) pH = 6.0, 0.1 M NaCl.

The clear deviation from linearity at pH = 6.0 for low concentrations of ammonium ferrous sulphate is most likely explained by the charge of the electroactive species in solution. In these solutions Fe(II) is present as hydrated Fe(II); $[Fe(H_2O)_6]^{2+}$. It is expected that, at high pH, where no protonation of the polymer backbone occurs, it is more difficult for a positively charged species to diffuse into the polymer layer. The polymer backbone is always positively charged so an electroactive species with a positive charge will have difficulties entering the polymer coating. The current-voltage curves obtained for these experiments were analysed using Koutecky-Levich plots. The intercepts of these plots were inversely proportional to the substrate concentration, indicating a mediating process that is controlled by the polymer layer (ref. 13).

Table 1 Rotating-disk electrode data for electrodes coated with polymer I. Electrolyte 0.1 M $NaClO_4$; rotation speed 600 rpm; Disk area 0.071 cm^2; Electrode I; $\Gamma_{Ru} = 3\times10^{-9}$ moles/cm^2 Electrode II; $\Gamma_{Ru} = 3\times10^{-8}$ moles/cm^2. Substrate $[Fe(CN)_6)]^{4-}$.

	i_{lim} (µA)		intercept (µA^{-1}cm^2)	
$[Fe(CN)_6]^{4-}$ (mM)	El.I	El.II	El.I	El.II
0.5	10.8	11.3	0	680
0.75	14.0	16.9	0	570
1.0	18.5	24.9	0	420
2.0	36.3	50.7	0	260

For the experiments with ferrocyanide, two electrodes coated with different amounts of polymer I were used (see Table 1). The linear range obtained was the same for both electrodes. The Koutecky-Levich plots obtained for both electrodes showed important differences. For the thin coating no intercept was obtained while for thicker coating an intercept was obtained which was inversely proportional to the substrate concentration. It seems clear that for thicker layers the mediating process is determined by the polymer layer, whereas for the thin layer the oxidation is limited by substrate diffusion from the bulk solution to the electrode surface.

CONCLUSIONS

The results obtained show that modified electrodes of the type described here can be used for the quantitative determination of Fe(II) using rotating disk techniques. The charge of the substrate is an important factor and a reduced response is obtained for the positively charged $[Fe(H_2O)_6]^{2+}$ species at high pH. In this study no detailed experiments concerned with possible interferences were carried out. Preliminary studies showed that the presence of Cu^{2+} (0.13 M) did not reduce the limiting current obtained for a 10^{-4} M $[Fe(H_2O)_6]^{2+}$ solution at pH = 2.0. It is furthermore worthwhile pointing out that a change of the pH from 2.0 to 6.0 does not change the response of the electrode very much. It is therefore concluded that in the presence of a suitable background electrolyte the analytical application of these electrodes is feasible.

ACKNOWLEDGEMENTS

We thank the National Board for Science and Technology for financial assistance.

REFERENCES

1   N.Oyama and F.C.Anson, J.Electrochem.Soc., 127, (1980), 640.
2   O.Haas and J.G.Vos, J.Electroanal.Chem., 113, (1980), 139.
3   O.Haas, M.Kriens and J.G.Vos, J.Am.Chem.Soc., 103, (1980), 1318.
     J.M.Calvert and J.J.Meyer, Inorg. Chem., 21, (1982), 3978.
5   J.M.Calvert, D.L.Peebles and R.J.Nowak, Inorg.Chem., 24, (1985), 3111 and references therein.
6   A.R.Guadalupe and H.D.Abruna Anal.Chem., 57, (1985), 142
7   L.M.Wier, A.R.Guadalupe and H.D.Abruna, Anal.Chem., 57, (1985), 2009.
8   D.M.T.O'Riordan and G.G.Wallace, Anal.Proc., 22, (1985), 199.
9   D.M.T.O'Riordan and G.G.Wallace, Anal Chem., 58, (1986), 128.
10   M.R.Szentirmag and C.R.Martin, Anal Chem, 56, (1984), 1898.
11a S.M.Geraty and J.G.Vos, J.Electroanal.Chem., 176, (1984), 389.
  b S.M.Geraty and J.G.Vos to be published.
12   J.M.Clear, J.M.Kelly, C.M.O'Connell and J.G.Vos, Chem.Res.M., (1981), 3039.

M.R. Smyth and J.G. Vos
Electrochemistry, Sensors and Analysis
© Elsevier Science Publishers B.V., Amsterdam — Printed in The Netherlands

POLYMER COATINGS ON SUSPENDED METAL MESH FIELD EFFECT TRANSISTORS

John Cassidy[1], John Foley[2], Stanley Pons[2] and Jiri Janata[3]

[1]Present address : School Of Chemical Sciences, National Institute of Higher Education, Dublin 9 (Ireland)
[2]Department of Chemistry, University Of Utah, Salt Lake City, Utah 84112 (U. S. A)
[3]Department of Bioengineering, University Of Utah, Salt Lake City, Utah 84112 (U. S. A)

ABSTRACT

The suspended gate field effect transistor has been modified with electrochemically deposited polymers for use as a neutral molecule sensor. Reversible responses were found for polar molecules and a simple reflectance experiment was carried out to examine the process.

INTRODUCTION

With the introduction of integrated circuitry, microtransducers for temperature, pressure and chemical composition are becoming more widely available (refs. 1-4). Neutral molecule sensors based on solid ionic conductors have been described (refs. 4-5) and sensors based on field effect transistors have been used to detect H_2S (refs. 7-8), CO (ref. 9), and NH_3 (refs. 10-11). The aim in this work is to extend the usefulness of the suspended gate field effect transistor, recently developed in the University of Utah, as a neutral molecule sensor. The growth in the detection of complex molecules in solution by means of enzymes and immunosensors (ref. 12) may conceivably be adapted for use in the gas phase by using polymers that form weak charge transfer complexes with gases. For example, Langmuir Blodgett films of phthalocyanines have been used in the detection of NO_2 (ref. 13). The sensing mechanism in this particular case was due to the change in conductance of the film from the charge transfer of the physisorbed NO_2 (ref. 14).

Electrochemically deposited polymers such as polypyrrole (refs. 15-16) and polyphenylene (refs. 17-18) have been recently been shown to be sensitive to neutral molecules. The electrochemical deposition and anion insertion into polymers such as polyphenylene (refs. 18, 21-23) and polyaniline (refs. 24-26) has been described. The use of these polymers in conjunction with the suspended gate field effect transistor was proposed since the device, coated electrochemically with palladium, was found to be reversibly sensitive to hydrogen.

310

THEORY

A detailed description of the device has already been given (ref. 20), aswell as its use as an alcohol sensor by electrochemically coating it with polypyrrole (refs. 15-16). A schematic view of the cross-section of the device is given in Fig. 1. The substrate is p type silicon with the source and drain terminals doped to form n type silicon. The gate is composed of a perforated metal mesh suspended above the silicon dioxide insulator. The device is similar to a metal insulator semiconductor field effect transistor for which the following potential current equations apply under saturation conditions ($V_d > V_g-V_t$)

$$I_d = \mu \; C_i W \; (V_g-V_t)^2/2L \tag{1}$$

where I_d is the current flowing between the drain and source. V_g is the applied gate potential, μ is the electron mobility, C_i the gate capacitance and L and W the length and width of the gate respectively. V_t is the threshold voltage which is a function of the metal semiconductor work function difference:

$$V_t = \Phi_{ms}+2\Phi_f-Q_b/C_i-Q_{ss}/C_i \tag{2}$$

where Φ_{ms} is the metal semiconductor work function difference. Φ_f is the fermi level, Q_b is the charge density in the space charge region and Q_{ss} is the surface state charge density. The change in the metal semiconductor work function difference Φ_{ms} on exposure to the compound of interest depends on the

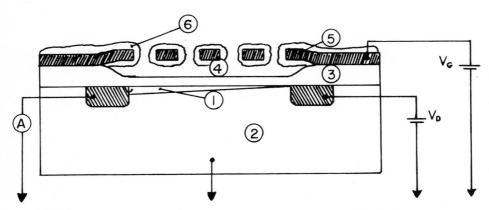

Fig. 1. Schematic of the suspended metal mesh field effect transistor. 1: the inversion channel, 2: Si substrate, 3: Si_3N_4/SiO_2 insulator, 4: the air gap open to gases, 5: the metal mesh, 6: the polymer coating.

change of electron work function of the electronic conductor adjacent to the gate insulator. Whether this change is due to a surface dipole or bulk effect is still a matter of investigation (ref. 16).

EXPERIMENTAL

The wafer of suspended gate field effect transistors (SGFET) was prepared at the Department of Bioengineering at the University of Utah (ref. 20). After scribing, the individual chips were attached to a gold Kovar TO4 header by heating to form a Si/Au eutectic. The individual chips were wirebonded and encapsulated leaving the sensing gate exposed. After encapsulation, characteristic curves (drain current against gate potential at constant drain potential) were plotted to verify that the chip was not damaged before and after electrochemical deposition. The mesh was used as a working electrode in a three electrode system for polymer deposition. The basic requirement is that the polymer be deposited on the inside of the mesh between the metal and the substrate. Polyphenylene was deposited from a solution of biphenyl in acetonitrile (10mM in 0.1M $LiAsF_6$). The polymer was deposited by scanning the potential from 0.0 to +1.7V with respect to a silver reference (Ag/Ag^+(0.01M) in 0.1M tetrabutylammonium tetrafluoroborate in acetonitrile). Polyaniline was deposited from an aqueous solution of aniline (0.1M aniline in 2M HCl) by scanning the potential from +0.2V to +0.8V with respect to a saturated calomel electrode.

The TO4 header containing the chip was mounted in a modified Carle gas chromatograph (model 211c) in a configuration parallel to a thermal conductivity detector (TCD). The effluent from the column (empty in this case) was split using a zero volume tee and directed into both detectors. Liquids were introduced by means of a syringe through the conventional septum inlet. The drain to source current of the SGFET was kept constant using a feedback system and the change in gate potential was monitored as a function of exposure to various compounds. The carrier was N_2 and the SGFET was held at a temperature of 90^{0}C. Reflectance spectra were taken using an IBM IR/98 fourier transform infra red spectrometer. Spectra of a polymer coated on a platinum electrode exposed to alcohol vapour were referenced against spectra of the same polymer in nitrogen.

RESULTS AND DISCUSSION

Deposition of polyphenylene from a solution of $LiAsF_6$ yields a polymer doped with AsF_6^- anions. The resulting polymer is stable in air unlike those doped in the gas phase (ref. 27). In a set of preliminary experiments the responses due to the exposure of polyphenylene modified meshes to various compounds were

312

recorded. Responses to methanol and acetone are shown in Fig 2(a, b). The response of a mesh coated with polyaniline is shown in Fig 2c. The rapid return to baseline is characteristic of the coated SGFET compared to the bare platinum mesh. Responses for the TCD are shown also as reference. In general, responses were found for polar compounds whereas for hexanes and toluene no response was obtained.

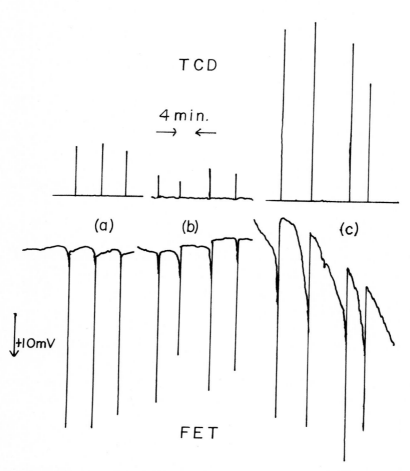

Fig. 2. Responses at a polyphenylene coated mesh in parallel with a TCD. The FET was in feedback mode and the change in gate potential is shown. Flow rate = 20cm^3/min. temperature = 90^{0}C. 1μL samples injected. V_{ds} = 1.23V, I_{ds} = .257mA (a) response for acetone, (b) response for methanol, (c) response for acetonitrile at a polyaniline coated mesh under similar conditions.

In an effort to make a preliminary examination of the interaction of polar molecules with the surface of the polymer, a simple reflectance study was carried out. A spectrum of electrochemically deposited polyphenylene on platinum in the presence of ethanol was referenced to the same polymer in nitrogen. The spectrum is shown in Fig. 3 and the peaks due to ethanol aswell as the peaks due to the interaction of the polymer with ethanol are present. The bands that are not due to the gaseous ethanol are those at 3300 cm^{-1} and 1666 cm^{-1}. In the spectrum of the polymer in nitrogen referenced to platinum there are peaks at 3250 cm^{-1} and 1670 cm^{-1} . It may therefore be conceivable that these bands are enhanced on exposure to ethanol although particular orientations of ethanol on the surface may be responsible. An identical experiment carried out with hexanes yielded no extra peaks due to any interactions. This is in agreement with the fact that no signal is obtained at the SGFET on exposure to hexanes. Thus it is probable that the interaction of the polar molecules with the polymer is responsible for the signal and not any diffusion through to the platinum.

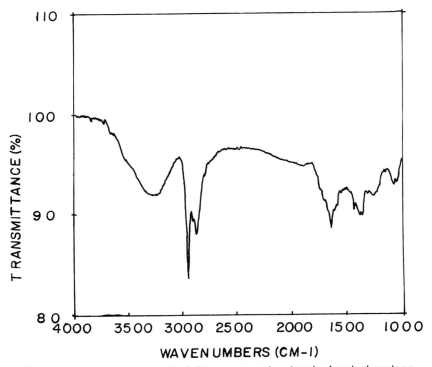

Fig. 3. The reflectance signal derived from a sample signal of polyphenylene exposed to ethanol against the reference signal of the polymer in N_2. The bands pointing down are due to the free ethanol along with those due to the interaction of ethanol with the polymer.

314

ACKNOWLEDGMENTS

The financial assistance of the Office of Naval Research and the SOHIO Oil
Company are appreciated.

REFERENCES
1   J. Janata in: J. Janata, R. J. Huber, (Eds.) Solid State Chemical Sensors,
    Academic Press, Orlando, FL 1985
2   H. Wohltjen, Anal. Chem.   56 (1984) 87A
3   T. Seiyama, K. Fueki, J. Shiokawa, (Eds.) Chemical Sensors: Proceedings of
    the International Meeting on Chemical Sensors, Fukuoka, Japan Elsevier,
    Amsterdam (1983)
4   J. Janata, J. Huber, in H. Freiser (Ed.), Ion Selective Electrodes in
    Analytical Chemistry, Plenum, NY(1980), Vol. 2
5   D. F. Shriver, G. C. Farrington, Chem. Eng. News 63 (1985) 47
6   D. E. Williams, D. McGeehan, in D. Pletcher (Senior Reporter) Eletrochemistry
    The Royal Society of Chemistry, London, 9 (1984) 246.
7   M. S. Shivaraman, J. Appl. Phys. 47 (1976) 3592.
8   J. P. Couput, B. Cornut, C. Chambra, F. Chauvet, in ref. 3 p408
9   K. Dobos, R. Strotman, G. Zimmer, Sens. Actuators, 4 (1983) 593
10 I. Lindquist, A. Spetz, M. Armgarth, C. Nylander, I. Lundstrom, Appl. Phys.
    Lett. 43 (1983) 839
11 C. Nylander, M. Armgarth, I. Lundstrom, in ref. 3 p203.
12 M. Aizuwa, in ref. 3 p. 683.
13 G. Roberts, Sens Actuators 4 (1983) 131
14 T. A. Jones, B. Bott, W. Hurst, B. Mann, in ref. 3 p. 90.
15 M. Josowicz, J. Janata, Anal. Chem. 58 (1986) 514.
16 M. Josowicz, J. Janata, Paper presented at the conference of the
    Electrochemical Society, Toronto, May 12-17 (1985)
17 M. Levy, J. Cassidy, S. Pons, J. Janata, Paper presented at FACSS meeting
    Philadelphia (1984)
18 J. F. McAleer, S. Pons, S. Bandyopadhyay, E. Eyring, (to be submitted)
19 J. F. Cassidy, S. Pons, J. Janata, Anal. Chem. (submitted)
20 G. F. Blackburn, M. Levy, J. Janata, Appl. Phys. Lett. 43 (1983) 700
21 I. Rubenstein, J. Polymer Sci. , 21 (1983) 3035
22 I. Rubenstein, J. Electrochem. Soc. 130 (1983) 1507
23 A. Pruss, F. Beck, J. Electroanal. Chem 172 (1984) 281
24 K. Kobayashi, H. Yoneyama, H. Tamura, J. Electroanal. Chem. 161 (1984) 419
25 K. Kobayashi, H. Yoneyama, H. Tamura, J. Electroanal. Chem. 177 (1984) 281
26 T. Oshaka, Y. Ohnuki, N. Oyama, G. Katagiri, K. Kamisako, J. Electroanal.
    Chem. 161 (1984) 399
27 G. B. Street, T. C. Clarke, I. B. M. J. Res. Develop. 25 (1981) 51

M.R. Smyth and J.G. Vos
Electrochemistry, Sensors and Analysis
© Elsevier Science Publishers B.V., Amsterdam — Printed in The Netherlands

THE ELECTROCHEMISTRY OF GASES OF MEDICAL INTEREST

P. Tebbutt[1], D. Clark[1], G. Robinson[1], C.E.W. Hahn[1] and W.J. Albery[2]

[1]Nuffield Department of Anaesthetics, Radcliffe Infirmary, Woodstock Road
Oxford OX2 6HE

[2]Chemistry Department, Imperial College, London SW7 2AZ

SUMMARY
    In this paper we present the results of preliminary investigations into
the effect of halothane on the reduction of oxygen in aqueous systems and the
effect of carbon dioxide on the reduction of oxygen in an aprotic solvent.

INTRODUCTION

    The continuous monitoring of anaesthetic and respiratory gases such as
nitrous oxide (N_2O), halothane ($CF_3CBrClH$), oxygen (O_2) and carbon dioxide (CO_2)
in the clinical environment is essential with respect to safety and desirable
for research purposes. The Clark-type oxygen sensor [ref.1] is widely used for
patient monitoring but other methods such as mass spectroscopic and infra-red
analyses are normally prohibitively expensive.

    In this paper we describe two aspects of our current electrochemical
research underlying the development of electrochemical sensors for the
determination of the above gases. Previous work in the department has led to
the development of a prototype sensor for the simultaneous determination of O_2
and N_2O [ref.2]. Interest is now focussed on the electrochemistry of O_2/CO_2
and O_2/halothane mixtures.

    The homogeneous chemical reaction between electrochemically generated
superoxide radical ($O_2^{\cdot-}$) and carbon dioxide in a number of solvents of low
proton availability has been reported [ref.3]. However, some uncertainty exists
about the exact mechanistic sequence and the electrochemical data is limited.
In this paper we present some additional data for dimethyl sulphoxide (DMSO) to
support the previously assigned mechanism [ref.3]. This investigation has been
motivated by the need to determine the respiratory gases on a breath-by-breath
timescale. A prototype detector, based on the Clark-type membrane electrode,
has been developed which exploits the nucleophilic attack on carbon dioxide by
the superoxide radical anion. It is hoped that this detector might provide a
solution to the "disconnect" problem, arguably one of the most pressing
problems facing modern surgery. This is the situation where the gas flow

system to the patient becomes partially or wholly disconnected - the disconnect problem. The inescapable conclusion of anaesthetists and technicians is that breath-by-breath analysis of output CO_2 from a patient during surgery is the only means of developing a fail-safe solution to the problem.

METHODS

The study of the mechanism of O_2 reduction on platinum in aqueous NaOH (0.1 mol dm^{-3}) in the presence and absence of halothane has been carried out using the rotating ring disc electrode (RRDE) method as described by Albery and Hitchman [ref.4].

The one-electron reduction characteristics of O_2 in DMSO have been studied in the presence and absence of dissolved CO_2 using cyclic voltammetry [ref.5]. Potential sweep rates used ranged from a few mVs^{-1} up to a few hundred Vs^{-1}, the upper limit being imposed by double-layer charging effects since the experiments were performed on a 1mm diameter platinum disc electrode. However, recent investigations have revealed that high quality data, uncomplicated by charging and uncompensated resistance effects, can be obtained at fast sweep rates by using ultramicroelectrodes [ref.6].

AnalaR grade materials were used throughout; all gases were medical grade; water was triply distilled and spectroscopic grade DMSO (BDH Chemicals Ltd) was purified using standard techniques [ref.7] and transferred to the cell under vacuum. The cell itself, described in detail in the literature [ref.8], facilitates further in-situ drying of the solvent. By rotating the whole assembly through 90° the reaction compartment contents are transferred to an upper chamber and can only return to the reaction compartment after passing through a column of activated alumina. The O_2/halothane mixtures were prepared by passing O_2 through a "Vapor" vaporiser (Draeger, West Germany).

RESULTS AND DISCUSSION

1. The effects of halothane on oxygen reduction

The simplest mechanism for the reduction of O_2 to H_2O that has been proposed is that as described by the two parallel reactions (1) and (2):

$$O_2 \xrightarrow{\text{4e,4H}^+} 2H_2O \tag{1}$$

$$O_2 \xrightarrow{\text{2e,2H}^+} H_2O_2 \xrightarrow{\text{2e,2H}^+} 2H_2O \tag{2}$$

Albery and Hitchman define the parameter, x,

$$x = \frac{\text{flux of } O_2 \text{ reduced direct to } H_2O}{\text{flux of } O_2 \text{ reduced to } H_2O_2} \tag{3}$$

which can be found by plotting the observed collection efficiency, i_D/i_R, as a function of the rotation speed, ω, of the RRDE.

i.e.

$$\frac{N_0 \cdot i_D}{i_R} = \frac{(1+2x) + 2(1+x) \cdot k \cdot D^{-2/3} \cdot v^{1/6} \cdot 0.643}{\omega^{1/2}} \tag{4}$$

where N_0 is the geometric collection efficiency and was found to be 0.16 for the electrode used. k is the rate constant for the reduction of H_2O_2, and D and v have their usual meanings [ref.4].

The ring was held at -0.3V (vs SCE) in order to reduce any H_2O_2 reaching it and both disc and ring currents were plotted against disc potential. (Fig. 1)

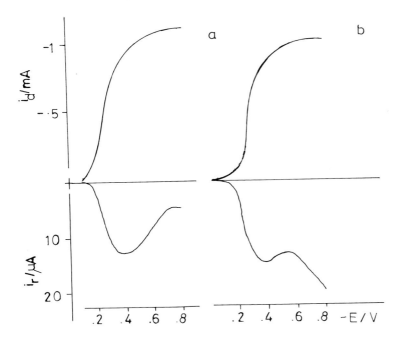

Fig. 1 Disc polarograms with corresponding i_R vs E_D plots for the reduction of O_2 on platinum, both in the absence, (a), and presence (b) of halothane.

From Fig. 1 it can immediately be seen that at potentials more negative than -0.5V much more H_2O_2 is produced when halothane is present in solution than in its absence. This is further reflected in plots of eqn.(4), shown in Fig. 2.

318

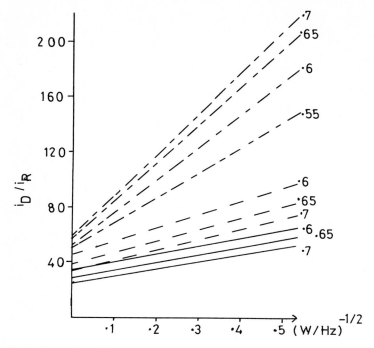

Fig. 2 Plots of eqn.(4) in the presence of; ——— 4%, ----- 1%, —--—- 0% halothane.

In the absence of halothane the results obtained were typical of those obtained by other workers who have studied this system [ref.4 and refs. therein] but when halothane was present the results showed distinctly different patterns. In the potential region of interest (negative of -0.5V) the intercepts of the plots in Fig. 2, at infinite rotation speed, increase with potential in the absence of halothane, but decrease with potential when halothane is present. Analysis of the slopes of these plots reveals that the rate constant, k , in the absence of halothane is about twice the value when it is present in a concentration of 1% and greater at 4%. However, this difference is not great and it has not yet been shown to be a quantitative measure of the amount of halothane present.

Halothane itself is electrochemically inert on platinum, though it can be reduced on both silver and gold. As the observed effect is essentially inhibitive in nature, and it is well known that the reduction of O_2 is affected by the surface preparation of the metal, it seems likely that the observed mechanistic change can be attributed to the absorption of halothane at the electrode. Competition for surface sites with O_2 inhibits the reduction of O_2, and relatively more H_2O_2 is produced. A more precise study is currently being carried out, the results of which will be the subject of future publications.

2. Cyclic Voltammetric studies of the superoxide/carbon dioxide titration reaction in dimethyl sulphoxide

Cyclic voltammetric studies of the one-electron reduction of O_2 to $O_2^{\cdot -}$ (eqn.(5)), at platinum (Pt) in DMSO reveal that it is a reversible process at low potential sweep rates but becomes irreversible at higher ones after passing through a quasi-reversible region at intermediate values (Fig. 3).

$$O_2 + e^- \longrightarrow O_2^{\cdot -} \tag{5}$$

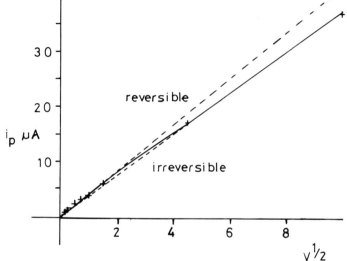

Fig. 3   A plot of the dependence of the peak current for the one-electron reduction of O_2 at Pt in DMSO at 37°C on the square root of the potential sweep rate.

$k^\ominus$ was determined by comparing experimental ΔE_p values with working curves constructed as a function of the variable ψ [ref.9] defined by

$$\psi = \frac{(RT)^{\frac{1}{2}} \cdot k^\ominus}{(nFD\pi\nu)^{\frac{1}{2}}} \tag{6}$$

where the symbols included in equation (6) have their usual meaning, and are defined in reference 9.

$k^\ominus = 1 \times 10^{-2}$ cm s^{-1}.   The diffusion coefficient, D_{O_2}, was taken as 3.0×10^{-5} cm^{2}s^{-1} as determined using the rotation-step method [ref.10].

Cyclic voltammetry is a very powerful technique for investigating coupled chemical reactions since it is ideally suited to initial mechanistic studies. The timescale of the experiment can be varied easily (and extensively) by changing the potential sweep rate, so that reaction intermediates can often be readily detected and identified.

In Fig. 4 the effect of dissolved CO_2 on the $O_2/O_2^{\cdot-}$ system is readily demonstrated. At lower sweep rates ($<500mVs^{-1}$) the reduction process is irreversible, and a broad oxidation peak appears, centred at +1.00V vs. Ag/AgCl. The peak current for the O_2 reduction is enhanced by 70% at 20mVs compared to the value obtained in the absence of CO_2 at the same sweep rate. At slower sweep rates the peak is replaced by a plateau. This evidence is consistent with a specific type of following chemical reaction mechanism, a catalytic mechanism, where the reactant (O_2) is regenerated chemically.

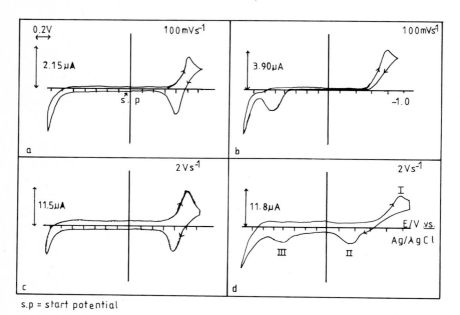

s.p = start potential

Fig. 4 Sample cyclic voltammograms for the O_2/CO_2 system at Pt in DMSO at 37°C. (a) & (c) $[O_2]=0.148mM$, no CO_2 present.
(b) & (d) $[O_2]=0.148mM$, $[CO_2]=0.074mM$.

At higher sweep rates ($>1Vs^{-1}$) an additional peak appears, with E_p = -0.20V vs. Ag/AgCl. This has not been reported previously. Moreover, this peak current increases with sweep rate but the second oxidation process at +1.00V vs. Ag/AgCl begins to diminish until at $10Vs^{-1}$ it is indistinguishable from background. It would be reasonable to attribute the process at -0.20V vs. Ag/AgCl to the oxidation of $O_2^{\cdot-}$ becoming accessible, as the timescale of the experiment is shortened. However, the qualitative shape of the voltammogram (Fig. 4(d)), the peak position with respect to the reduction process, and the effect on the second oxidation process suggest that it might be due to an intermediate, possibly the radical anion $CO_4^{\cdot-}$. This would be consistent with the previously proposed mechanism [ref.3]. Thus, it is likely that the overall mechanism is (with reference to Fig. 4(d)):

$$O_2 + e^- \longrightarrow O_2^{\cdot-} \qquad \text{(electrode)} \qquad (5)$$

$$O_2^{\cdot-} + CO_2 \longrightarrow CO_4^{\cdot-} \qquad \text{(solution)} \qquad (7)$$

$$CO_4^{\cdot-} + CO_2 \longrightarrow C_2O_6^{\cdot-} \qquad \text{(solution)} \qquad (8)$$

$$C_2O_6^{\cdot-} + O_2^{\cdot-} \longrightarrow C_2O_6^{2-} + O_2 \quad \text{(solution)} \qquad (9)$$

'percarbonate'

The rate constant for the $O_2^{\cdot-}/CO_2$ reaction has been determined from the limiting current plateau [ref.9] for the reduction process at $10\,mVs^{-1}$, under second-order conditions, to be $1.35 \times 10^3 m^{-1} 1\, s^{-1}$. This is in good agreement with the value reported by Sawyer et al. [ref.3], and with our own previous

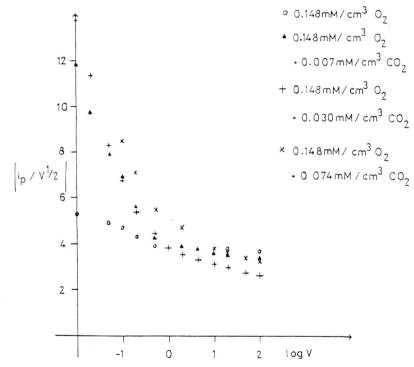

$\circ$ 0.148mM/cm^3 O_2

$\blacktriangle$ 0.148mM/cm^3 O_2

$\cdot$ 0.007mM/cm^3 CO_2

$+$ 0.148mM/cm^3 O_2

$\cdot$ 0.030mM/cm^3 CO_2

$\times$ 0.148mM/cm^3 O_2

$\cdot$ 0 074mM/cm^3 CO_2

Fig.5 Confirmation of the catalytic reaction mechanism for the $O_2^{\cdot-}/CO_2$ system.

evaluations, measured using the rotating ring-disc electrode titration method as described by Albery and Hitchman [ref.4]. The catalytic mechanism is confirmed by Fig. 5. The marked increase of the $i_p/\nu^{\frac{1}{2}}$ value (where i_p is the reduction peak current and ν is the potential sweep rate) on decreasing the sweep rate readily identifies the mechanism.

A controlled potential electrolysis, involving the generation of $O_2^{\cdot-}$ at a gold (Au) disc in DMSO containing 0.2 mol dm^{-3} tetraethylammonium perchlorate (TEAP) and an excess of CO_2, resulted in the formation of a solid at the electrode surface. The Laser-Raman spectrum obtained for this solid is consistent with $(TEA^+)_2C_2O_6$ [ref.11].

These results will be discussed more fully in a future publication.

ACKNOWLEDGEMENT

P.T. and D.C. wish to thank the M.R.C. of Great Britain for financial support during the course of this work. D.C. also wishes to acknowledge the considerable practical assistance given by Dr. J. Heinze and his group at the University of Freiburg, West Germany.

REFERENCES

1   L.C. Clark Jr., Trans.Am.Soc.Art.Int.Org. 2(1956) 41-45.
2   W.N. Brooks, D.Phil. Thesis, University of Oxford, (1980).
    P. Maynard, M.Sc. Thesis, Polytechnic of Central London, (1983).
    E. Hall, Journal of Microcomputer Applications, 7(1984) 319-327.
3   J.L. Roberts, T.S. Calderwood and D.T. Sawyer, J.Am.Chem.Soc., 106(1984) 4667-4670.
4   W.J. Albery and M.L. Hitchman, Ring Disc Electrodes, Pergamon Press, 1971.
5   A.J. Bard and L.R. Faulkner, Electrochemical Methods, John Wiley and Sons, 1980.
6   D. Clark, J. Mortenson and J. Heinze, Unpublished results.
7   C.K. Mann, J.Electroanal.Chem., 4(1969) 57.
8   H. Kiesele, Anal.Chem., 53(12), (1981) 1953.
9   Southampton Electrochemistry Group, Instrumental Methods in Electrochemistry, Ellis Howard Series in Physical Chemistry, Ed. Prof. T.J. Kemp.
10 P. Barron, Ph.D. Thesis, Imperial College, University of London, 1982.
11 D. Clark and A.R. Mount, Unpublished results.

M.R. Smyth and J.G. Vos
Electrochemistry, Sensors and Analysis
© Elsevier Science Publishers B.V., Amsterdam — Printed in The Netherlands

THEORY OF 2 DIMENSIONAL DIFFUSION IN A MEMBRANE ENCLOSED AMPEROMETRIC GAS SENSOR

J. M. Hale
Orbisphere Labs, 144 route de Thonon, CH-1222 Vesenaz, Geneva, Switzerland

SUMMARY
   Equations are derived for the steady state current at a membrane enclosed amperometric gas sensor having a disc-shaped detecting electrode, with explicit account taken of diffusion of gas to the edge of the disc.
   The relative importance of the edge current depends only on the ratio of electrode radius to membrane thickness. When this ratio is below 88, the edge current exceeds 1 % of the "disc current" (proportional to electrode area). For large values of this ratio the edge current is proportional to the length of the circumference of the disc.

INTRODUCTION

   Membrane enclosed amperometric gas sensors, especially oxygen detectors, represent one of the most important applications of electroanalytical chemistry. Control of many industrial processes in power generation, food and drug processing, oil extraction and refining, sewage degradation, combustion and plastic manufacturing, cited as examples, depends upon the sensing of oxygen. Hence, a thorough understanding of the details of the mechanism and the operation of these sensors is needed, for improvement and optimisation of their performance.

   Although theoretical models of such sensors have been discussed by many authors (1-5), almost all these works had in common a complete neglect of effects introduced by the edge of the working electrode. The common assumption is that concentration variations need only be accounted for in the direction normal to the electrode, whence the partial differential equation describing diffusion of electroactive gas through the membrane and internal electrolyte layer contains only one spatial variable. This will be referred to herein as the one dimensional (1-D) theory.

   Careful experimental studies (6,7) have revealed some disagreements between the predictions of this theory, and recorded performance, which have been ascribed to edge effects. The most striking example of such a disagreement, is seen in the shape of the transient current following a sudden "step" change in the partial pressure of the electroactive gas being sensed. This transient is well described by the 1-D theory provided that the ratio of initial to final partial pressures is fairly close to unity, but is distinctly different in the sense that the measured current exceeds the predicted current in the later part of the transient, when the pressure ratio is greater than about ten. It has been suggested that (6,7) this effect comes about because the edge current is not really negligible, and has a longer time constant than the current component due to diffusion along the normal to the electrode surface. No quantitative theory of this effect has been published, though some authors have made illustrative calculations after introducing some simplifying approximations. Recently, for example, Vacek, Linek and Sinkule (8) have studied the transient diffusion through the sample, membrane and electrolyte layers toward an electrode, invoking hemispherical geometry. Jensen, Jacobsen and Thomsen (9) treated the case of lateral diffusion to the edge only through an electrolyte layer, and assumed diffusion through the membrane only in the normal direction.

Another disagreement between theory and practice which has been noticed previously, concerns the non linear dependence of the steady state current from the sensor upon electrode area, at small electrode dimensions. Again it has been concluded that the microelectrode collects "additional" electroactive gas, as compared to the one dimensional model, from a zone surrounding the edge of the electrode and this extra contribution becomes increasingly significant as the electrode size is reduced. This problem is conceptually and mathematically simpler than the transient one because the time variable is omitted (essentially put equal to infinity). It was stated in mathematical terms by Siu and Cobbold (10), though their solution was unfortunately incorrect.

This steady state problem is reconsidered below, and is formally solved in terms of an integral equation. Some asymptotic equations are also derived for the limiting cases of small, and large electrode dimensions, with a simplifying restriction to diffusion in a single layer - the membrane. In

practice it is fairly easy to make the electrolyte layer so thin
as to become negligible, and to increase the stirring of a
liquid sample to such an extent that diffusion within it can be
ignored. A cylindrically symmetrical system is treated with a
disc shaped working electrode, and a uniformly thick membrane.
In this case the diffusion equation contains two spatial
variables, the distance y normal to the electrode, and the
distance from the central axis. We therefore refer to this
treatment as a two-dimensional (2-D) theory. The problem is
mathematically identical to that of an electrostatically charged
disc placed midway between and parallel to two infinite grounded
planes, and has received considerable attention in the
mathematical literature (11).

The most interesting result obtained below, comes from the
asymptote for thin membrane/large radius electrodes. It is shown
that the current may be resolved into two components: the usual
1-D component proportional to electrode area, and an additional
component proportional to electrode diameter but independent of
membrane thickness. The ratio of these terms depends only on the
ratio of electrode diameter to membrane thickness. Thus it is
shown that the edge current is greater than 1 % of the usual 1-D
current for working electrode radii less than about 90 membrane
thicknesses. It is therefore predicted that most commercially
available sensors must exhibit a significant edge current.

THE BOUNDARY VALUE PROBLEM

Let p represent the fugacity of the electroactive gas at any
point y, ϱ within the membrane, where y is the distance
coordinate along the normal from the electrode, and ϱ is the
radial distance from the central axis of cylindrical symmetry.
At the outside of the membrane, the fugacity (partial pressure)
of electroactive gas is assumed equal to that in the sample, p_s.
The diffusion coefficient and solubility of the electroactive
gas in the membrane material are represented by D_m and S_m
respectively. Let the radius of the electrode be a, and the
membrane thickness be b. Furthermore, define the dimensionless
quantities :

$V = (p_s - p) / p_s$; $x = y / a$; $r = \varrho / a$; $h = b / a$:

$F = \text{Flux} / (D_m \, S_m \, a \, p_s)$

Then the boundary value problem to be solved is :

$$\frac{\partial^2 V}{\partial x^2} + \frac{\partial^2 V}{\partial r^2} + \frac{1}{r}\frac{\partial V}{\partial r} = 0 \tag{1}$$

$$V = 1 \text{ at } x = 0, \ 0 \leq r \leq 1 \tag{2}$$

$$V = 0 \text{ at } x = h, \ 0 \leq r < \infty \tag{3}$$

$$\frac{\partial V}{\partial x} = 0 \text{ at } x = 0, \ r > 1 \tag{4}$$

It is required to calculate the flux of electroactive gas to the electrode,

$$\text{Flux} = \int_0^a 2\pi \varrho \ D_m \ S_m \left(\frac{\partial p}{\partial y}\right)_{y=0} d\varrho \tag{5}$$

Hence,

$$F = -\int_0^1 2\pi r \left(\frac{\partial V}{\partial x}\right)_{x=0} dr \tag{6}$$

is the quantity to be calculated, proportional to the electrode current.
The usual 1-D approximation is to write $\dfrac{\partial V}{\partial r} = 0$

whence

$$(\partial V / \partial x)_{x=0} = -1 / h$$
and

$$F = \pi a / b \tag{7}$$

A formal solution, most suitable for thick membranes

A full solution may be obtained by using the infinite Hankel integral transform. This is defined by :

$$\overline{V}(p,x) = \int_0^\infty rV(r,x) J_0(pr) dr \tag{8}$$

The transformed problem has solution :

$$\overline{V} = B(h, p) [e^{-px} - e^{-p(x-2h)}] \tag{9}$$

so that the solution for the original problem may be written formally, using the inversion formula for the Hankel transform :

$$V(r,x) = \int_0^\infty p B(h, p) J_0(rp) [e^{-px} - e^{-p(x-2h)}] dp \tag{10}$$

Now using the boundary conditions in the plane $x = 0$, we derive two simultaneous integral equations which specify the solution to the problem :

For $r \leq 1$:

$$\int_0^\infty pB \left[1 - e^{-2hp} \right] J_0 \ (rp) \ dp = 1 \tag{11}$$

and for $r > 1$:

$$\int_0^\infty p^2 B \left[1 - e^{-2hp} \right] J_0 \ (rp) \ dp = 0 \tag{12}$$

Tranter (11), following Williams (12) and Cooke (13), shows how these equations may be reformulated as a standard Fredholm integral equation of the second kind.

Using this method it is found that

$$V(r, \ x) = \int_0^\infty U \ (h, \ p) \ \frac{\sinh \ [p \ (h-x)]}{\cosh \ [hp]} \ J_0 \ (rp) \ dp \tag{13}$$

where

$$U \ (h, \ p) = \int_0^1 H \ (z, \ h) \ \cos \ (pz) \ dz \tag{14}$$

$$H \ (z, \ h) = \frac{2}{\pi} - \int_0^1 H \ (t, \ h) \ K \ (z, \ t) \ dt \tag{15}$$

and

$$K \ (z,t) = \frac{2}{\pi} \int_0^\infty [\tanh \ (hp) - 1] \ \cos \ (tp) \ \cos \ (zp) \ dp \tag{16}$$

Equation (15) for $H(z,h)$ is a Fredholm integral equation. For the purpose of numerical solution of the problem it is useful to combine equations (14) to (16) into a single integral equation.

Substituting equation (16) into (15) :

$$U \ (h, \ p) = \frac{2}{\pi} \ \frac{\sin p}{p} - \int_0^1 H(t,h) \ [\int_0^1 \cos(pz) \ K \ (z,t) \ dz] \ dt \tag{17}$$

Multiplying equation (16) by $\cos pz$, integrating over z from 0 to 1, and substituting the result into equation (17) it is found that :

$$U(h,p) = \frac{2}{\pi} \ \frac{\sin p}{p} + \frac{1}{2}\int_0^\infty U(h,q) \ [1-\tanh(h,q)]$$
$$[\sin(p-q) \ / \ (p-q) + \sin(p+q) \ / \ (p+q)] \ dq \tag{18}$$

Equations (13) and (18) specify a complete solution to the original problem. The integral for the flux in equation (6) becomes :

$$F = 2\pi \int_0^\infty J_1 (p) \; U (h,p) \; dp \tag{19}$$

In the limit of infinite h, that is, diffusion to a disc electrode from a semi-infinite medium, we find :

$$U(\infty,p) = \frac{2}{\pi} \frac{\sin p}{p} \tag{20}$$

$$V(r,o) = \frac{2}{\pi} \int_0^\infty \frac{\sin p}{p} J_0 (rp) \; dp \tag{21}$$

that is:

$$V(r,o) = 1 \text{ for } r > 1 \tag{22}$$

$$= \frac{2}{\pi} \arcsin (1/r) \text{ for } r > 1 \tag{23}$$

$$(\partial V / \partial x)_{x=0} = 0 \quad \text{for } r > 1 \tag{24}$$

$$= -(2 / \pi) (1 - r^2)^{-\frac{1}{2}} \text{ for } 0 < r \leq 1 \tag{25}$$

from equation (6) the total flux in this case is :

$$F = 4 \tag{26}$$

Equations (22) and (24) confirm that the solution derived satisfies the original boundary conditions in the plane x = 0. Equation (26) specifies the total steady state flux to a circular disc electrode in a semi-infinite medium, and agrees with a result obtained by Siu and Cobbold (10).

In the case of h large though finite, it can be shown that

$$U(h,p) = N(h) \sin p / (2\pi p) \tag{27}$$

where

$$N(h) = 4 \left\{ 1 - \frac{2}{\pi} \int_0^\infty \left[\frac{\sin q}{q} \right]^2 [1 - \tanh (hq)] \; dq \right\}^{-1} \tag{28}$$

Then the flux can be calculated explicitly with the result :

$$F = N(h) \tag{29}$$

That N(h) approaches the value 4 for h→∞ follows by inspection of equation (28). In the limit of very large h, the tanh function in equation (28) can be approximated by :

$$\tanh(hq) = 1 + 2 \sum_{n=1}^{\infty} (-1)^n e^{-2nhq} \qquad (30)$$

This function rapidly approaches unity, for h very large, whence $(\sin q / q)$ may be replaced by unity in equation (28) and the integral evaluated. Then :

$$F = 4 [1 + 2 \ln2 / (\pi h)] \qquad (31)$$

This is an explicit equation for the asymptotic dependence of the flux upon h as h approaches infinity.

An approximate solution valid for thin membranes (14)

In the previous section, a complete solution to the diffusion problem was obtained which, however, is awkward to use for the derivation of numerical results when:

$$a >> b$$

where b is the membrane thickness and a the radius of the working electrode. In the case of a >> b, that is of thin membranes and for macroscopically large electrode diameters, it is possible to ignore the circular-disc shape of the electrode and to explore the details of the diffusion near the edge of a semi-infinite plane electrode. The boundary value problem for this case is :

$$\frac{\partial^2 p}{\partial x_1^2} + \frac{\partial^2 p}{\partial x_2^2} = 0 \text{ for } -\infty < x_2 < \infty, \ -\infty < x_1 < \infty \qquad (32)$$

where

$$p = 0 \text{ at } x_1 = 0, \ 0 < x_2 < \infty \qquad (33)$$

$$p = p_s \text{ at } x_1 = b \qquad (34)$$

$$\frac{\partial p}{\partial x} = 0 \text{ at } x_1 = 0, \ -\infty < x_2 < 0 \qquad (35)$$

Of course the total flux of electroactive gas reaching the semi-infinite electrode must be infinite, but as will be seen below, it is easy to distinguish the incremental current arising from the local edge effect.

Equation (32) is Laplace's equation in 2-dimensions and may be treated by the powerful method of complex conjugate functions (15). We define

$$V = \pi^2 (p-p_s) / 2 p_s, \quad x = \pi x_1 / 2b, \quad y = \pi x_2 / 2b \tag{36}$$

$$z = y + ix \quad \text{where } i = \sqrt{-1} \tag{37}$$

and

$$W = V + iU = \int_0^\infty \ln [\tanh ((-iz -t) / 2)] \, \varrho(t)dt \tag{38}$$

U is known as the stream function, obeys Laplace's equation, and is related to V through the Cauchy-Riemann relations.

$$\frac{\partial V}{\partial y} = \frac{\partial U}{\partial x} , \quad \frac{\partial V}{\partial x} = - \frac{\partial U}{\partial y}$$

In order to understand the form of equation (38) it is most convenient to refer to the electrical analogue of the diffusion problem, which, as mentioned in the introduction is the determination of the electrostatic potential V in the vicinity of a semi-infinite charged plane at potential $-\pi^2/2$ placed midway between and parallel to two infinite grounded planes located at $x = \pm\pi/ 2$. In that problem the analogue of the fugacity gradient at the electrode, needed for calculation of the flux of electroactive gas, is the electrostatic charge density $\varrho(y)$ on the semi-infinite plane. This charge density, with alternating signs, appears on all the image planes at the same coordinate y. The potential at any point x,y is obtained by summing the potentials generated by all of the infinitesimal charges ϱdy, leading to the ln tanh term in equation (38), and then by integrating over y from 0 to ∞.

In order that the potential (which is the real part of W, the result of the integration) take the values specified in the boundary conditions on the boundary planes, the charge density must have the form :

$$\varrho(y) = [1 - \exp(-2y)]^{-\frac{1}{2}} \quad 0 < y < \infty \tag{39}$$

Substituting this function into equation (38) and evaluating the integral yields :

$$W = -(\pi^2/2) + \pi x + i\pi y + i\pi \ln [1 + \sqrt{1- (\cos 2x - i\sin 2x)e^{-2y}}] \tag{40}$$

It was verified that the real part of this function satisfies all boundary conditions.

To calculate the additional current arising from the "edge effect", we integrate the fugacity gradient $\varrho(y)$ from the edge of the electrode to some point $L = l\pi / 2b$ where the fugacity distribution would be practically indistinguishable from that on a "1-dimensional" electrode (where $\varrho \rightarrow 1$)

$$\frac{2b}{\pi} \int_0^L (1-e^{-2Y})^{-\frac{1}{2}} \, dy = \frac{2b}{\pi} \ln [e^L + \sqrt{e^{2L} - 1} \,]$$

$$\approx 1 + 2b \ln 2 / \pi \tag{41}$$

since l is considered large. In the absence of the edge effect the result of this integral would be 1, so the effect of the edge is to increase the width of the electrode by $2b \ln 2/\pi$

Returning now to the case of a circular disc, it follows that the total flux of electoactive gas should be calculated as if the electrode had an effective "1-D" area :

$$A = \pi a^2 + 2\pi a \quad . \quad 2b \ln 2/\pi \tag{42}$$

But the flux in the 1-D approximation is

$$Flux = A \, D_m \, S_m \, P_S \, / \, b$$

$$= \pi a^2 \, D_m \, S_m \, P_S \, / \, b + 4 \ln 2 \, D_m \, S_m \, P_S \, a \tag{43}$$

or

$$F = Flux \, / \, (D_m \, S_m \, P_S \, a) = \pi a \, / \, b + 4 \ln 2 \tag{44}$$

The first term on the right side of equation (43) is the usual flux term calculated by means of the 1-D approximation, and the second term is the incremental flux arising from the edge effect. By comparing these terms we can derive the critical radius a_C at which the edge term becomes significant, eg one percent of the 1-D flux :

$$a_C \approx 100 \quad (4b \, / \, \pi) \ln 2 = 88.25 \, b$$

Electrodes having radii less than this threshold value have significant edge currents, and those having $a > a_C$ should have insignificant edge effects.

DISCUSSION

All of the forgoing results can be summarised in the form of a graph of the relationship between the calculated total flux of electroactive gas to the electrode, and the ratio of electrode

radius a to membrane thickness b. It is particularly convenient
to plot this graph with the dimensionless quantity

$F = \text{Flux} / (D_m\, S_m\, P_s\, a)$

on the ordinate. Figure 1 presents the graph in this form.

First, the 1-D approach of equation (7) gives a straight line
with slope π and zero intercept at a/b = 0. The latter condition
corresponds to a membrane of infinite thickness and the zero
intercept means that there is a negligible rate of penetration
of the gas through such a membrane, in the one-dimensional
system which has a one-to-one correspondance between the area of
the source and the area of the electrode.

In contrast equation (26) predicts a non-zero steady state flux
for such a system, when diffusion to a finite sized electrode is
properly accounted for with a 2-D theory. The reason is that the
source area is effectively infinite in this case and compensates
for the infinite length of the diffusion path.

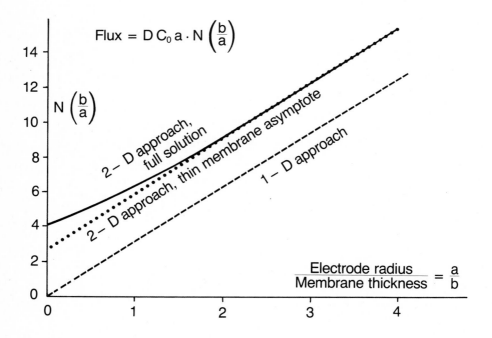

Fig. 1. Comparison of estimates of the steady state diffusion
controlled flux to a disc electrode, made by various
approximations.

Equation (31) shows that a membrane is effectively infinitely thick, if its thickness exceeds about 44.12 electrode radii.

Equation (44) predicts a straight line parallel to that calculated from the 1-D case but separated from it by the amount 4 ln 2. This constant shift reflects the fact that the edge effect gives rise to an additional term in the equation for the flux, compared to that derived from a 1-D treatment, proportional to the length of the circumference of the electrode.

Finally, the flux in systems intermediate between the thin membrane asymptote region ($a/b \geq 4$) and the infinitely thick membrane case ($a/b = 0$), was calculated from equation (19), using equation (18) to compute the unknown function U (h,p) by an iterative method. The curvature of this plot at small a/b reflects the increasing importance of the edge effect as the electrode dimensions are reduced relative to the membrane thickness.

REFERENCES

1. H. K. Mancy, D. A. Okun and C. N. Reilley, J. Electroanal. Chem. **4**, 65-92 (1962).
2. V. Linek, J. Sinkule and V. Vacek, "Dissolved Oxygen Probes" in Comprehensive Biotechnology, M. Moo-Young Editor, Oxford 1985, vol. 4, chapter 19, 363-394.
3. M. L. Hitchman, "Measurement of Dissolved Oxygen" Wiley, New York (1978).
4. E. Gnaiger and H. Forstner, "Polarographic Oxygen Sensors" Springer Verlag, Berlin (1983).
5. Y. H. Lee and G. T. Tsao, Adv. Biochem. Eng. **13**, 36-86 (1979)
6. C. E. W. Hahn, A. H. Davies and W. J. Albery, Resp. Physiol. **25**, 109-133 (1975).
7. J. M. Hale and M. L. Hitchman, J. Electroanal. Chem. **107**, 281-294 (1980).
8. V. Vacek, V. Linek and J. Sinkule, J. Electrochem. Soc., **133**, 540-547 (1986).
9. O. J. Jensen, T. Jacobsen and K. Thomsen, J. Electroanal. Chem. **87**, 203-211 (1978).
10. W. Siu and R. S. C. Cobbold, Med. Biol. Eng. **14**, 109-121 (1976).
11. C. J. Tranter, "Integral Transforms in Mathematical Physics", Chapman & Hall, (1971) p. 117.
12. W. E. Williams, Proc. Edinburgh Math. Soc. **12**, 213-216 (1961).
13. J. C. Cooke, Quart. J. Mech. Appl. Math. **IX**, 103-110 (1956).
14. R. P. Feynman, Private Communication (1985).
15. W. R. Smythe, "Static and Dynamic Electricity", Mc Graw Hill (1939) Problem 26, p. 103.

M.R. Smyth and J.G. Vos
Electrochemistry, Sensors and Analysis
© Elsevier Science Publishers B.V., Amsterdam — Printed in The Netherlands

USE OF PROTEIN COATINGS ON PIEZOELECTRIC CRYSTALS FOR ASSAY OF GASEOUS POLLUTANTS

G. GUILBAULT, J. NGEH-NGWAINBI, P. FOLEY and J. JORDAN

Chemistry Department, University of New Orleans, New Orleans, Louisiana   70148

SUMMARY

The use of protein coatings (antibodies and enzymes) on piezoelectric crystals is described, for the assay of gaseous pollutants. With cholinesterase as the substrate, organophosphorus compounds can be detected; using parathion antibodies, the antigen parathion is easily measured. Sensitivity is parts per billion, the response time is fast (< 1 min), and the stability is excellent.

INTRODUCTION

The piezoelectric crystal is based on the principle that the frequency of vibration of an oscillating quartz crystal is decreased by the deposition of a small mass on the electrode's surface.

By putting a highly selective coating on the crystal, a response indicates the presence of the molecule for which the coating is selective. Quantitatively, the gaseous pollutants are adsorbed or partitioned by the coating of the electrode, thereby increasing the mass on the crystal and decreasing its frequency, according to the Sauerbrey equation:

$$\Delta F = - 2.24 \times 10^6 \; \frac{F_o^2}{A} \; \Delta Ms$$

where ΔF = the change in frequency of the crystal

  F_o = the basic frequency of the crystal

  A = area of the electrode

  ΔMs = mass deposited by the adsorption or partitioning

The magnitude of the mass change, Ms, is linearly related to the concentration of substance present in the atmosphere.

The piezoelectric effect was first described by Pierre and Jacques Curie in 1880. Cady built the first oscillator in 1911, but the first analytical papers were not published until 1964, by King, who proposed the detector for gas chromatography, and by Guilbault and co-workers, who used it as a detector for organophosphorus compounds in air. King's detector was marketed first by Esso Research, and now Dupont, as a moisture analyzer, and also by a now defunct Florida company, as a GC instrument utilizing the PZ detector. Over a hundred papers have been since published, on the use of the PZ detector for assay of atmospheric pollutants.

Preliminary studies utilizing biological substrates, such as enzymes and antibodies, have proven succesful as coatings for a piezoelectric crystal detector. Immobilization of cholinesterase and parathion antibodies allows for the detection of organophosphorus pesticides at the ppb level with the cholinesterase coatings exhibiting a general response to all organophosphorus pesticides and the antibody coatings exhibiting a more specific response to parathion. Excellent reproducibilities, coating lifetimes, response times and selectivities are observed. This research is the first use of proteins as coatings for the direct assay of gaseous compounds, thus making this method an attractive alternative to some conventional techniques currently in use.

EXPERIMENTAL

The experimental set-up is shown in **Fig. 1.** Before measurements of the

Schematic of SET Up.

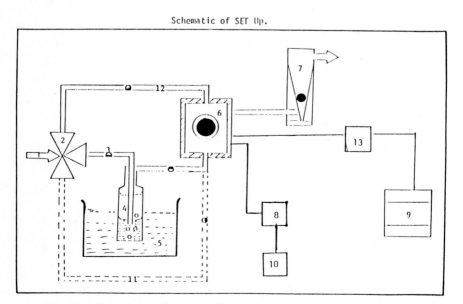

Fig. 1. Experimental Set-Up Used.

response of a coated crystal to a particular pesticide, were tested, the gas delivery system was purged for at least 24 hours with the substrate saturated carrier gas. This ensured that only the pesticide of interest was in the delivery system when measurements were recorded. The gas permeation tubes allow entry of a fixed and known concentration of pesticide into the detection chamber.

In each case, the electrode portion of the crystals were coated by first applying equal amounts of protein to both sides of the crystal using a microliter syringe. The protein could then be immobilized using appropriate ratios of glutaraldehyde and bovine serum albumin, applied to the crystal in the same manner as the protein. Excess glutaraldehyde was washed off using a 0.1 $\underline{M}$ phosphate buffer. The crystal was then placed in a desiccator for at lease one hour before testing to allow the coating to dry.

The coated crystals were then placed in the test cell. The pure carrier gas was used to establish a stable baseline. Once this was done, the test gas was introduced to the cell using a four-way valve and the subsequent decrease in frequency was recorded as the response of that coating to the pesticide being tested. Generally, four measurements of this type were made in succession in order to evaluate the reproducibility of each response, as shown for a cholinesterase coating with pesticide in **Fig. 2.** This was continued throughout the lifetime of each coated crystal.

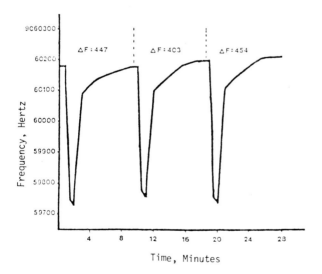

Fig. 2. Response to repeated exposures of Malathion.

The PZ detector is mounted for analytical work as shown in Fig 1. The inlet from the atmosphere is split into a T junction, to allow gas to pass on both sides of the coated crystal. A set frequency, usually 9 or 14 MHz, is applied, and the frequency of oscillation, which indicates the amount of pollutant present, is recorded using a frequency counter and recorder.

In order to generate a known concentration of the test gas at very low part-per-billion to part-per-trillion concentrations, a gas permeation tube is used, as shown in Fig. 1. Pure carrier gas mixes with the diffused test gas to generate a known concentration.

A PZ 101 instrument, which has been built with a separate NiCd power supply for 8 hour field operation, air pump with variable intake, electronic gate for varying sensitivity and readout of the concentration or frequency, is sold commercially by Universal Sensors (P. O. Box 736, New Orleans, La. 70148).

RESULTS AND DISCUSSION

The reaction sequence for organophosphorus detection using an enzyme coating is shown in below.

Malathion + Acetylcholinesterase ======= E-I
    I   E ΔF (reversible response)

Typical response times were on the order of 30-60 seconds with complete recovery in 5-10 minutes, as shown in Fig. 2.

Calibration curves for acetylcholinesterase to Malathion and DIMP = (Diisopropyl Methyl Phosphonate) are shown in **Fig. 3.** Excellent results were obtained.

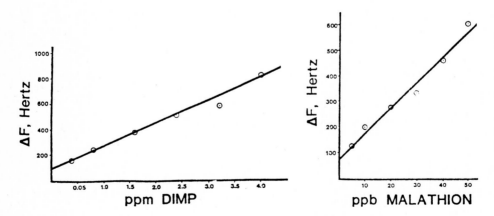

Fig. 3. Malathion and DIMP calibration curves.

Next, let us consider the use of antibodies as coatings for the detection of gas antigens.

A typical response curve for the antibody coating to parathion is shown in **Fig. 4.** Note the fast response and good reproducibility. Recovery time was on the order of 1-2 minutes while responses were complete in 2-3 minutes.

The response of the antibody coating in the gas phase to other similarly structured pesticides is shown in Table 1. Comparison is made to the response of the antibody to these same compounds in solution. The similarity in response shows that indeed an antigen-antibody reaction is being monitored in the gas phase.

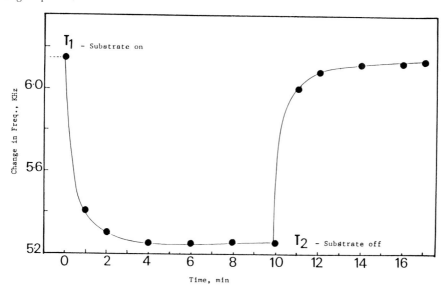

Fig. 4. Response of Detection to 35 ppb of Parathion.

A reproducibility of better than 4% for all concentrations of antibody tested was achieved, and the calibration plot for parathion is linear in the part-per-billion to part-per-million range.

No interferences from humidity were observed as long as it remained constant. A larger response was recorded at higher humidities than at relatively low humidities, hence calibration curves at different humidities are first required.

The enzyme coated crystals had a lifetime of approximately 40 days when stored in a desiccator at room temperature while the antibody coatings had a lifetime of approximately 18 days when stored in the same manner.   The stability of the antibody coated crystal is shown in **Fig. 5.**

These preliminary results make the future use of piezoelectric crystals as immunochemical sensors very promising.   This technique is also potentially useful and convenient in the area of environmental analysis,   and an alternative to radioimmunoassay.   Use of this technique for the assay of substrates,   such as formaldehyde in air with formaldehyde dehydrogenase,   for detection of contraband drugs with cocaine or morphine antibodies,   or explosives in airport security with antibodies specific for those compounds, are only a few of the applications we have developed.

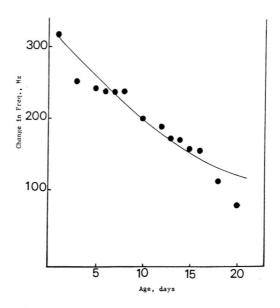

Fig. 5.   Long Term Stability Studies on the Response of a Parathion Antibody Coating with Age.

TABLE 1

Amount of pesticide causing 50% inhibition of parathion binding to antiserum in solution (ppb) needed to produce a 400 Hz change in the gas phase.

Pesticide	Amt. (gs) in solution[*]	Amt. (ppb) in gas phase
Parathion	50	36
Malathion	70,000	106
Paraoxon	1,850	--
Methyl Parathion	15,000	158
p-Nitrophenol	10,000	680
Disulfoton	--	560
Ethion	--	102

[*]Ercegovich et al. J. Agric. Food Chem. 1981, 29, 559-563.

ELECTROANALYTICAL METHODS IN CLINICAL

AND PHARMACEUTICAL CHEMISTRY

M.R. Smyth and J.G. Vos
Electrochemistry, Sensors and Analysis
© Elsevier Science Publishers B.V., Amsterdam — Printed in The Netherlands

RECENT APPLICATIONS OF ELECTROANALYSIS IN THE BASIC MEDICAL SCIENCES

WILLIAM R. HEINEMAN, EDWARD DEUTSCH AND H. BRIAN HALSALL

Department of Chemistry, Biomedical Chemistry Research Center, University of Cincinnati, Cincinnati, OH   45221 (U.S.A.)

SUMMARY

The concept of enzyme immunoassay with electrochemical detection has been demonstrated with assays developed for analytes of medical importance. Detection is by thin layer hydrodynamic amperometry in conjunction with liquid chromatography. Homogeneous assays for phenytoin and digoxin and heterogeneous assays for digoxin, α_1-acid glycoprotein and IgG have been developed. A detection limit of 10 pg/mL has been achieved with a sandwich assay for IgG.

Technetium radiopharmaceuticals for skeletal imaging have been formulated by electrochemical reduction of TcO_4^- in the presence of a diphosphonate ligand. The electrode potential influences the distribution of technetium diphosphonate complexes formed and can therefore be used to generate specific complexes with the best imaging properties.

INTRODUCTION

Some electroanalytical techniques have been used in conjunction with the basic medical sciences. Perhaps the most commonly used device is the Clark oxygen electrode, which is routinely employed to measure dissolved oxygen levels in patients. Electrochemical sensors for the in vivo monitoring of other analytes of importance in medical care are being developed. Some electroanalytical techniques are also used in clinical laboratories. For example, glucose is detected by an amperometric enzyme electrode and pH, CO_2 and electrolytes such as K^+, Na^+, Ca^{2+}, and Cl^- are measured by ion selective electrodes. This paper describes recent developments in the use of electroanalytical techniques in conjunction with immunoassay and nuclear medicine.

ENZYME IMMUNOASSAY WITH ELECTROCHEMICAL DETECTION

Immunoassay methods are used routinely for the low level (μg/mL-pg/mL) determination of compounds of clinical interest in a variety of biological matrices. Radioimmunoassay is the traditionally employed immunoassay method; however, enzyme immunoassay is competing with radioimmunoassay as a fast, reliable and accurate method that does not involve the use of radioisotopes. This technique combines the selectivity of the antigen/antibody reaction with the inherent chemical amplification ability of an enzyme label to achieve low

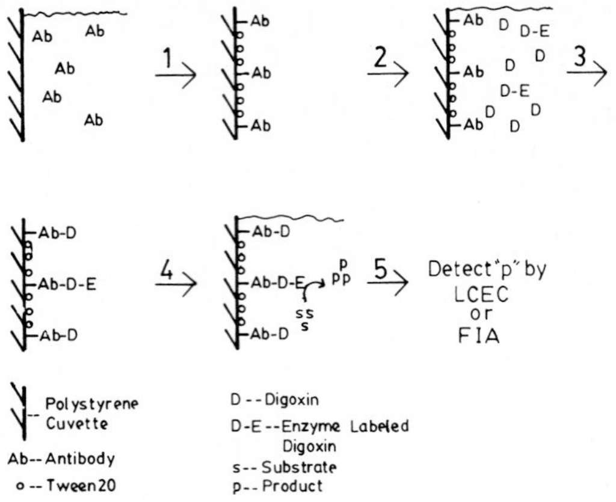

Fig. 1. Competitive heterogeneous enzyme immunoassay with electrochemical detection of "p".

detection levels. Homogeneous and heterogeneous enzyme assays have been developed for a wide variety of substances including drugs, hormones, steroids and proteins.

Electrochemical enzyme immunoassay is based on antigen labeled with an enzyme that catalyzes the production of an electroactive product. Hydrodynamic amperometry has proved to be very effective as the detection technique for enzyme immunoassay.

Heterogeneous Immunoassay

Heterogeneous enzyme immunoassays with electrochemical detection are based on the Enzyme Linked Immunosorbent Assay (ELISA) technique in which antibody is typically immobilized on the walls of small volume (ca. 500 μL) plastic cuvettes. The general procedure is outlined in Figure 1 for the determination of drug D.

Reagent cuvettes are first prepared by attaching specific antibody to the inside walls by passive adsorption or covalent bonding. The cuvettes are then rinsed with a non-ionic surfactant to cover any exposed surface between

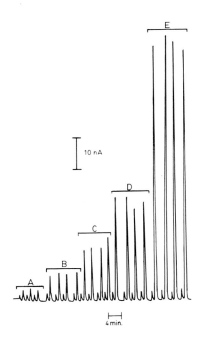

Fig. 2. Heterogeneous enzyme immunoassay with LCEC detection of enzyme-generated phenol for a series of digoxin standards in plasma solutions. Concentration of digoxin in plasma samples (A) 5.0, (B) 2.0, (C) 1.0, (D) 0.5, and (E) 0.0 ng/mL. [Reprinted from (ref. 1) with permission. Copyright 1986, American Chemical Society.]

antibody molecules. Reagent cuvettes of this type can be prepared in advance and stored. Enzyme-labeled drug (D-E) and samples or standards containing D are added to the tubes for competitive equilibration with the antibody. In the competition between D and D-E for a limited number of antibody binding sites, a greater concentration of D in the sample results in less D-E being bound to adsorbed Ab. Unbound D and D-E are then rinsed from the tubes and substrate (s) is added. At a fixed time, sample is withdrawn and analyzed for the electroactive product (p). In a typical standard plot of current from the electrochemical measurement of p vs. concentration of D standard, the electrochemical signal diminishes with increasing D concentration since less

348

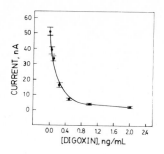

Fig. 3. Standard curve for dioxin standards in pooled human serum. [Reprinted from (ref. 1) with permission. Copyright 1986, American Chemical Society.]

D-E is able to interact with Ab and hence is rinsed out of the cuvette.

An enzyme immunoassay of this type has been demonstrated for digoxin (ref. 1), a stimulant commonly used for the treatment of chronic heart disease. The therapeutic range is both low and very narrow (0.8 to 2.0 ng/mL in blood). Alkaline phosphatase, which catalyzes the hydrolysis of the phenyl phosphate ester to phenol and phosphate, is used as the enzyme label. The enzyme-generated phenol is easily detectable by liquid chromatography with electrochemical detection (LCEC), whereas phenyl phosphate is electroinactive and hence noninterfering (ref. 2). Injected sample (typically 20 μL) flows through a thin-layer electrochemical cell where phenol is oxidized at a carbon-paste working electrode at +870mV vs. Ag/AgCl. An LCEC chromatogram showing a series of injections from an immunoassay is shown in Figure 2, and a standard plot, which is typically non-linear in a competitive assay of this type, is shown in Figure 3. A detection limit of 50 pg/mL has been achieved for digoxin, which illustrates the low detection level capability of the technique. Results of analyses on patient samples are in good agreement with results by an accepted radioimmunoassay procedure. Assays have also been developed for α_1-acid glycoprotein (ref. 3) and for immunoglobulin G (ref. 4).

The heterogeneous enzyme immunoassay with electrochemical detection has several advantageous features. The detection limit, as demonstrated for digoxin and IgG, is typically in the low pg/mL range, and was a function of the antigen-antibody binding constant, rather than the ability to detect phenol by LCEC. Consequently, even lower limits should be attainable. Since the sample is rinsed out of the reagent tubes before adding substrate, problems with interferences by electroactive constituents in the sample and possible fouling of the electrode by the adsorption of protein films are eliminated. These concepts were demonstrated with research instrumentation for LCEC. Automation of many steps in the procedure and the treatment of data is needed for its ready application to the routine analysis of many samples.

Homogeneous Immunoassay.

Immunoassays of the homogeneous type rely on a change in the intensity of the label signal that occurs when D-E binds with D to form Ab:D-E. The general equilibrium scheme for a homogeneous immunoassay is shown below

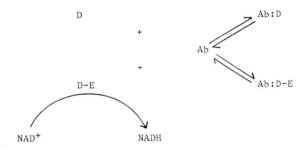

where E is glucose-6-phosphate dehydrogenase. The ability to distinguish between D-E and Ab:D-E enables the immunoassay to be done without a separation step. The immunoassay relies on reduction in the rate of enzyme catalysis when the antibody binds to the enzyme-labeled antigen. This enables free D-E to be distinguished from Ab:D-E on the basis of the rate of production of NADH. The catalytic activity of Ab:D-E is substantially diminished as a consequence of steric hindrance and limited substrate access. Since a larger concentration of D results in more D-E being in the unbound and, consequently, more active form, a calibration plot can be based on the rate of NADH formation versus D concentration. An assay of this type is called an Enzyme Multiplied Immunoassay Technique (EMIT^{TM}) and is commonly used with spectrophotometric detection of NADH (ref. 5).

Phenytoin, an antiepileptic drug, has been used as an initial test system for developing enzyme immunoassays with amperometric detection of NADH (ref. 6). This immunoassay scheme is amenable to the electrochemical detection of NADH by LCEC. The concentration of NADH is monitored by oxidation at 0.75 V vs Ag/AgCl at a glassy carbon electrode in a flow-through thin layer cell. Antibody and proteins in the sample are removed by a reversed-phase C-18 liquid chromatography precolumn to prevent adsorption on the electrode and consequent fouling. Assays of serum samples from patients on phenytoin treatment gave results that were in good agreement with a standard spectro-photometric procedure at therapeutic levels (10-20 μg/mL).

Enzyme labels of this type can take advantage of enzyme amplification as a means of achieving a low detection limit. The electrochemical technique can be extended to the determination of species at concentrations in the low ng/mL range, as has been demonstrated with digoxin (ref. 7).

. The main advantage of the homogeneous immunoassay, compared to the hetero-geneous assay, is the absence of a step for separating the antibody-bound antigen from the free antigen. This usually translates into a simpler assay procedure. However, homogeneous assays are usually restricted to higher detection levels than heterogeneous assays and are more susceptible to inter-ferences from other sample constituents which would otherwise be removed in a separation step.

ELECTROCHEMICAL GENERATION OF RADIOPHARMACEUTICALS

Radiopharmaceuticals have been used extensively in recent years as organ imaging agents in nuclear medicine (refs. 8-9). Noninvasive procedures have been developed for imaging the skeleton, heart, brain, kidney and lungs. These procedures are based on the tendency of the body to concentrate some chemical form of a radioisotope (primarily ^{99m}Tc) at the organ of interest (ref. 10). An image of the organ can then be obtained by a gamma scan of the appropriate portion of the body.

Diphosphonate complexes of ^{99m}Tc are widely used as skeletal imaging agents. A typical skeletal imaging agent is formulated on the day of use by mixing a solution of TcO_4^- (containing $^{99m}TcO_4^-$ and its daughter $^{99}TcO_4^-$) with a reducing agent, a diphosphonate ligand, and an antioxidant stabilizer such as ascorbic acid. The chemistry of the formulation involves reduction of TcO_4^- to lower oxidation states [Tc(V), Tc(IV), Tc(III)] with concomitant formation of Tc-diphosphonate complexes. This reaction mixture is injected into the patient without purification.

Recently, we have developed an anion exchange high performance liquid chromatography (HPLC) method for the separation of the component complexes in Tc-diphosphonate radiopharmaceuticals for skeletal imaging (ref. 11). This methodology has enabled us to explore the complicated chemistry involved in the formulation of skeletal imaging agents. Separation of ^{99m}Tc(NaBH$_4$)-HEDP reaction mixtures prepared by NaBH₄ reduction of pertechnetate in the presence of HEDP (hydroxyethylidine diphosphonate) has shown the presence of at least seven Tc-containing species in "carrier-added" preparations and three components in "no-carrier added" preparations (ref. 11). The distribution of complexes within these mixtures can be varied by controlling pH, the concentration of HEDP, the concentration of Tc, and the presence of air (ref. 12). Suitable control of these formulation conditions can yield mixtures which consist of essentially (85%) one component. The chromatographically separated Tc-HEDP components exhibit distinctly different biodistributions (ref. 13). Similar results have been obtained with methylene diphosphonate MDP (refs. 14-15) and dimethylamino diphosphonate DMAD (refs. 16-17). Thus, it appears that optimum imaging efficacy can be achieved by injection of a single Tc-complex, rather than a radiopharmaceutical "mixture" of complexes.

As a part of our systematic study of variables that affect the generation of Tc-diphosphonate complexes, we have investigated inert electrodes as reductants. Radiopharmaceuticals were prepared by the reduction of pertechnetate in the presence of HEDP, ascorbic acid and sodium acetate by controlled potential electrolysis. The resulting Tc-HEDP radiopharmaceuticals were analyzed by HPLC to determine the distribution of the component Tc-HEDP complexes. Chromatograms of radiopharmaceuticals prepared at pH 5.6 at several electrode potentials are shown in Figure 4. Five major Tc-HEDP components can be identified on the basis of chromatographic retention time. These five components have been observed in previous studies on NaBH₄-reduced preparations of Tc-HEDP (ref. 12). Apparently, electrochemical reduction generates no species that are different from those obtained by chemical reduction with NaBH₄. The applied potential has a dramatic effect on the distribution of Tc among the five complexes. At -0.8V only complexes c and g are present in any great amount. Decreasing the potential to -0.9, -1.0 and -1.2V causes complexes c and f to predominate and complex g to become only a minor component. Complex e is a major component at -1.0V as evidenced by the substantial shoulder on f. Shifting the potential to -1.5V results in substantial evolution of hydrogen gas from the mercury electrode. Complex d is now a major component in addition to c, e and f. None of component g, which was the major component at -0.8V, is detectable at -1.5 V.

The electrochemical reduction of pertechnetate in the presence of HEDP, under the proper conditions, yields a very high percentage of the complex

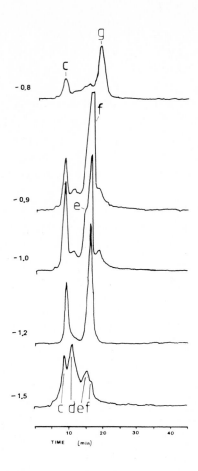

Fig. 4. Chromatogram of radiopharmaceutical prepared by electrochemical reduction of TcO₄⁻ at various reduction potentials. 1.0 mM TcO₄⁻, 0.85 M sodium acetate, 67 mM HEDP, 85 mM ascorbic acid, pH = 5.6.

labeled c, which has been shown to have exceptionally good bone uptake (ref. 13). This particular complex is produced in only low yields when prepared by chemical reduction of ^{99m}TcO₄-. Thus, electrochemistry shows promise in aiding the development of a more efficacious bone imaging agent by allowing selective generation of those individual ^{99m}Tc-HEDP complexes with the best imaging properties.

REFERENCES

1 K.R. Wehmeyer, H.B. Halsall, W.R. Heineman, C.P. Volle, I. Chen, Competitive heterogeneous enzyme immunoassay for digoxin with electro-chemical detection, Anal. Chem., 58 (1986) 135-139.
2 K.R. Wehmeyer, M.J. Doyle, D.S. Wright, H.M. Eggers, H.B. Halsall, W.R. Heineman, Liquid chromatography with electrochemical detection of phenol and NADH for enzyme immunoassay, J. Liq. Chromatogr., 6 (1983) 2141-2156.
3 M.J. Doyle, H.B. Halsall, W.R. Heineman, Enzyme-linked immunoadsorbent assay with electrochemical detection for α_1-acid glycoprotein, Anal. Chem., 45 (1984) 2355-2360.
4 K.R. Wehmeyer, H.B. Halsall, W.R. Heineman, Heterogeneous enzyme immunoassay with electrochemical detection: competitive and "sandwich"-type immunoassays, Clin, Chem., 31 (1985) 1546-1549.
5 A.F. Rosenthal, M.G. Vargas, C.S. Klass, Evaluation of Enzyme-Multiplied Immunoassay Technique (EMIT) for Determination of Serum Digoxin, Clin. Chem., 22 (1976) 1899-1902.
6 H.M. Eggers, H.B. Halsall, W.R. Heineman, Enzyme Immunoassay with Flow-amperometric detection of NADH, Clin. Chem., 28 (1982) 1848-1851.
7 D.S. Wright, H.B. Halsall, W.R. Heineman, Enzyme immunoassay with amperometric detection, The Pittsburgh Conference and Exposition on Analytical Chemistry and Applied Spectroscopy, Atlantic City, New Jersey, March 5-9, 1984, No. 533.
8 G. Subramanian, B.A. Rhodes, J.F. Cooper, V.J. Sodd (Eds.), Radiopharma-ceuticals, The Society of Nuclear Medicine, New York, 1975.
9 B.A. Rhodes, B.Y. Craft, in: Basics of Radiopharmacy, C.V. Mosby Co. St. Louis, Missouri, 1978.
10 N.D. Heindel, H.D. Burns, T. Honda, L.W. Brady (Eds.), The Chemistry of Radiopharmaeuticals, Masson, New York, 1978, Chapter 17.
11 T.C. Pinkerton, W.R. Heineman, E. Deutsch, Separation of technetium hydroxyethylidene diphosphonate complexes by anion-exchange high performance liquid chromatography, Anal. Chem., 52 (1980) 1106-1110.
12 J.P. Zodda, S. Tanabe, W.R. Heineman, E. Deutsch, The effect of formulation conditions on the distribution of $^{99m}Tc(NaBH_4)$-HEDP components separated by anion exchange HPLC, Int. J. Appl. Radiat. Isot., in press.
13 T.C. Pinkerton, D.L. Ferguson, E. Deutsch, W.R. Heineman, K. Libson, In vivo distributions of some component fractions of $Tc(NaBH_4)$-HEDP mixtures separated by anion exchange high performance liquid chromatography, Int. J. Radiat. Isot., 33 (1982) 907-915.
14 S. Tanabe, J.P. Zodda, E. Deutsch, W.R. Heineman, Effect of pH on the formation of $Tc(NaBH_4)$-MDP radiopharmaceutical analogues, Int. J. Radiat. Isot., 34 (1983) 1577-1584.
15 S. Tanabe, J.P. Zodda, K. Libson, E. Deutsch, W.R. Heineman, The biological distributions of some technetium-MDP components isolated by anion exchange high performance liquid chromatography, Int. J. Radiat. Isot., 34 (1983) 1585-1592.
16 M.E. Holland, W.R. Heineman, E. Deutsch, The effect of formulation conditions on the distribution of $^{99m}Tc(NaBH_4)$-DMAD radiopharmaceutical analogues separated by anion exchange high performance liquid chromatography, manuscript in preparation.
17 M.E. Holland, J. Bugaj, W.R. Heineman, E. Deutsch, The biological distributions in normal rats and in an osteogenic rat model of some technetium-DMAD components isolated by anion exchange high performance liquid chromatography, manuscript in preparation.

M.R. Smyth and J.G. Vos
Electrochemistry, Sensors and Analysis
© Elsevier Science Publishers B.V., Amsterdam — Printed in The Netherlands

APPLICATIONS OF MODERN VOLTAMMETRIC TECHNIQUES IN CLINICAL CHEMISTRY

J.P. HART

Division of Cellular Biology, Kennedy Institute of Rheumatology, Bute Gardens, Hammersmith, London, W6 7DW.

SUMMARY

Voltammetry/electrochemistry, following high-performance liquid chromatography (LCEC), can be a very powerful analytical tool in clinical chemistry; this is due to the high selectivity and sensitivity offered by the technique. A brief description is given of several different types of electrochemical flow-through cell based on the principle of either amperometric, or coulometric, detection. The application of these detectors to the analysis of endogenous levels of some vitamins and coenzymes in body fluids is discussed.

INTRODUCTION

The analysis of vitamins and coenzymes, at physiological concentrations in biological fluids, is of obvious clinical interest. However, these determinations present difficult analytical problems. There are several reasons for this: the biological compounds are often present at very low levels e.g. the mean circulating concentration of vitamin K_1 was found to be 335 pgml^{-1} for normal subjects and only 98 pgml^{-1} for patients suffering with osteoporosis and a fractured neck of femur (refs. 1, 2)- in addition the vitamin, or coenzyme, may exist as a structurally related group e.g. vitamin B_6 is a group of compounds consisting of pyridoxal, pyridoxine, pyridoxamine and the 5'-phosphate esters of these three compounds (ref. 3); there is also the possibility of interference from other naturally occurring compounds. Therefore, there is a need for very sensitive and selective methods for measuring endogenous levels of these substances in body fluids.

One of the most powerful techniques to emerge in recent years involves voltammetric detection, or as it is generally known, electrochemical detection, following high-performance liquid chromatography (LCEC) (refs. 4, 5). This technique has been shown to possess great potential for the analysis of trace amounts of electroactive organic compounds in complex matrices; concentrations down to the picogram per millilitre level have been readily measured (ref. 4). Many vitamins and coenzymes are electroactive (refs. 6, 7); consequently their analysis using LCEC should be possible. It is only recently that this has been realised.

This paper describes several different types of electrochemical detection systems used in LCEC; the application of these to the determination of some vitamins and coenzymes in body fluids is discussed.

ELECTROCHEMICAL DETECTORS
Amperometric detectors

In amperometric detection an electrochemical reaction occurs at the electrode surface which electrolyses much less than 100% of the electroactive species (typically 1%) (ref. 8).

(i) Single-electrode cells. Perhaps the most popular of the single working electrode arrangements for amperometric detection, is the thin layer cell. There are several variations of this cell type (ref. 9); in one of our studies we used a Model TL5 containing a glassy carbon working electrode (area ≈0.07 cm^2) which was obtained from Bioanalytical Systems, West Lafayette. In this particular cell the working electrode is situated in one half of the cell and is parallel to the flowing stream of liquid. The counter electrode is made from stainless steel tubing and forms the cell outlet; the reference electrode is silver/silver chloride. Both of these electrodes are situated in a separate compartment in the other half of the cell; the cell halves are separated by a gasket which may be one of several thicknesses, but is typically 0.002 to 0.005 cm. The equation for the current produced in a thin layer cell has been reported by Elbicki et al (ref. 10) and is given in equation 1.

$$i = 1.47 \ n \ F \ C \ \left(\frac{DA}{b}\right)^{2/3} \overline{U}^{1/3} \tag{1}$$

where i is the current density in μAcm^{-2}; n is the number of electrons transferred per mole; F is Faraday's constant; C is the concentration of analyte in mmoles L^{-1}; D is the diffusion coefficient $cm^2 s^{-1}$; A is the electrode area in cm^2; b is the gasket thickness in cm; $\overline{U}$ is the average volume flow rate in $cm^3 s^{-1}$. Clearly, thinner gaskets are to be prefered; in fact we obtained an increase in signal height for vitamin K_1, in agreement with equation (1), when the gasket thickness was decreased from 0.005 cm to 0.002 cm (ref. 11). As is also obvious from equation (1) current increases with flow-rate and electrode area, but in practice the values of these parameters are limited. This is due to the increase in background current, which accompanies the analyte signal enhancement, and which causes an increase in the noise level. Since it is the signal to noise ratio which is of importance in LCEC, samll background currents are desirable. These are mainly due to electroactive impurities in the mobile phase (ref. 8); therefore it is important to use pure reagents to prepare the mobile phase.

Several other important single-electrode detectors have been described (ref. 8). These include cells containing tubular, and disc, solid electrodes, as well as mercury electrodes in the opposite, parallel and normal configurations.

A variety of new electrode materials for use in amperometric detectors have been reviewed recently (ref. 12).

(ii) Dual-electrode cells. At the present time the two most popular arrangements of dual electrodes in thin-layer cells are the parallel and series configuration (ref. 13).

In the parallel configuration two identical electrodes are arranged side-by-side so that the eluent meets both electrodes simultaneously (Fig. 1a). In one mode of operation potentials are chosen at two points on the voltammetric wave e.g. E1 and E2, Fig. 1c. The ratio of the resulting currents i_1/i_2 is calculated and is compared to a standard of a pure compound for confirmation of peak identity. In another mode it is possible to monitor both oxidisable, and reducible, species simultaneously i.e. in the same injection. In this case the potential of one elctrode is held at a value which is sufficiently positive to initiate an oxidation process (E1, Fig. 1d); the other is held at a potential where reduction occurs (E2, Fig. 1d).

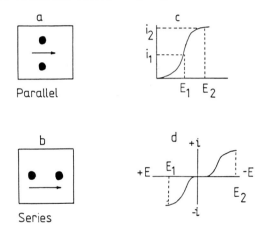

Fig. 1. Dual-electrodes in thin-layer cells (for explanation of letters a-d, see text.

In the series configuration one electrode is place upstream of the other (Fig. 1b). These may be operated in what is called the "redox" mode when a reversible or quasi-reversible reaction is involved. The upstream electrode (designated the "generator" electrode) is set at a potential where, for example, reduction occurs (E2 Fig. 1d); the product is then detected downstream at the second electrode (designated the "detector" electrode) which is held at a more positive potential, where re-oxidation occurs (E1 Fig. 1d). This mode offers several advantages over the reductive mode: the magnitude of the detector electrode is usually such that the background current is low and consequently the noise minimal; in addition, if the oxidation potential is sufficiently positive, no interference from dissolved oxygen occurs.

Coulometric detectors

In coulometric detection an electrochemical reaction occurs as the electrode surface which electrolyses 100% of the electroactive species (ref. 8).

Both single-electrode and dual-electrode coulometric detectors have been recently reviewed (ref. 12).

We have investigated the latter type of detector in several of our studies (refs. 14, 15). In these studies we used a commercially available cell containing dual porous graphite electrodes in the series configuration (this was a Model 5011 electrochemical cell and was obtained from Environmental Science Associates). The "generator" electrode and "detector" electrode areas are about 5 cm^2 and 0.6 cm^2 respectively. With this arrangement it is possible to use the redox mode, in the manner described above, but in this case total electrolysis occurs at the upstream electrode. Therefore, better sensitivity is to be expected with this type of cell than with the dual-amperometric detectors.

It is also possible to use another mode with the series dual-electrode and this has been designated the "screen mode". The upstream electrode potential is set at a value close to the foot of the analyte wave, where only the interferences are electrolysed. This effectively filters out the potential interference and improves the selectivity of the technique; this generally improves the sensitivity because higher gains may be employed.

APPLICATIONS
Fat-soluble vitamins

Vitamin K$_1$ (phylloquinone) has recently been shown to be essential in the reaction which converts peptide glutamate to γ-carboxyglutamate (Gla) (ref. 16). These Gla residues act as bidentate ligands for Ca^{2+} and are consequently of importance in calcification. It was for this reason that we were interested in measuring circulating vitamin K$_1$ levels in normal subjects and patients with disease of bone.

At the out-set of our studies the only method capable of measuring endogenous phylloquinone levels was that developed by Shearer et al (ref. 17); this involved H.P.L.C. with ultraviolet detection (LCUV). Since LCEC is generally known to be more sensitive than LCUV, for electroactive compounds, we decided to investigate the former for the development of assays for circulating vitamin K$_1$ levels.

Initially, we optimised the conditions for the mobile phase for use with an

octyl column and found that 95% MeOH-0.05 M acetate buffer pH 3.0 was suitable
for retention fo vitamin K_1. Using this medium, the electrochemical behaviour
of the vitamin was studied using cyclic voltammetry with a planar glassy carbon
electrode (ref. 11). Only one peak was obtained on the forward scan which is
consistent with the reduction of the quinone to the hydroquinone form of the
vitamin. The reverse scan also showed one peak which is due to the re-oxidation
of the hydroquinone back to the quinone. Since the separation between the two
peaks was greater than 30 mV the process is quasi-reversible. In addition, a
plot of the current function, $i_p/CV^{\frac{1}{2}}$ vs. $V^{\frac{1}{2}}$ (where i_p is the peak current in
μA, C is the concentration in mM and V is the scan rate in mVs^{-1}) indicated
that the hydroquinone, but not the quinone, was adsorbed onto the electrode
surface. The overall mechanism is shown below:

Quinone   Hydroquinone
                 (adsorbed)

$$R= -CH_2CH=CCH_2 \left[CH_2CH_2CHCH_2 \right]_3 H$$

This electrochemical process can be exploited, using two different
electrochemical detection systems, for the assay of vitamin K_1. First, the
reduction process may be used; this can be performed with a single working
electrode cell. In addition, since the reaction is quasi-reversible it is
possible to use the redox mode technique with series dual-electrodes. We have
investigated both of these detectors.

In our first method, we used a thin-layer cell containing a glassy carbon
working electrode which was operated in the reductive mode (ref. 11). A
solvent extraction procedure involving hexane and ethanol was used to separate
the vitamin from plasma (10 ml). After further purification using silica
cartridges, and a semi-preparative step with normal phase liquid chromatography,
the residue was analysed by LCEC. The reverse-phase (octyl) column was used
in conjunction with the mobile phase described above which was set at a flow-
rate of 1 mlmin^{-1}; the applied potential was -1.0 V vs. Ag/AgCl. This
procedure was found to be about three times more sensitive than the original
LCUV method and was suitable for the determination of endogenous phylloquinone
in the plasma of normal subjects. However, we needed to improve the
sensitivity of the method in order to measure depressed circulating levels of
the vitamin and also to reduce the volume of plasma required for the assay.

We achieved the required sensitivity by simply replacing the thin-layer cell with a coulometric detector containing dual porous graphite electrodes in series (ref. 14). These were operated in the redox mode: the upstream electrode was held at -1.3 V to reduce the quinone to the hydroquinone which was detected downstream at the second electrode, which was held at 0 V, where re-oxidation to the quinone occurred. The limit of detection of this method is about 50 pg for vitamin K_1, which was suitable for measuring depressed circulating levels in patients with osteoporosis (Fig. 2). Our present method is about 30 times more sensitive than the original LCUV method (ref. 17).

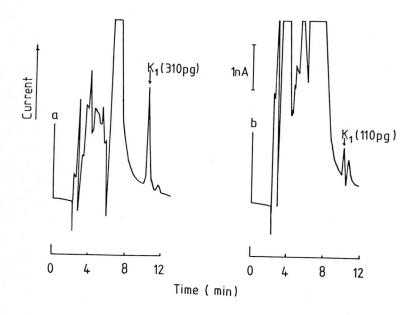

Fig. 2. Chromatograms of plasma samples obtained by LCEC in the redox mode for: a) a normal subject (endogenous concentrations of 240 pgml^{-1} in plasma); and b) a patient with osteoporosis and fractured neck of femur (endogenous concentration of 60 pgml^{-1} in plasma). (Reprinted with permission from ref. 14).

Vitamin E comprises a group of compounds known as tocopherols. The structure of these are shown below:

R_1, R_2 and R_3 are CH_3 in α-tocopherol.

R_1 and R_3 are CH_3, R_2 is H in β-tocopherol.

R_1 is H, R_2 and R_3 are CH_3 in γ-tocopherol.

R_1 and R_2 are H, R_3 is CH_3 in δ-tocopherol.

This group is of clinical interest because a deficiency can cause neuro-pathological and neuromuscular abnormalities in children; one possible cause of deficinecy is intestinal malabsorption (ref. 18). Therefore, reliable methods for the assay of the vitamin, preferably on small samples of blood, are required.

Since the four tocopherols possess a phenolic hydroxy group they all undergo oxidation at this moiety. Since δ-tocopheorl does not appear to circulate at measurable levels it may conveniently be used as an internal standard. Chou et al, (ref. 18) described a method for the determination of α-tocopherol, together with β- plus α-tocopherol, which required only 25 µl of serum. The vitamers were extracted with 100 µl of ethanol and 100 µl of heptane. A 50 µl aliquot of the heptane phase was evaporated to dryness and the residue dissolved in 50 µl of methanol for injection onto an octadecylsilyl (ODS) column. In this case the mobile phase consisted of 95% MeOH-acetate buffer pH 5.0; the thin-layer cell contained a glassy carbon working electrode which was operated in the oxidative mode at a potential of +1.0 V. The predominant form of the vitamin in normal subjects was found to be α-tocopherol; the limit of detection for the three circulating forms of the vitamin was below 0.1 µgml^{-1} which was well below the normal circulating values (4.3 - 9.7 µgml^{-1} for α-tocopherol; 1.8 - 3.9 µgml^{-1} for β- and γ-tocopherol). A similar procedure has also recently been described by Vandewoude et al (ref. 19). Both of these LCEC methods produced chromatograms that were surprisingly free of interference considering the minimal clean-up used.

Water-soluble vitamins

As mentioned earlier vitamin B_6 is a group consisting of six compounds and all of these contain the 3-hydroxy pyridine nucleus. We are currently interested in the measurement of circulating levels of two of these vitamers, namely pyridoxal (PL) and pyridoxal 5'-phosphate (PLP); the structure of these is shown below:

Pyridoxal   Pyridoxal phosphate

Pyridoxal 5'-phosphate is the predominant form in human blood and it is the coenzyme for many biological processes. Pyridoxal circulates at lower levels, but is the form required for transport into cells. Recently, it has been discovered that PLP increases the rate of formation of Gla residues (ref. 20) which suggests that this could be another essential cofactor in the process of calcification. Therefore, we are interested in measuring circulating levels of both PL and PLP in normal subjects and patients with diseases of bone, particularly osteoporosis. We have carried out some preliminary studies towards an assay for PL and PLP in plasma (ref. 15).

Initial studies showed that PL could be separated from PLP, and that these were also resolved from PM, PMP and PN, on a reverse phase column (ODS) using 0.05 M acetate buffer pH 6.25 as mobile phase. Using this medium the electro-chemical behaviour of PL and PLP was studied at a glassy carbon electrode using cyclic voltammetry. PLP showed one irreversible oxidative peak only at about +1.0 V vs. S.C.E. However, PL showed one oxidative peak at +1 V, which was also the result of an irreversible reaction, together with a reductive peak at about +0.3 V. On the second forward scan a further oxidative peak occurred, at about +0.4 V, as well as the initial oxidation peak at +1 V. The quasi-reversible couple was considered to be the result of a follow-up reaction involving a dimerisation process (ref. 15).

The coulometric cell, containing dual porous graphite electrodes in series was used in conjunction with an ODS column to investigate the possibility of measuring circulating levels of PL and PLP. From our initial studies it was shown that the maximum current for both vitamers occurred at +1 V. It was also shown that the onset of oxidation did not occur until +0.7 V; therefore, it was possible to use the cell in the oxidative screen-mode with the upstream electrode at +0.6 V and downstream electrode at +1.0 V. Using this detection system, and the conditions described above, the response with concentration was linear over three orders of magnitude (i.e. 200 pg - 100 ng).

It was not possible to measure circulating PL and PLP levels on plasma simply deproteinized with TCA due to interference from naturally occurring compounds. Therefore, the possibility of using a "clean-up" step with a strong cation exchange (SCX) cartridges was investigated. Pyridoxal was retained at pH 4.0 and could be eluted with mobile phase, but unfortunately PLP could not be retained under any of the acidic conditions investigated. Therefore, we hydrolysed PLP to PL with acid-phosphatase. Fig. 3 shows the chromatograms obtained for a normal subject, and as is clear a well-defined peak appeared at 13 min for PL. The pre-peak, due to an unidentified naturally occurring compound, could be removed by simply reducing the potential, at the downstream electrode, to +0.9 V; this also allowed us to increase the sensitivity by a factor of 5. We have found several anomalies in our preliminary studies which we are currently investigating. When these have been resolved we intend carrying out studies on circulating PL and PLP levels in normal subjects and patients with diseases of bone.

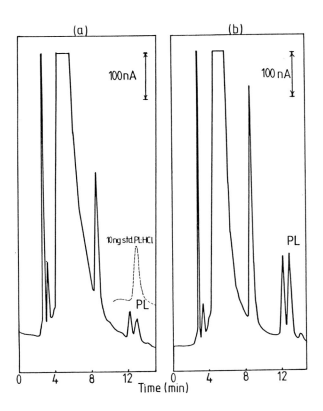

Fig. 3. Chromatograms of a normal plasma sample by LCEC in the oxidative-screen mode for: a) plasma without acid phosphatase; and b) with acid phosphatase. Electrode-1, +0.6 V; Electrode-2, +1.0 V; guard cell, +1.5 V. (Reprinted with permission from ref. 15).

Vitamin C (ascorbic acid) deficiency results in the well-known disease know as scurvy. This compound is an ene-diol and undergoes an electrochemical oxidation reaction to produce dehydroascorbic acid which then undergoes an irreversible solvation reaction as shown below:

$$\text{HO-C} \overset{O}{\underset{\text{HO-C}}{\parallel}} \quad \xrightarrow{-2e- \ 2H^+} \quad \overset{O}{\underset{\text{O=C}}{\parallel}} \quad \xrightarrow{H_2O} \quad \text{HO-C} \overset{OH \ O}{}$$

Reverse-phase columns have generally been used to separate the vitamin from interferences in biological fluids (refs. 21, 22). Iriyama et al (ref. 21) carried out a preliminary protein precipitation step on body fluids with meta phosphoric acid-EDTA. In order to measure ascorbic acid in serum or cebrospinal fluid a 0.5 ml aliquot of the sample was taken, whereas 5 µl were taken for urine analysis. After filtration, a 10 µl volume was injected onto the column. Detection was carried out with a glassy carbon working electrode held at +0.8 V vs. Ag/AgCl; the mobile phase consisted of 0.1 M dipotassium phosphate-citric acid buffer pH 4.6 containing 1 mM EDTA. These authors showed that uric acid did not interfere with vitamin C measurements and, in fact, could be measured simultaneously. The concentration of vitamin C in human serum, urine and cerebrospinal fluid were 12.8, 36.1, and 9.01 µgml^{-1} respectively.

Coenzymes

The coenzyme group known as pterins are of clinical interest because of their involvement in several diseases such as kidney dysfunction, Parkinsons' disease and senile dementia. They may also be involved in the biosynthesis of neurotransmitters (ref. 23).

This group of eoenzymes can exist naturally as the fully oxidized species, as the 7,8-dihydropterins, and as the 5,6,7,8-tetrahydropterins. It is therefore a difficult problem to analyse the different species, and these three possible oxidation states simultaneously. However, an LCEC method has recently been described by Lunte and Kissinger (ref. 23) which is capable of carrying out this type of assay on urine samples. These authors used a thin-layer electrochemical cell containing dual glassy carbon electrodes in the parallel configuration. One electrode was held at a potential of +0.8 V, to measure the reduced forms of the coenzymes and the other electrode was held at -0.7 V, to monitor the oxidised species. Peak identification in urine samples was confirmed by poising the two electrodes at different positions on the voltammetric wave and measuring the current ratios as described earlier, e.g. for the tetrahydropterins the electrode potentials were set at +0.3 V and +0.5 V. The peak current ratios for tetrahydroneopterin and tetrahydrobiopterin in urine were 0.59 and 0.63 respectively; in standards of the same compounds, the ratios were 0.61 and 0.61 respectively. Therefore good agreement was obtained between these two sets of results, which confirmed the identity of the peaks.

Another coenzyme of clinical interst is glutathione (α-glutamylcysteinyl-glycine); it exists in the oxidised form as the disulphide (GSSG) and in the reduced state as the thiol (GSH). One of its functions is to act as a reducing substance to keep red cell haemoglobin in the reduced state. It is therefore of clinical interest, to known the relative concentrations of these forms in whole blood. Until recently, the oxidised form needed to be reduced to the thiol by rather tedious wet chemical methods before quantitative analysis. However, Allison and Shoup (ref. 24) recently described an LCEC method for the simultaneous determination of GSSG and GSH using dual mercury/gold amalgam electrodes in series and operated in the redox mode.

The principle of operation is shown below:

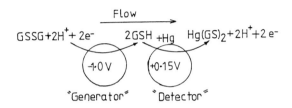

The first electrode ("generator" electrode) is held at −1.0 V to reduce the disulphide (GSSG) to the thiol (GSH) with the consumption of two electrons and two protons. The thiol is then detected downstream at the second electrode which is held at +0.15 V. At this electrode mercury is oxidised in the presence of the thiol to produce the mercury complex $(Hg(GS)_2)$ and two protons and two electrons. Obviously endogenous thiol does not react at the first electrode, but is detected at the second electrode. The two signals were separated because of the difference in the retention times of GSSG and GSH on the reverse-phase (ODS) column. The mobile phase used in this procedure consisted of 99% 0.1 M monochloro-acetate, pH 3.0/1% methanol; the limits of dectection for GSH and GSSG were 3.5 pmole and 5.7 pmole respectively. Whole blood was analysed after the addition of Na_2EDTA and $HClO_4$, followed by centrifugation and filtration; the concentrations of GSH and GSSG were 1.1 mM and 1.8 µM. Therefore, the limits of detection of the two compounds, determined by the dual-electrode technique, were well below the normal circulating levels.

CONCLUSION

This paper has described a few applications of LCEC to the determination of vitamins and coenzymes in body fluids. These examples clearly demonstrate how powerful the technique can be in this particularly difficlut area of clinical analysis. It is feasible that most of the vitamins and many coenzymes could be measured by the appropriate LCEC method. When optimising the procedure careful consideration should be given to the electrode material, electrode configuration and mode of operation, as well as chromatographic factors.

ACKNOWLEDGMENTS

The author wishes to thank Dr. J. Chayen and Dr. L. Bitensky for helpful discussions and the Arthritis and Rheumatism Council for Research for financial assistance.

REFERENCES

1   J.P. Hart, A. Catterall, R.A. Dodds, L. Klenerman, M.J. Shearer, L. Bitensky and J. Chayen, The Lancet, 2 (1984) 283.
2   J.P. Hart, M.J. Shearer, L. Klenerman, A. Catterall, J. Reeve, P.N. Sambrook, R.A. Dodds, L. Bitensky and J. Chayen, J. Clin. Endocrinol. Metab., 60 (1985) 1268-1269.
3   J.E. Leklem and R.D. Reynolds, Methods in Vitamin B-6 Nutrition, Plenum Press New York and London, 1981.
4   K. Bratin, C.L. Blank, I.S. Krull, C.E. Lunte and R.E. Shoup, Internat. Lab., 14 (1984) 24-43.
5   S.A. McClintock and W.C. Purdy, Internat. Lab., 14 (1984) 70-77.
6   J.P. Hart, in J. Chayen and L. Bitensky (Eds.), Investigative Microtechniques in Medicine and Biology, Vol. I, Marcel Dekker Inc., New York, 1984, Ch. 5.
7   J.P. Hart, Trends in Anal. Chem., 5 (1986) 20-25.
8   H.B. Hanekamp, P. Bos and R.W. Frei, Trends in Anal. Chem., 1 (1982) 135-140.
9   K. Bratin, P.T. Kissinger and C.S. Bruntlett, J.Liq. Chromatogr., 4 (1981) 1777-1795.
10 J.M. Elbicki, D.M. Morgan and S.G. Weber, Anal. Chem., 56 (1984) 978-985.
11 J.P. Hart, M.J. Shearer, P.T. McCarthy and S. Rahim, Analyst, 109 (1984) 477-481.
12 H.G. Barth, W.E. Arber, C.H. Lochmuller, R.E. Majors and F.E. Regnier, Anal. Chem., 58 (1986) 211R-250R.
13 C.E. Lunte, P.T. Kissinger and R.E. Shoup, Anal. Chem., 57 (1985) 1541-1546.
14 J.P. Hart, M.J. Shearer and P.T. McCarthy, Analyst, 110 (1985) 1181-1183.
15 J.P. Hart and P.J. Hayler, Anal. Proceedings (in press).
16 P.V. Haushcka, J.B. Lian and P.M. Gallop, Trends Biochem. Sci., 3 (1978) 75-78.
17 M.J. Shearer, P. Barkhan, S. Rahim and L. Stimmler, The Lancet, 2 (1982) 460-463.
18 P.P. Chou, P.K. Jaynes and J.L. Bailey, Clin. Chem., 31 (1985) 880-882.
19 M. Vandewoude, M. Clayes and I. De Leeuw, J. Liq. Chromatogr., 311 (1984) 176-182.
20 W.K. Kappel and R.E. Olson, Arch. Biochem. Biophys., 235 (1984) 521-528.
21 K. Iriyama, M. Yoshiura and T. Iwamoto, J. Liq. Chromatogr., 8 (1985) 333-344.
22 K. Iriyama, M. Yoshiura, T. Iwamoto and Y. Ozaki, Anal. Biochem., 141 (1984) 238-243.
23 C.E. Lunte and P.T. Kissinger, Anal. Chem., 55 (1983) 1458-1462.
24 L.A. Allison and R.E. Shoup, Anal. Chem., 55 (1983) 8-12.

M.R. Smyth and J.G. Vos
Electrochemistry, Sensors and Analysis
© Elsevier Science Publishers B.V., Amsterdam — Printed in The Netherlands

ELECTROCHEMICAL BEHAVIOUR OF PLATINUM (II) ANTITUMOUR DRUGS AT PLATINUM AND CARBON PASTE ELECTRODES

J-M. KAUFFMANN, G.J. PATRIARCHE and F. MEBSOUT

Free University of Brussels (U.L.B.), Institute of Pharmacy, Campus Plaine 205/6, Boulevard du Triomphe, B-1050 Brussels, (Belgium).

SUMMARY

The electrochemical behaviour of platinum (II) complexes used in cancer chemotherapy (cis-platin, carboplatin) has been investigated using cyclic voltammetry at carbon paste (C.P.E.) and platinum electrodes. Measurements performed in neutral aqueous media as a function of free chloride ions have permitted to establish the nature of the oxidative products and to point out a marked different redox behaviour between the two platinum (II) species studied. Platinum (II) derivatives possessing no chloride ligand are not electroreducible at the C.P.E. and their oxidation occurs at more positive potentials. Use of the carbon paste electrode has permitted to establish that the reduction of the platinum (IV) oxidized products of carboplatin (JM 8) does not regenerate the starting molecule but gives rise to the appearance of cis-platin structures, and the greater the number of halide ligands in the oxidized products, the easier will be their electroreduction.

INTRODUCTION

Following its widespread introduction into clinical practice in the early 1970s, the drug cis-platin (cis-diammine dichloroplatinum (II), CDDP) has received considerable interest in order to understand its mechanism of action as an antitumour drug. Although a second generation platinum (II) analogue of reduced toxicity but comparable therapeutic activity has emerged, cis-platin still plays the leading part among the metal-containing antitumour drugs (refs. 1-3).

Electrochemistry of platinum (II) complexes has already been extensively investigated by Hubbard and coworkers (ref. 4) but working electrodes used were exclusively platinum based. As platinum surfaces are highly reactive electrochemically, giving rise to perturbing surface oxides at positive potentials and rapid hydrogen adsorption - desorption phenomena at negative potentials, and as free chloride ions greatly affect the electrode response due to the halide bridged pathway (ref. 4), it was of interest to study the electrochemical behaviour of platinum (II) anti-cancer drugs at carbon based electrodes.

The present report will be devoted to the redox behaviour of cis-platin and of a new developed molecule, JM 8 [diammine (1,1-cyclobutanedicarboxylato)-

platinum (II)] which, in contrast to the former, contains no halide ligand in its structure as illustrated in figure 1.

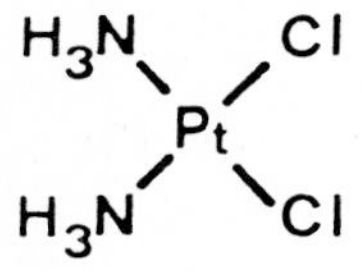

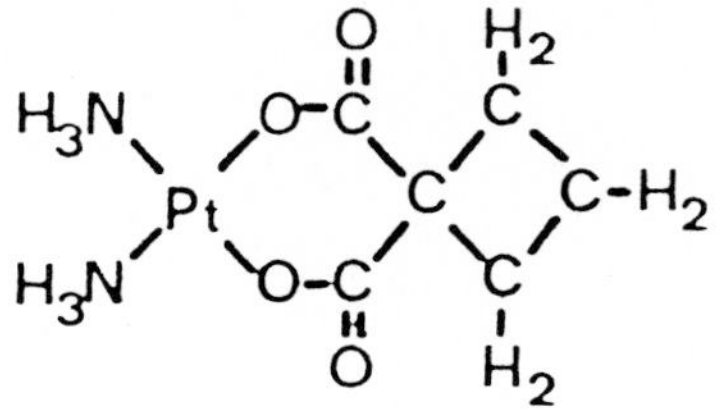

cis-platin
(CDDP)

carboplatin
(JM 8)

Fig. 1. Structure of the platinum complexes investigated.

The experiments were performed comparatively at the carbon paste and at the platinum electrodes using cyclic voltammetry. Measurements have been conducted in the absence and in the presence of free chloride ions, as they play a prominent part both biologically (ref. 5) and chemically (ref. 4) in the reactivity of the molecules.

EXPERIMENTAL

Instrumentation

Voltammetric measurements were carried out with a Model 175 Universal programmer / Model 174 potentiostat (Princeton Applied Research) and a PAR RE 0074 recorder.

Measurements were performed using a conventional three electrode design at $20.0 \pm 0.1°C$. The reference electrode , a saturated calomel electrode (S.C.E.), was placed into a Purley tube filled with the electrolyte used for the investigation. Glassy carbon was the auxiliary electrode. Working electrodes were a carbon paste electrode (Metrohm EA 267) prepared from standard paste (Metrohm EA 207 C) of spectroscopic grade carbon powder and Uvasol liquid paraffin ; the other working electrode was a platinum electrode (Radiometer P101) geometric area = 0.16 cm^2. All potentials are referred against a saturated calomel electrode (S.C.E.) Tacussel type C 10.

The experiments were carried out at pH 6, the pH of the solutions being measured using a Tacussel 60 pH-meter.

Reagents and solutions

The authors are indebted to the BRISTOL-MYERS International Corp. (Brussels, Belgium) for generous gifts of JM 8 and CDDP.

To minimize hydrolysis (ref. 6) of the molecules, care was taken to freshly prepare the solutions to be analyzed (1.10^{-3} M) by direct dissolution in double-distilled water. The measurements were carried out as soon as the solution was ready.

Supporting electrolyte was 0.1 M sodium perchlorate and was prepared with pure grade reagents (Merck P.A.). Traces of oxygen were removed from the solution to be analyzed by passing purified oxygen-free nitrogen through the cell.

Cleaning and activation of the platinum electrode

Before every set of experiments, the platinum electrode was cleaned by dipping it into diluted and boiling nitric acid and finally rinsed with double-distilled water.

Activation was essential in order to improve the electrode response and to yield reproducible current - potential curves. Following steps were applied:
- cyclic voltammetry (CV) experimentation (scan rate = 20 mV.s^{-1}) was performed during 10 minutes in order to deposit platinum particles (platinization) onto the electrode surface from a bath containing 1.10^{-3} M potassium hexachloro-platinate (IV) and 1 M sodium chloride. The potential scan range was performed between -0.5 and +1.0 V.
- the platinized platinum electrode prepared was then carefully washed with double distilled water.

Before each measurement, adsorbed chloride was removed by polarization steps (CV) during six cycles in 0.5 M H_2SO_4 with potential ranges from 0.0 to +1.5 V (a rest period was then observed at +0.45 V during 2 minutes). Then, oxidation of the surface was realized by dipping the platinized electrode during 5 min. into a sulfo-chromic solution. Finally, the electrode was carefully rinsed with double-distilled water and with supporting electrolyte solution and used as soon as ready.

RESULTS AND DISCUSSION

Cyclic voltammetry of JM 8 at the C.P.E.

Investigating the cathodic potentials with an initial potential of +0.5 V, reduction of 1.10^{-3} M JM 8 is not observed at the electrode, in the absence or in the presence of free chloride ions.

Considering the oxidation side, with the same initial potential, one anodic peak E_{pa} is detected at high positive potentials near solvent decay, the reaction being diffusion controlled in the range of scan rates investigated, 5 to 500 mV.s^{-1}. Influence of chloride ions on anodic peak shape is observed when concentration exceeds 1.10^{-2} M, resulting in a progressive peak intensity diminishing as shown in table 1.

TABLE 1
Analysis of the cyclic voltammograms of 1.10^{-3} M carboplatin in 0.1 M NaClO$_4$ at the C.P.E. as a function of chloride ion concentration.

$\begin{bmatrix} Cl^- \end{bmatrix}$ (M)	E_{p_a} V	α	ia μA	E_{p_c} V	ic μA
				$- 0.55^{(I)}$	10
0	+ 1.55	0.24	120		
				$- 0.80^{(0)}$	---
				$- 0.20^{(II)}$	6
1.10^{-3}	+ 1.55	0.20	120	$- 0.55^{(I)}$	---
				$- 0.75^{(0)}$	---
				$- 0.20^{(II)}$	10
1.10^{-2}	+ 1.52	0.20	120	$- 0.45^{(I)}$	---
				$- 0.70^{(0)}$	---
				$- 0.05^{(III)}$	4
1.10^{-1}	+ 1.50	0.20	100	$- 0.25^{(II)}$	---
				$- 0.60^{(C)}$	---
				$- 0.05^{(III)}$	9
1	+ 1.48	0.19	85	$- 0.25^{(II)}$	---
				$- 0.60^{(0)}$	---

E_{pa}: anodic peaks potential

: transfer coefficient calculated using eqn $n = 0.048(E_{pa} - E_{pa/2})$

i_a : anodic peak intensity

E_{pc}: cathodic peaks potential

i_c : cathodic peaks intensity

---: detected but not measurable.

Reversing the scan direction after E_{pa}, corresponding cathodic peaks are obtained at very negative potentials, illustrating the high degree of irreversibility of the process. The presence of chloride ions markedly modifies the nature of the oxidation products obtained at E_{pa}, as suspected by the appearance of additional cathodic peaks (figure 2 and table 1). In the absence of chloride ions, two cathodic peaks are obtained: $E_{pc}(I)$ and $E_{pc}(0)$, the last one giving concomittant gas evolution maintaining the potential at $E_{pc}(0)$. At 1.10^{-3} M in chloride, a second reduction peak, $E_{pc}(II)$, is developed whose intensity increases with an increase in chloride ions till 1.10^{-2} M, then diminishes on behalf of a new peak $E_{pc}(III)$. Simultaneousely, $E_{pc}(I)$ is diminishing and is no more detected at 1.10^{-1} M in chloride. Peak $E_{pc}(0)$ is continuously present with an increasing intensity till 1.10^{-1} M in chloride,

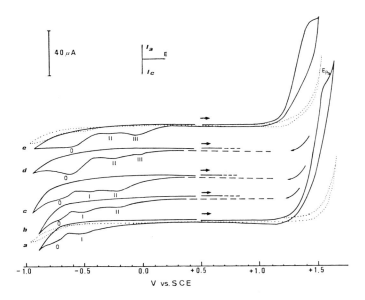

Fig. 2. Cyclic voltammograms of 1.10^{-3} M JM 8 in 0.1 M NaClO$_4$ at the C.P.E.
Scan rate = 20 mV.s^{-1}; starting potential = +0.5 V; initial anodic
scan: a) $[Cl^-]$ = 0 , b) $[Cl^-]$ = 1.10^{-3} M , c) $[Cl^-]$ = 1.10^{-2} M , d) $[Cl^-]$ = 1.10^{-1} M , e) $[Cl^-]$ = 1 M. Surface renewed after each cycling; dotted lines = supporting electrolyte.

after which it diminishes. Cyclic voltammograms, realized as a function of the switching potential $(E\lambda)$, show the irreversible nature of these cathodic peaks in the range of scan rates 5 - 500 mV.s^{-1} as illustrated during four successive cycles on the same surface in figure 3.

Changing the sweep direction at $E\lambda_1$, $E\lambda_2$ and $E\lambda_3$ (cycles 1, 2 and 3) produces no new anodic peak beside E_{pa}. However, the scanning till $E\lambda_3$ has induced a drastic surface modification which is observed ultimately. Indeed during the fourth and subsequent cycles (fig. 3, curve c, dashed lines), when reversing the sweep at $E\lambda_1$ or $E\lambda_2$, formation of a new anodic peak $(E_{pa}2^*)$, corresponding to the reoxidation of reductants generated during peaks $E_{pc}(II)$ or $E_{pc}(III)$, occurs. However, peak $E_{pa}2^*$ is never detected when reversing at $E\lambda_3$, we may only detect a small anodic peak $E_{pa}1^*$. Intensity of peaks $E_{pa}1^*$ and $E_{pa}2^*$ gradually increases while cycling, attaining a limiting value after 15 minutes. Thus reduction of the oxidized species occurs in a common final electronic step, $E_{pc}(0)$, during which a modification of the C.P.E. occurs, attributable to platinum particles electrodeposition, giving rise to hydrogen evolution (ref. 7) and to the appearance of new anodic peaks: $E_{pa}1^*$ and $E_{pa}2^*$.

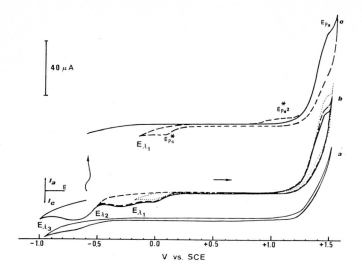

Fig. 3. Cyclic voltammograms of 1.10^{-3} M JM 8 in 0.1 M NaClO$_4$ and 1 M NaCl
at the C.P.E. as a function of inversion potential Eλ. Scan rate =
20 mV.s^{-1}, starting potential = +0.5 V, initial anodic scan. Several cycles
on the same surface. a) supporting electrolyte; b) $\cdots$ first cycle ;
--- second cycle, —— third cycle, c) fourth cycle.

Peak $E_{pa}1^*$ corresponds to the oxidation of the platinum particles deposited,
as proved by cycling the modified C.P.E. in a supporting electrolyte in the
absence of JM 8; moreover the formation of $E_{pa}1^*$ is inhibited in presence of
chloride ions. Because of the presence of electrodeposited platinum particles,
we may observe the catalytic reoxidation of the reductants generated during
peaks $E_{pc}(II)$ and $E_{pc}(III)$.

Comparing these processes with the reduction and oxidation peaks of
cis-platin on C.P.E. and upon consideration of the works of A.T. Hubbard (ref.8)
relative to the electrooxidation of platinum (II) complexes, we may interpret
the redox behaviour of JM 8 at the C.P.E. We may deduce the presence of three
oxidized structures: structure I in the absence of chloride, structures I and
II between 1.10^{-3} and 1.10^{-2} M in chloride and structures II and III between
1.10^{-2} and 1 M in chloride. Taking into account that perchlorate is a non-
complexing ion (ref. 9) and assuming that the oxidation of the compound
presumably involves a single two-electron step to platinum (IV), we may suggest
the structures represented in figure 4, the number and the respective concentration ratio of these forms being directly related to chloride concentration.

The reduction of the oxidized forms of JM 8, which may occur in step $E_{pc}(I)$,
$E_{pc}(II)$ or $E_{pc}(III)$ gives rise to new structures different than JM 8.

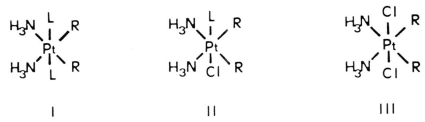

Fig. 4. Schematic structures of the three oxidized compounds.
L = H₂O or OH ; R = organic ligand.

Indeed $E_{pa}2^*$ doesn't correspond to the catalytic oxidation of JM 8 nor $E_{pc}(0)$ to the reduction of the species. For these reasons, and due to the fact that organic ligands are easily leaving groups compared to NH_3 (ref. 9), and taking into account the redox behaviour of cis-platin at the C.P.E., we may assume that the reduction of structures I, II and III gives rise to the products represented in figure 5, with elimination of the organic ligand.

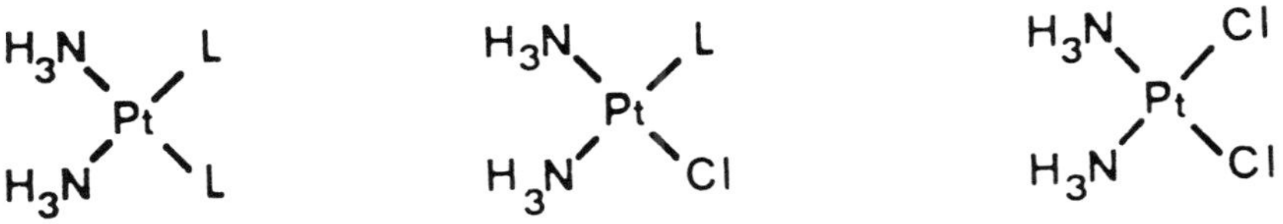

Fig. 5. Schematic structures of the three reduced compounds.

Compounds II' and III' having one or two halide ligands are catalytically oxidizable $(E_{pa}2^*)$ due to the in-plane bridging effect (ref. 8), contrary to compound I'whose reoxidation is not detected at the modified C.P.E. The further reduction of structures I', II' or III' involves the common step $E_{pc}(0)$ giving rise to platinum electrodeposition, which is similar to the behaviour of cis-platin at the C.P.E.

Cyclic voltammetry of JM 8 at the platinum electrode

The study of JM 8 at a smooth platinum electrode confirms the high degree of irreversibility of the redox behaviour of platinum (II) complexes possessing no halide ligands (ref. 8). Addition of free chloride ions in the solution improves slightly the processes due to the axial bridging effect (ref. 8). Like at the C.P.E., reduction at the platinum electrode is not detected, but oxidation and subsequent reduction give rise to cis-platin structures as mentioned above. They may be further reduced with platinum electrodeposition and appearance of the corresponding catalytic properties of a platinized platinum electrode. At the activated electrode (fig. 7 I), the formation of one or two reduction peaks after E_{pa}^* as a function of chloride ions confirms the presence of species II and III. Reduction of species I is not detected due to the irreversible nature of the process which is masked by a rapid hydrogen evolution at the platinized platinum electrode. Similarly with the results obtained at the C.P.E. we may conclude that at 1.10^{-3} M in chloride, species II appears whose concentration increases till 1.10^{-2} M in chloride, then diminishes on behalf of species III, the most abundant form in 1 M chloride solution.

Cyclic voltammetry of cis-platin at the C.P.E.

Reduction of cis-platin is observed at the electrode giving a sharp irreversible reduction peak $E_{pc}(0)$ (fig. 6 II) whose intensity diminishes at scan rates higher than 10 mV.s^{-1}. Reversing the scan direction after $E_{pc}(0)$

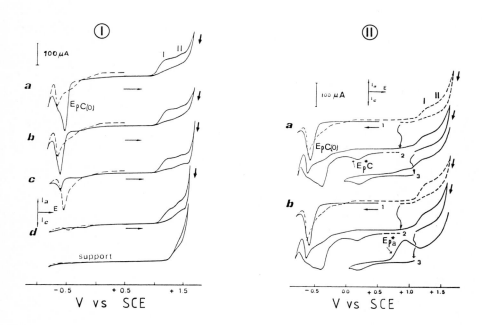

Fig. 6. Cyclic voltammograms of 1.10^{-3} M cis-platin in 0.1 M NaClO$_4$ at the C.P.E. as a function of chloride ions. Starting potential = +0.5 V, scan rate = 20 mV.s^{-1}. a) [Cl$^-$] = 0 ; b) [Cl$^-$] = 4.10^{-3} M ; c) [Cl$^-$] = 1.10^{-1} **M** ; d) [Cl$^-$] = 1 M.
I : initial anodic scan ; II : initial cathodic scan.

produces overlapping curves due to the reactivity of the deposited platinum particles. In the presence of chloride ions, peak potential remains constant but intensity decreases progressively.

Oxidation of 1.10^{-3} M cis-platin, with a starting potential of +0.5 V, gives two anodic peaks $E_{pa}(I)$ and $E_{pa}(II)$ whose intensities and potentials are independent of chloride ions concentration, the process being diffusion controlled in the range of scan rates 5 - 500 $mV.s^{-1}$ (fig. 6 I). Reversing the scan direction after the anodic peaks, we do not observe well-defined corresponding reduction steps; just a slight and broad current starting at +0.2 V preceeding a sharp reduction peak located at potentials corresponding to cis-platin reduction are obtained. The presence of chloride ions produces only a slight increase of the first reduction step but drastically limits $E_{pc}(0)$ formation at concentration higher than 1.10^{-1} M. Several cycles conducted on the same surface (fig. 6 II, cycles 1, 2, 3) produce the same electrode modification process as observed for JM 8, but to a greater extent. The shape of the quasi-reversible couple E_{pa}^{*}/E_{pc}^{*} is particularly well-marked in the presence of chloride ions (fig. 6 II, curve b). Continuing cycling, the peaks are increasing as well as the reversibility of the reaction and available potential ranges are becoming restricted.

Contrary to JM 8, cis-platin oxidation occurs at the platinum particle deposit, a phenomenon which may be related to the in-plane bridging effect.

During ageing of the 1.10^{-3} M cis-platin and 0.1 M NaClO$_4$ solution, pH decreases till 4 and we observe a progressive peak $E_{pa}(I)$ diminution and a peak $E_{pa}(II)$ increase . As the pH has no influence on anodic peak shape, $E_{pa}(II)$ may correspond to the oxidation of cis-platin hydrated species (ref. 10) Moreover reduction of a cis-platin aged solution is not feasible at the C.P.E. in the different media investigated.

Cyclic voltammetry of cis-platin at the platinized platinum electrode

In figure 7 II are reported the voltammograms of 1.10^{-3} M cis-platin in 0.1 M NaClO$_4$ as a function of chloride ions concentration. By analogy with JM 8, peaks $E_{pa}(I')^{*}$ and $E_{pc}(I')^{*}$ may correspond to a non-catalyzed redox couple. Peak $E_{pa}(I)^{*}$ formation may be assigned to catalyzed oxidation by the free chloride ions liberated during partial cis-platin hydrolysis.

From the appearance of two distinct reduction peaks as a function of chloride ions and by taking into account the results of JM 8 and those of Hubbard (ref. 8), we may suggest the presence of two cis-platin oxidized structures reduced at $E_{pc}(II)^{*}$ and $E_{pc}(III)^{*}$: structure I between 1.10^{-3} M and 1.10^{-1} M in chloride ions which disappears progressively on behalf of structure II, the most abundant form in 1 M chloride ions solution.

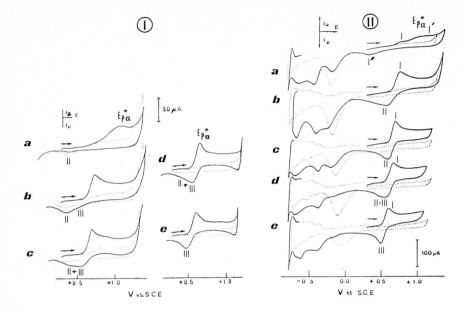

Fig. 7. Cyclic voltammograms of 1.10^{-3} M JM 8 (I) and 1.10^{-3} M cis-platin (II) in 0.1 M NaClO$_4$ at the platinized platinum electrode as a function of chloride ions concentration. Scan rate = 20 mV.s^{-1} , initial anodic scan. Dotted lines = supporting electrolyte.
a) $[Cl^-]$ = 0 ; b) $[Cl^-]$ = 1.10^{-3} M ; c) $[Cl^-]$ = 1.10^{-2} M ; d) $[Cl^-]$ = 1.10^{-1} M
e) $[Cl^-]$ = 1 M.

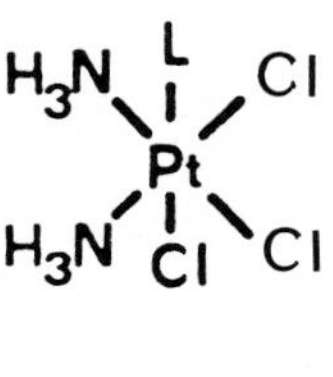

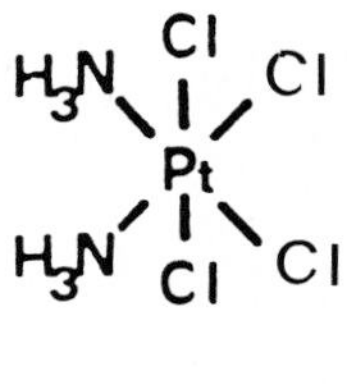

Fig. 8. Schematic structures of cis-platin oxidized species.

CONCLUSION

Cyclic voltammetric experiments, conducted with a carbon paste electrode offer further insights into the redox properties as well as into the aqueous stability of platinum (II) complexes. Use of the carbon paste electrode furnishes several interesting advantages, when studying platinum (II)

derivatives as a function of free chloride ions, in comparison with the
platinum electrodes. Indeed anodic and cathodic available potentials are
extended and perturbing oxide surface phenomena are negligible. Moreover in
presence of free halides, the electrochemical oxidation processes are only
slightly affected.

The comparative study of two platinum (II) anti-cancer drugs differing in
their ligand nature has permitted to point out a marked different redox
behaviour owing to the use of the C.P.E.. That is, non chloride ligands,
in this case aquo or organic, tend to stabilize the platinum (II) species as
well   towards oxidation   as towards reduction. Moreover the use of the
C.P.E. has permitted to suggest the nature of the oxidation products of JM 8
in the presence and in the absence of free chloride ions and that the
subsequent reduction of these oxidized forms gives rise to platinum deposits.

Finally the C.P.E. permits the quantitative study of cis-platin hydrolysis
as a function of time and its use should be of great interest for further
investigations oriented towards platinum (II) complexes - biological molecules
interaction studies (ref. 11).

ACKNOWLEDGEMENTS

Thanks are expressed to the Bristol Company for providing samples
generousely and to the "Fonds National de la Recherche Scientifique"
(F.N.R.S. Belgium) for support to one of us (G.J.P.).

REFERENCES

1 R. Dagani, Science,219 (1985) 20 - 22.
2 K. Harrap, Platinum Metals Rev., 28 (1984) 14 - 19.
3 S.J. Lippard, in Platinum, Gold and other Metal Chemotherapeutic Agents,
   S.J. Lippard (Ed), A.C.S. Symposia Series, (1982).
4 C.N. Lai, A.T. Hubbard, Inorg. Chem., 11 (1972) 2081 - 2091.
5 S.J. Lippard, Science, 218 (1982) 1075 - 1082.
6 I.S. Krull, X.D. Ding, S. Braverman, C. Selavka, F. Hochberg,
   L.A. Sternson, J. Chromatog., 21 (1983) 166 - 173.
7 R.D. Giles, J.A. Harrison, H.R. Thirsk, J. Electroanal. Chem., 20 (1969)
   47 - 60.
8 J.P. Cushing, A.T. Hubbard, J. Electroanal. Chem., 23 (1969) 183 - 203.
9 M.E. Howe-Grant, S.J. Lippard, Aqueous platinum (II) chemistry; binding to
   biological molecules, in H. Siegel (Ed), Metal ions in biological systems,
   vol 11, M. Dekker, New York, 1980, pp 63 - 125.
10 R.B. Martin, Hydrolytic Equilibria of Cis-diammine dichloroplatinum (II),
   in S.J. Lippard (Ed), Platinum, Gold and other Metal Chemotherapeutic Agents
   A.C.S. Symposia Series, 1982, pp 231 - 244.
11 J-M. Kauffmann, G.J. Patriarche, Conference at the "Journées d'Electrochimie
   Florence, Italy, 1985.

M.R. Smyth and J.G. Vos
Electrochemistry, Sensors and Analysis
© Elsevier Science Publishers B.V., Amsterdam — Printed in The Netherlands

POLAROGRAPHIC DETERMINATION OF THE SULPHOXIDE METABOLITES OF S-CARBOXYMETHYL-L-CYSTEINE

A.D. WOOLFSON, J.S. MILLERSHIP AND E.I. KARIM
Pharmacy Department, The Queen's University of Belfast, Belfast BT9 7BL, U.K.

SUMMARY
   Polarography has been investigated as a possible analytical method for the sulphoxide metabolites of S-carboxymethyl-L-cysteine. Although the compounds themselves were found to be polarographically inactive, acid hydrolysis yielded electro-reducible components subsequently identified as cysteine (SCMC-sulphoxide) and methylmethane thiosulphate/pyruvic acid (SMC-sulphoxide). The hydrolysis reaction was quantitative and the differential pulse polarographic peaks due to cysteine and pyruvic acid could be used for the analysis of the corresponding sulphoxide metabolites.

INTRODUCTION
   Metabolism, in humans, of the mucolytic drug S-carboxymethyl-L-cysteine(SCMC) has been shown to proceed via sulphoxidation, decarboxylation and N-acetylation. Sulphoxidation is the major metabolic route and SCMC-sulphoxide, N-acetyl-SCMC-sulphoxide, S-methyl-cysteine (SMC)- sulphoxide and N-acetyl-SMC-sulphoxide have all been identified in urine samples following administration of the drug (refs. 1-3). Investigations of the metabolism of SCMC have, however, indicated a wide variation in the amount of the drug excreted in the urine as sulphoxide metabolites. It has therefore been suggested (ref. 4) that a polymorphic distribution of sulphoxidation capacity may exist within a given population. Unfortunately, studies to date have been hindered by the lack of a reliable analytical method for these compounds. Therefore a study of the polarographic behaviour of the four sulphoxides was initiated. Although sulphoxides in general are non-reducible at the dropping mercury electrode (ref. 5), the reported chemistry of the compounds (refs. 6-9) suggested the possibility of generation of an electroactive species via an initial hydrolysis reaction.

METHODS
Materials

   SCMC, cysteine and cystine were all obtained from Sigma Chemical Co., U.K. SMC, methyl disulphide and pyruvic acid were obtained from Aldrich, U.K. N-acetyl-SCMC and N-acetyl SMC were prepared by the method of Zbarsky and Young (ref. 10). All the sulphoxides were prepared by oxidation of the corresponding sulphide with 30% hydrogen peroxide (ref. 7). Methyl methane thiosulphate was prepared by peroxide oxidation of methyldisulphide (ref. 11).

All polarograms were recorded in Britton-Robinson buffer adjusted to constant ionic strength. De-aeration of solutions was by passage of oxygen-free nitrogen for 5 min.

Polarography

Polarograms were obtained with a PAR 174A polarographic analyser (EG&G Princeton Applied Research, U.K.) and a conventional three-electrode system comprising a dropping mercury electrode (DME) equipped with a drop timer, saturated calomel reference electrode (SCE) and platinum auxiliary. The polarographic cell was maintained at 20° and polarograms were recorded on an EW-11 X-Y plotter (Rikadenki-Mitsui U.K.).

Polarographic conditions in the differential pulse mode were as follows:- initial potential zero Volts; scan rate 5mV s^{-1}; pulse amplitude 50 mV; drop time 1s; sensitivity 0.2-1.0 µamps full-scale as required.

Cyclic voltammograms were obtained with a Metrohm E506 Polarecord and E612 scanner (Metrohm, Switzerland) and recorded with a Gould-Advance 2001 digital storage oscilloscope/ X-Y plotter.

Hydrolysis procedures for sulphoxides

A sulphoxide solution (0.1mg ml^{-1}, 1ml) was mixed with 1M H_2SO_4 (1ml) in a sealed test tube. The solution was heated for 30 min. in a boiling water bath. The recorded peak currents for the hydrolysates remained constant for heating periods beyond 30 min.

Calibration graphs

For SCMC and SMC sulphoxides, stock aqueous solutions (0.05 mg ml^{-1}) were freshly prepared daily. Aliquots from 50-250 µl were mixed with equal quantities of 1M H_2SO_4, hydrolysed, cooled and diluted to 20 ml with supporting electrolyte prior to transfer to the polarographic cell. Within this range the peak current/concentration plots were linear with correlation coefficients >0.99. The lowest detectable amount of each sulphoxide was 1.25 µg.

RESULTS

Polarograms of SCMC, N-acetyl-SCMC and their corresponding sulphoxides showed no electro-activity throughout the entire pH range of the supporting electrolyte. Polarograms of these compounds after hydrolysis were recorded in pH4 buffer and are presented in Fig 1. Similarly, SMC and its derivatives were polarographically inactive. Following hydrolysis, electroactivity was apparent only with the sulphoxides (Fig. 2). Cyclic voltammograms of

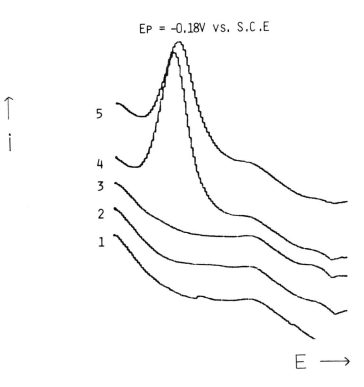

EP = -0.18V vs. S.C.E

5

4

3

2

1

↑
i

E ⟶

Fig. 1   Differential pulse polarograms obtained following acid hydrolysis of
         SCMC(2), N-acetyl-SCMC(3), SCMC-sulphoxide (4) and N-acetyl-SCMC-
         sulphoxide (5) in pH4 Britton-Robinson buffer (1).

SCMC-sulphoxide (hydrolysate) and L-cysteine indicated that both reductions
were quasi-reversible whereas those of SMC-sulphoxide (hydrolysate) and
pyruvic acid gave no signal on the reverse sweep.

    Following acid hydrolysis, the number of electrons (n) in the reduction
reaction was obtained from the slope of a log (peak current) vs. time plot.
Values of n were:- SCMC-sulphoxide 0.73 (cysteine 0.80), SMC-sulphoxide
1.75 (pyruvic acid 1.82).

    Peak potential/pH plots for both SCMC-sulphoxide (pH range 2-10) and
SMC-sulphoxide (pH range 2-6 at Ep approx.-0.99V vs. S.C.E.) were linear and
yielded slopes, respectively, of 59 and 71 mV.

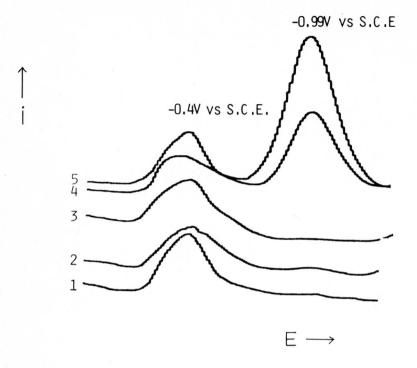

$\uparrow$
i

−0.99V vs S.C.E

−0.4V vs S.C.E.

5
4

3

2

1

E $\longrightarrow$

Fig. 2   Differential pulse polarograms obtained following acid hydrolysis of
SMC(2), N-acetyl-SMC(3), N-acetyl-SMC-sulphoxide(4) and SMC-
sulphoxide(5) in pH4 Britton-Robinson buffer (1).

DISCUSSION

Chemical investigation of the acid hydrolysis of SCMC-sulphoxide indicated
that the major products were cysteine and cystine (ref. 6)., with cleavage
of the sulphoxide bond attributed to the presence of the carboxymethyl group.
Cystine itself hydrolyses on standing to produce cysteine (refs. 12-13).
Acid hydrolysis of SMC-sulphoxide, however, has been shown to yield the
sulphide itself plus pyruvic acid, methyldisulphide and methyl methanethiosulphonate (refs. 7-9). SCMC and SMC are both resistant to acid hydrolysis.

In the present study initial investigations confirmed the polarographic
inactivity of all the sulphides and sulphoxides. Following acid hydrolysis of
SCMC-sulphoxide a well-defined peak at E_p −0.18V vs. S.C.E. was observed in

the differential pulse polarogram at pH 4. At pH values >5 this peak gradually broadened and decreased in intensity. Identical polarographic behaviour was observed following acid hydrolysis of N-acetyl-SCMC-sulphoxide. Under identical conditions a similar peak at -0.18V vs. S.C.E. was also recorded for cysteine. Addition of cysteine to SCMC hydrolysate at pH4 in the polarographic cell produced a linear increase in peak current. Under the pertaining conditions cysteine reacts with metallic mercury to form mercuric cystinate which is then rapidly reduced at the electrode (ref. 14). One electron and one proton are transferred per molecule. The slope of the peak potential/pH graph for SCMC-hydrolysate, together with an experimental determination of n, were in agreement with the values expected for the reduction of cysteine. Cyclic voltammograms of both cysteine and the hydrolysate confirmed that the well-defined peak at pH 4 following acid hydrolysis of SCMC-sulphoxide or its acetylated derivative corresponded to reduction of cysteine.

The polarographic behaviour of SMC-sulphoxide and its acetylated derivative was found to be different from, and more complex than, the SCMC derivatives. In pH 4 buffer two peaks were recorded at E_p-0.4V(broad) and E_p-0.99V vs. S.C.E. (well-defined). At pH values >6 this latter peak current diminishes rapidly and a subsequent peak appears at a more negative potential. Under the same polarographic conditions (pH 4), methyl disulphide was polarographically inactive, methylmethane thiosulphonate gave two peaks (E_p-0.4V vs. S.C.E., broad: E_p-0.93V vs. S.C.E., very weak) and pyruvic acid gave a strong peak at -0.99V vs. S.C.E. On investigation, the hydrolysate peak at -0.99V vs. S.C.E. corresponded to a two proton, two electron reduction, consistent with the reduction mechanism of undissociated pyruvic acid. The change in the polarogram above pH 6 was therefore due to the pyruvate anion. Finally, a cyclic voltammogram of the hydrolysate at pH 4 was consistent with that of pyruvic acid. For both SMC-sulphoxide and its acetylated derivative, acid hydrolysis therefore yielded two polarographically active components. However, only the peak recorded in pH 4 buffer and due to undissociated pyruvic acid was found to be of analytical use.

The hydrolysis procedure reported for the sulphoxides was both simple and rapid, resulting in the generation of polarographically active components. In all cases, 30 minutes heating with acid in a boiling water bath and subsequent dilution of the cooled hydrolysate to 20 ml (pH 4 Britton-Robinson buffer) yielded a maximum peak current which remained constant if the heating time was increased. Investigation of the use of this indirect polarographic method for human metabolic studies on S-carboxymethyl-L-cysteine is continuing.

REFERENCES

1. R.H. Waring, The metabolism of S-carboxymethyl-L-cysteine in rodents, marmasets and humans, Xenobiotica, 8 (1978) 265-270.
2. R.H. Waring, Variation in human metabolism of S-carboxymethyl-L-cysteine, Eur. J. Drug. Metab. Pharmacokinet., 5 (1980) 49-52.
3. R.H. Waring and S.C. Mitchell, The metabolism and elimination of S-carboxymethyl-L-cysteine in man, Drug. Metab. Dispos., 10 (1982) 61-62.
4. R.H. Waring, S.C. Mitchell, J.R. Idle and R.L. Smyth, Polymorphic sulphoxidation of S-carboxymethyl-L-cysteine in human, Biochem. Pharmacol., 31 (1982) 3151-3154.
5. K.G. Stone, Polarographic behaviour of some organic compounds, J. Amer. Chem. Soc., 69 (1947) 1832-1833.
6. K. Takahashi, Products of performic acid oxidation and hydrolysis of S-carboxymethyl-L-cysteine and related compounds, J. Biochem., 74 (1973) 1083-1089.
7. R.L.M. Synge and J.L. Wood, S-methyl-L-cysteine oxide in cabbage, Bichem. J., 64 (1956) 252-259.
8. C.J. Morris and J.F. Thompson, The identification of S-methyl-L-cysteine oxide in plants, J. Amer. Chem. Soc., 78 (1956) 1605-1608.
9. F. Ostermayer and D.S. Tarbell, Products of acid hydrolysis of S-methyl-L-cysteine oxide; the isolation of methyl methanethiosulphonate and mechanism of hydrolysis, J. Amer. Chem. Soc., 82 (1960) 3752-3755.
10. S.H. Zbarsky and L. Young, Mercapturic acids I: synthesis of phenyl-L-cysteine and L-phenylmercapturic acid, J. Biol. Chem., 151 (1943) 211-215.
11. H.J. Backer and H. Kloosterziel, Esters thiosulphoniques, Rec. Trav. Chim Chim., 73 (1954) 129-132.
12. K. Shinohara and M. Kilpatrick, The stability of cysteine in acid solution, J. Biol. Chem., 105 (1934) 241-245.
13. J.I. Routh, Decomposition of cysteine in aqueous solutions, J. Biol. Chem., 126 (1939) 297-304.
14. I.M. Kolthoff, W. Stricks and N. Tanaka, The anodic waves of cysteine at the convection and dropping mercury electrodes, J. Amer. Chem. Soc., 77 (1955) 5211-5215.

M.R. Smyth and J.G. Vos
Electrochemistry, Sensors and Analysis
© Elsevier Science Publishers B.V., Amsterdam — Printed in The Netherlands

DETERMINATION OF BENZODIAZEPINES BY ADSORPTIVE STRIPPING VOLTAMMETRY

L. HERNANDEZ, A. ZAPARDIEL, J.A. PEREZ LOPEZ and V. RODRIGUEZ
Department of Analytical Chemistry, Autonoma University, 28049 Madrid, SPAIN

SUMMARY
    Adsorptive stripping voltammetry of benzodiazepines and thienodiazepines is discussed. These compounds are adsorbed at the HMDE at positive potentials below the peak potential obtained with differential pulse polarography. Concentrations of 10^{-7}-10^{-9} M are usually measurable with the relative standard deviation ranging from 2 to 6%.

INTRODUCTION

    The determination of compounds with the azomethine group has been carried out by differential pulse polarography (refs. 1-3), but it is possible to get lower detection limits for the determination of this kind of compounds.

    An electrochemical study of benzodiazepines and thienodiazepines at the mercury electrode indicates, in most of the cases studied, the existence of adsorption processes at the electrode surface during the reduction of these molecules. The electrocapillary and the capacity-potential curves indicate also the adsorption of the species in their oxidized form, and that it is possible to use the technique of adsorptive stripping voltammetry in the determination of these drugs.

    Kalvoda (ref. 4) has determined diazepam and nitrazepam using adsorptive stripping voltammetry. Wang (refs. 5-6) has outlined the advantages and disadvantages of this technique and reviewed the published applications.

    In this paper we describe the determination of the following benzodiazepine compounds; BrTDO (I): 7-bromo-5-(2-chlorophenyl)-1,3-dihydro-2H-thieno-2,3-e-1,4-diazepam-2-one; BROMAZEPAM (II): 7-bromo-1,3-dihydro-5-pyrid-2-yl-2H-1,4-benzodiazepin-2-one; PRAZEPAM (III): 7-chloro-1-cyclopropylmethyl-1,3-dihydro-5-phenyl-2H-1,4-benzodiazepin-2-one; CLORAZEPATE (IV): potassium 7-chloro-2,3-dihydro-2-oxo-5- phenyl-1H-1,4-benzodiazepine-3-carboxylate; MEDAZEPAM (V): 7-chloro-2,3-dihydro-1-methyl-5-phenyl-1H-1,4-benzodiazepine.

    The adsorptive stripping analysis of these compounds has been carried out using a hanging mercury drop electrode, and the accumulation step has been carried out at a potential where no reduction of the molecules occurred. The electrochemical determination was made by differential pulse voltammetry.

EXPERIMENTAL

Equipment

    A METROHM 646 VA Processor and a 647 VA Stand were used. The working electrode was a hanging mercury drop electrode (surface 0.44 mm^2), the reference electrode was   Ag/AgCl/KCl (3M) and the auxiliary electrode was a glassy carbon rod (2x65 mm).

Reagents

Stock solutions (1.0x10^{-2}M) of the pure compounds (Hoffman-La-Roche) were prepared daily by dissolving the compounds in methanol and/or deionized water. The solutions were stored in the dark at 4°C.

Other reagents used were acetate(Ac), Britton-Robinson(B-R), ammonium acetate, phosphate and boric acid/borate buffers and sodium hydroxide and sulphuric acid solutions of concentrations ranging from 0.02 to 0.40 M.

All solutions were prepared from analytical grade reagents using deionized water (Milli Q and Milli RO - MILLIPORE).

Procedures

Working solutions were prepared by dilution of the stock solution in methanol, with the exception in the case of clorazepate that it was dissolved in water, with different concentrations of the supporting electrolytes. In this way, the ionic strength corresponding to each solution ranged between 0.4 and 0.02 M.

The solution was degassed with nitrogen for 20 minutes (including an additional 30 seconds before each adsorptive stripping cycle), and a potential E_a (accumulation potential) applied. This potential is less negative than the potential corresponding to the peak potential obtained in the study of the reduction of every compound by differential pulse polarography. The solution in the cell was stirred continously during the accumulation time, t_a. The solution was then maintained at rest during the time t_r (rest time). The determination step was carried out using differential pulse polarography (DPP) starting at the initial potential E_a and scanning to a potential above the peak potential. The scan was finished at -1.20 V and the adsorptive stripping cycle was repeated using a new drop. All data were obtained at 20°C.

RESULTS AND DISCUSSION

In order to optimise accumulation conditions, there are several factors which must be investigated: nature of solvent, ionic strength, mass transport, temperature, pH, potential, time. Several of these parameters, and also some instrumental ones affect directly the voltammetric response, mainly in the form and reproducibility of the waves.

The results obtained in the study of the influence of the electrolyte and pH indicate that, in general, the best electrolytes for analytical purposes are Britton-Robinson buffers of pH between 5 and 7 with a concentration of 0.04 M, and acetate buffers of pH between 5 - 6 and concentrations between 0.04 and 0.25 M (see Table I). Very acidic and very alkaline solutions are not adequate for analytical purposes, since either the adsorption is decreased or hydrolysis of several drugs are observed.

Figure 1 summarizes the values of i_p obtained for BrTDO at different pH's and using different electrolytes. Figure 2 indicates the effect of the accumulation time (t_a) on the i_p values obtained for clorazepate at different concentrations of acetate buffers of pH 5.0 .

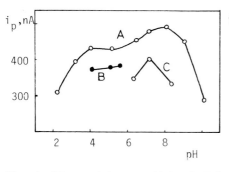 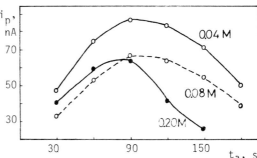

Fig. 1. Plots of ip vs. pH for BrTDO (4.5x10^{-7}M) in A) B-R buffer pH=2-10; B) acetate buffer pH=4-6; C) phosphate buffer pH=6.0-8.5

Fig. 2. Plots of i_p vs. accumulation time of clorazepate (1.2x10^{-7}M) in acetate buffer of various concentrations. E_a = -0.5 V ; t_r= 15 s .

The effect of the accumulation time on i_p was observed on increasing t_a from zero seconds to the time where a maximum adsorption is observed (in Figure 3 the recordings obtained with prazepam are given); above the time of the maximum, i_p decreases with increasing t_a (Figure 4).

In Table I, the optimised accumulation times for every compound are summarized. These times range usually between 50 and 170 seconds.

In the study of the variation of i_p with the accumulation potential (Figure 5) it was observed that for every drug there exists an optimum range of E_a for the adsorption process to be produced; this range corresponded usually to potentials about 300-320 mV more positive that the peak potential, with the exception of bromazepam where the difference is 150 mV (see Table I).

The half-peak widths observed in the voltammograms of solutions of the studied compounds, at 10^{-7} M concentration levels and under optimised conditions, are always lower than 100 mV.

It has been observed that i_p increases linearly with the surface of the drop, remaining practically constant on changing the stirring speed from 1220 to 2620 rpm. The best conditions resulted to be 1920 rpm and 0.44 mm^2 drop size.

Experiments were carried out to investigate the effect of rest time on i_p. These experiments indicated that i_p remains practically constant with the rest time.

Our studies indicated also that peak heigths changed linearly with the ampli-

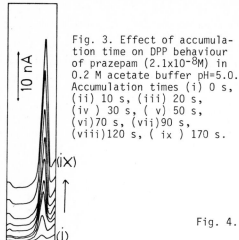

Fig. 3. Effect of accumulation time on DPP behaviour of prazepam (2.1×10^{-8}M) in 0.2 M acetate buffer pH=5.0. Accumulation times (i) 0 s, (ii) 10 s, (iii) 20 s, (iv) 30 s, (v) 50 s, (vi) 70 s, (vii) 90 s, (viii) 120 s, (ix) 170 s.

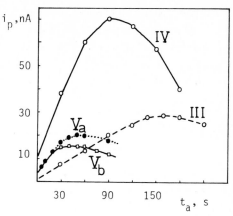

Fig. 4. Plots of peak current vs. accumulation time for III (2.1×10^{-8}M; E_a=-0.55 V); IV (1.0×10^{-7}M; E_a=-0.50 V); V_a (1.5×10^{-7}M; E_a=-0.55 V); V_b (1.5×10^{-7}M; E_a=-0.40 V).

tude of the applied pulse, up to -100 mV. If the increase in the peak heigth and the peak width with increases pulse amplitude is considered, the results obtained indicated that it is not recommended, in general, to use amplitudes lower than -50 mV nor greater than -80 mV. In order to get an adequate analytical response, the following parameters were chosen: scan speed, 10-20 mV/s; scan amplitudes, lower than 0.6-0.7 V; pulse repetition, 0.4 seconds.

From the voltammograms recorded at different concentrations for each drug (Figure 6), linear calibrations were obtained at several accumulation times (Fig. 7).

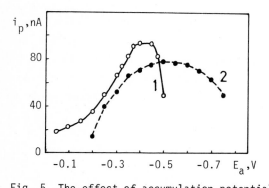

Fig. 6. DPP curves for (i) 0, (ii) 1.99×10^{-8}M, (iii) 3.97×10^{-8}M, (iv) 5.93×10^{-8}M, (v) 7.87×10^{-8}M clorazepate in 0.04 M acetate buffer pH=5.0 (t_a=90 s; E_a= -0.50 V ; t_r=15 s).

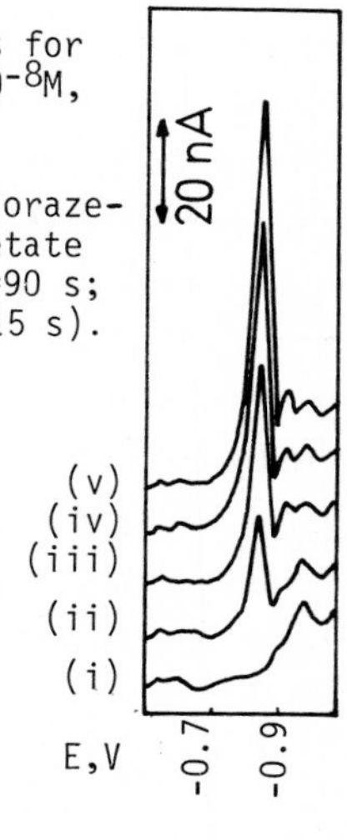

Fig. 5. The effect of accumulation potential E_a on peak currents for 1. Bromazepam (1.0×10^{-7}M); B-R 0.04M; pH=5.0; 2. Clorazepate (1.0×10^{-7}M); acetate 0.04M; pH=5.0.

For the optimum conditions for the analytical determination, the corresponding slopes of the calibration plots were obtained and these values and the corresponding correlation coefficients are given in Table I.

According to the calibration plots, it is possible to determine each drug in
the concentration range indicated. For greater concentrations it is also possible to apply this method, but it is necessary to change the working conditions.
It is also possible to use the method of standard addition for these determinations.

The detection limits usually increase with the accumulation time (i.e. for
prazepam for t_a=0 seconds, the detection limit was found to be 6.7×10^{-9}M and
for t_a=120 seconds, the detection limit was 1.0×10^{-9}M). For the optimum accumulation times considered in Table I, the detection limits ranged around 5.0×10^{-10}M.

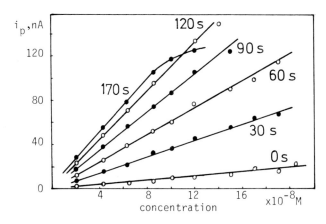

Fig. 7. Calibration cur-
ves for prazepam obtained
for various accumulation
times in 0.2M acetate bu-
ffer pH=5.0 (E_a=-0.55 V ;
t_r=15 s).

The reproducibility of the results is a direct function of the instrumental
conditions, attributed to the automatic control provided by the VA 646 Processor.
In all the cases studied and for the concentration ranges indicated above, the
typical standard deviation is lower than 6% (2% for 5×10^{-8}M).

The determination of these drugs was also carried out in artificial serum
with good results, but the resulting detection limits were about 5×10^{-9}M.

Chloride ion does not seem to affect the adsorption of these drugs, but adsorption is affected by ionic strength; usually adsorption increases on increasing ionic strength up to μ=0.3.

It is possible to conclude that adsorptive stripping voltammetry at the hanging mercury drop electrode is suitable for trace measurements of various benzodiazepines and thienodiazepines, as they are adsorbed according to a fast and
spontaneous process. The sensitivity is significantly enhanced over that obtained by conventional solution phase pulse voltammetry.

TABLE I

Compounds	I		II	III	IV	V
Electrolytes	B-R(0.04M)	Ac(0.25M)	B-R(0.04M)	Ac(0.20M)	Ac(0.04M)	B-R(0.04M)
pH	7.0	5.6	5.0	5.0	5.0	5.0
E_a, V (*)	-0.740	-0.625	-0.400	-0.550	-0.500	-0.550
t_a, s	80	80	90	170	90	60
E_p, V (*)	-1.040	-0.937	-0.545	-0.873	-0.850	-0.860
Linear response, M	$2\text{-}80 \times 10^{-9}$	$2\text{-}50 \times 10^{-9}$	$1\text{-}8 \times 10^{-8}$	$2\text{-}9 \times 10^{-8}$	$1\text{-}9 \times 10^{-8}$	$4\text{-}9 \times 10^{-8}$
Sensitivity, nA.l.mol^{-1}	2×10^9	1×10^9	1×10^9	1×10^9	9×10^8	2×10^8
Correlation coefficients	-0.991	-0.998	-0.999	-0.999	-0.999	-0.995
Detection limits, M	2×10^{-10}	3×10^{-9}	1×10^{-9}	4×10^{-9}	3×10^{-9}	2×10^{-9}

(*) vs Ag/AgCl/KCl (3M)

ACKNOWLEDGEMENTS
The authors wish to thank CAICYT for economic support of this project.

REFERENCES
1. M.A. Brooks, in F. Smyth (Ed.), Polarography of Molecules of Biological Significance, Academic Press Inc., London, 1979, Ch. 3, pp. 85-109.
2. H. Schutz, Benzodiazepines a Handbook, Springer Verlag, Heidelberg, 1982,
   p. 266.
3. M.A. Fernandez Arciniega, Doctoral Thesis, Autonoma University of Madrid,
   1984, Ch. IV, pp. 159-288.
4. R. Kalvoda, Anal. Chim. Acta, 162 (1984) 197-205.
5. J. Wang, Stripping Analysis: Principles, Instrumentation and Applications,
   Verlag Chemie, Dearfield Beach, 1985.
6. J. Wang, International Laboratory, October (1985) 68-76.

M.R. Smyth and J.G. Vos
Electrochemistry, Sensors and Analysis
© Elsevier Science Publishers B.V., Amsterdam — Printed in The Netherlands

REDUCTION MECHANISM OF TRIAZOLAM

R.M. JIMENEZ[1], R.M. ALONSO[1] and L. HERNANDEZ[2]

[1]Chemistry Department. Science Faculty of UPV. 48080 Bilbao (Spain)

[2]Department of Analytical Chemistry. Autonoma University. 28034 Madrid (Spain)

SUMMARY

The polarographic behaviour of Triazolam in aqueous methanol over a pH range $1 < pH < 10.5$ has been studied. The number of electrons, the charge transfer coefficients, α, and the number of hydrogen ions involved in the reaction process have been calculated. Various mechanisms of the reduction reaction have been proposed, which explain the polarographic behaviour of the species of Triazolam existing in solutions at the different pH studied.

INTRODUCTION

One of the most important groups of psychotropic drugs is that of benzodiazepines (ref.1). In the last years a great effort has been made in developing new drugs belonging to this group in order to achieve novel drugs with interesting pharmacological activity.

Triazolobenzodiazepines (ref.2) are one of the newer benzodiazepine derivatives, which have a triazolo nucleus bound to the diazepinic ring. Triazolam belongs to this family and is used as a minor tranquilizer with hypnotic properties (refs. 3,4).

The existence of the electroactive azomethine group, $>C=N-$, which can be reduced at a dropping mercury electrode (DME) allows the study of this compound by polarography.

Oelschager et al. (ref.5) studied the polarographic reduction of Triazolam in 10% dimethylformamide, while Triballet et al. (ref.7) carried out a study of the acid-base equilibrium and the electrochemical reduction in absolute ethanol.

The aim of this work is a detailed study of the reduction of Triazolam in aqueous methanol over a wide pH range 1 – 10.5, and to postulate the possible reaction mechanism for the electrochemical reduction of the compound.

EXPERIMENTAL

Apparatus

Princenton Applied Research Corporation (PAR) model 174 polarographic analyzer equipped with a PAR model 303 SMDE cell and a Houston Omnigraphic X-Y recorder was used.

A three electrode semi-micro cell with a working volume of 10.0 ml was employed containing a dropping mercury electrode (capillary tube of inside diameter

0.015 mm) as the indicator electrode, saturated Ag/AgCl/KCl as the reference
electrode and a platinum wire as the auxiliary electrode.

Polarography was performed in the sampled dc and the differential polarographic modes. A pulse of 25 mV was applied using a drop-time of 0.5 s, the electrode area was 0.0096 cm^2. Scans were recorded from −0.30 to −1.40 V (vs.Ag/AgCl
electrode) at a scan rate of 5 mV s^{-1} with a full range of 1.5 V.

Cyclic voltammetry was performed with an AMEL model 4488 oscillographic polarograph. The electrode assembly, in this case, consisted of a hanging mercury
drop electrode as the working electrode, a saturated calomel electrode as the
reference and a platinum as the auxiliary electrode.

The number of electrons was determined by microcoulometry using a AMEL 461
polarograph.

All experiments were performed at 20 ± 1°C. Dissolved air was removed from the
solutions by degassing with oxygen-free nitrogen for 8 minutes.

pH-meaurements were carried out with a Radiometer pHM64 digital pH-voltimeter
using a combined glass/calomel electrode Ingold gK 2301C.

Reagents

Triazolam was kindly supplied by Upjohn Co. A stock solution of the drug
(2.94×10^{-3} M) was prepared in Merck p.a. methanol and stored in the dark under
refrigeration. The pH was varied using a stock Britton-Robinson buffer (pH=1.88).
All the other reagents used in this work were of analytical grade.

RESULTS AND DISCUSSION

Polarograms of Triazolam in 8% methanol in 0.5 M KCl and Britton-Robinson
buffers as supporting electrolytes recorded at different pH values, showed a
single peak over the whole pH range.

At acidic medium, pH < pK_a, the initial polarographic peak shifts with time to
more negative potential (Fig.1). This behaviour can be explained by the acidic
hydrolysis of the compound (ref.7,8), which has been studied by spectroscopic
methods (ref.9). The acidic hydrolysis of Triazolam is shown in the Scheme 1.

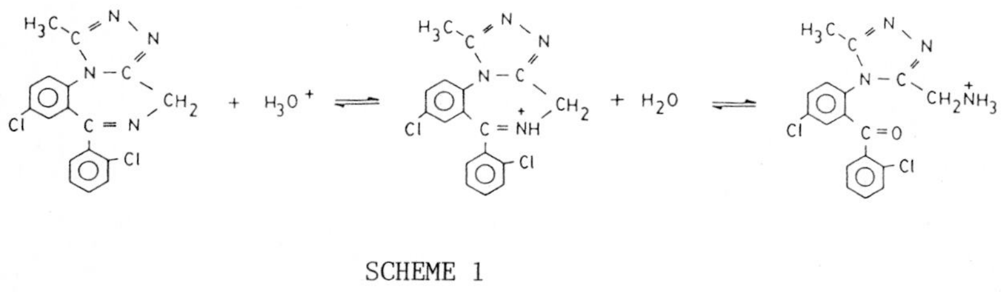

SCHEME 1

The existence of a single peak at the whole pH range can be explained by the
reduction of every species of Triazolam (open-ring, protonated and unprotonated
form) in solution through the same electroactive group $>$C=N−.

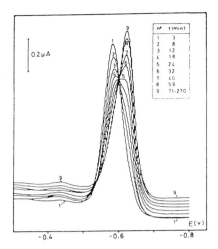

Fig.1. DPP polarogram variation of Triazolam at pH=1.3 as a function of time.

The variation of peak potential E_p with pH (Fig.2) showed for pH < pK_a two straight time-dependant lines. The peak potential was measured both at t=3 min and at t= ∞

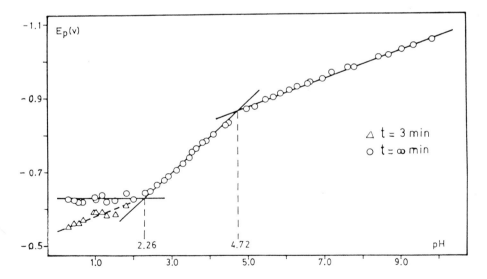

Fig.2. Influence of pH on the peak potentials of Triazolam

The number of hydrogen ions taking part in the electrode reaction, m, was calculated from the plot E_p vs pH according to the expression:
$$\Delta E_p / \Delta pH = (0.059/\alpha n) m.$$
The number of electrons, n, was determined by microcoulometry and was found to be 2 over the whole pH range. The charge transfer coefficient, α, was calcul-

ated from plots of $\log(i_d-i)/i$ vs pH.

Various mechanisms can be proposed which explain the polarographic behaviour of Triazolam in the different pH zones studied.

pH < 2.3

I) at t = 3 min. When the hydrolysis reaction does not occur.

m = 1 and n = 2

$$H_3C-C \cdots N\,H^+ \quad N-C-CH_2 \cdots + H^+ + 2e \;\rightleftharpoons\; H_3C-C \cdots N\,H^+ \quad N-C-CH_2 \;\; CH-NH$$

II) at t = ∞ . When the hydrolysis equilibrium is reached.

m = 0 and n = 2

$$H_3C-C \cdots N\,H^+ \quad N-C-CH_2\overset{+}{N}H_3 \;\; C=\overset{+}{O}H \;\rightleftharpoons\; H_3C-C \cdots N\,H^+ \quad N-C \;\; C=\overset{+}{N}H \;\; CH_2 + H_3O^+$$

$$\Big\Updownarrow 2\,e$$

$$H_3C-C \cdots N\,H^+ \quad N-C \;\; CH-NH \;\; CH_2 + H_2O$$

The reduction mechanism proposed for this pH zone at time=∞ is a CE mechanism. The hydrolysis equilibrium is shifted from the open-ring form to the closed-ring one at the electrode surface and this latter form is reduced. This explains the increased irreversibility of the process at these pH values.

The reversibility of the process has been studied by different polarographic techniques (ref.10).

The fact that both amino and ketonic groups (in the open-ring form) can be protonated explains why the peak potential is pH independent.

2.3 < pH < 4.7

m = 2 and n = 2. The mechanism involves the reduction of azomethine group to the corresponding amine (ref.11).

<u>pH > 4.7</u>

$m = 1$ and $n = 2$. The decrease of the plot E_p vs. pH slope for these pH values can be explained by an EC mechanism, where the Triazolam is first reduced at the electrode surface and then an acid-base equilibrium occurs; the resonance proton of the triazolo ring takes part in that equilibrium, so that the overall reduction process involves only one proton.

REFERENCES

1 H. Schutz, Benzodiazepines. A Handbook, Springer-Verlag, 1982.
2 M. Gall, J.B. Hester, Ger. Offen. DE 3,413,709 (Cl. C07D407/04),25 Oct.(1984).
3 M. Gall, B.V. Kamdar & R.J. Collins, J.Med.Chem., 21(12) (1978) 1290.
4 R.I.H. Wang & S.L. Stockdale, J.Int.Med.Resp., 1 (1973) 600.
5 H. Oelschager, F.I. Sengun & A. Kruskopf,Fresenius Z.Anal.Chem., 315 (1983)53.
6 C. Triballet, P. Boucly & M. Guernet, Bull. Soc. Chim. France,2(3) (1981) 113.
7 M. Konishi, K. Hirai & Y. Mori, J. Pharm. Sci., 71 (1982) 1328.
8 R.M. Jimenez, R.M. Alonso, F.Vicente & L.Hernandez, submitted for publication.
9 R.M. Jimenez, E. Dominguez, L. Badia, R.M.Alonso & F.Vicente,submitted for
   publication.
10 R.M. Jimenez, Doctoral Thesis, Bilbao 1985.
11 H. Oelschger, Bioelectrochem. Bioenerg., 10 (1983) 25.

FLOW INJECTION AMPEROMETRIC DETERMINATION OF CLOTIAZEPAM AT A GLASSY CARBON
ELECTRODE

R.M. ALONSO[1], R.M.JIMENEZ[1], M.A. FERNANDEZ-ARCINIEGA[2] and L. HERNANDEZ[2]
[1]Chemistry Department,Science Faculty of UPV, 48080 Bilbao (Spain).
[2]Department of Analytical Chemistry, Autonoma University, 28034 Madrid (Spain).

SUMMARY

Clotiazepam has been oxidized at a glassy carbon electrode in alkaline medium, which allows its determination by flow injection analysis with amperometric detection. Removal of oxygen is unnecessary and the rectilinear concentration range is found to be between 0.5 and 20 ppm, with coefficients of variations on ten determinations less than 2%.

INTRODUCTION

Benzodiazepines are a group of compounds of considerable therapeutic importance as tranquilizers and anxyolitic agents (ref. 1).

1-methyl-5-o-chlorophenyl-7-ethyl-1,2-dihydro-3H-thieno |2,3-e||1,4| diazepin-2-one, Clotiazepam, belongs to the Thienodiazepine family; its only difference with the Benzodiazepines is that a thiophene ring instead of a benzene ring is bound to the diazepinic moiety. Nakanishi et al.(ref.2) studied the pharmacological properties of this compound, in comparison with those of Benzodiazepines, showing that its properties were similar and that the compound is a well-tolerated and excellent antianxiety agent.

Flow injection analysis (FIA), introduced in 1975 by Ruzicka and Hansen (ref. 3) as a simple and convenient appoach to continous flow automatic analysis, offers great advantages, such as flexibility, speed and high sampling rates, which allows its application in many fields.

The electroactive group in Clotiazepam,the $C=N-$ group,is reducible at a dropping mercury electrode, and its polarographic determination has been carried out (ref. 4).

In this work, a study on the oxidation of Clotiazepam at a glassy carbon electrode in different media as well as the development of a determination method of the compound by FIA system with amperometric detection have been carried out.

EXPERIMENTAL

Apparatus

Voltammograms were obtained using a PAR 174 polarographic analyser (Prince - ton Applied Research) with three-electrode operation(glassy carbon electrode, platinum counter electrode and calomel reference electrode). In all experiments a Houston Instruments Omnigraphic 2000 X-Y recorder was used.

For linear sweep voltammetry (LSV) a sweep rate of 10 mV s^{-1} and a low pass filter of 1swas used.

In the flow injection experiment, an Ismatec Mini-S peristaltic pump was used to maintain the flow of eluent and injections were made with a Rheodyne injection valve (5020) with an injection volume of 75 μl; the injection valve was connected to the detector cell by means of 1 meter of 0.58 mm bore tubing.

The detector cell used in the flow injection analysis experiments consisted of a home-made simple wall-jet detector, which is used partially immersed in electrolyte solution (the same as the carrier solution) in order to allow electrical contact to the reference electrode. The cell system was throughly described by Fogg et al. (ref. 5).

It was found to be unnecessary to de-gas the eluent.

Reagents

Clotiazepam was kindly supplied by ESTEVE Laboratories, Barcelona.

A standard Clotiazepam solution of 3.13x10^{-3} M was prepared in Merck p.a. methanol and kept under refrigeration to avoid degradation.

A 1 M stock solution of sodium hydroxide was prepared from doubly-distilled water.

All the reagents used in this study were of analytical grade.

RESULTS

Linear sweep voltammetry

Linear sweep voltammetry studies of Clotiazepam showed that at a glassy carbon electrode, the compound was only oxidized in alkaline medium. No electro-activity was observed at other pH values studied. This fact could be due to the hydrolysis of the compound in alkaline medium (ref.6).

Clotiazepam has a single peak in alkaline medium at 0.16 V, which offers the possibility of developing a procedure, based on flow injection analysis with amperometric detection , for a rapid determination of the drug.

Previous experiments showed that the hydrolysis reaction of the compound appears at pH values greater than 10 and is a kinetic process, depending on the OH$^-$ concentration. This reaction is practically instantaneous at a pH value of 12; therefore to increase the measurement speed,all the experiments have been carried out at this pH.

Determination of Clotiazepam by flow injection analysis

As said earlier, the compound was electroactive only in alkaline medium (see Fig.1), and therefore sodium hydroxide was chosen as the medium for the flow injection determination of Clotiazepam. 0.05 M NaOH was used as carrier solution and all the drug samples were prepared in this medium.

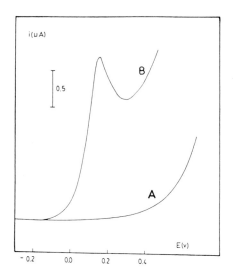

Fig. 1. Linear sweep voltammogram of Clotiazepam at a glassy carbon electrode in alkaline medium. (A): supporting electrolyte 0.05 M NaOH, (B) : Clotiazepam

A hydrodynamic voltammogram was obtained by injecting the Clotiazepam solution in 0.05 M NaOH, recording the signals obtained at different potentials and plotting the peak current vs. potential (see Fig.2). From this plot, a potential of + 0.05 V was chosen for the quantitative determination of the compound.

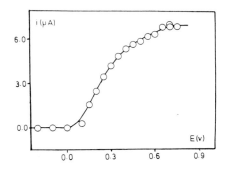

Fig. 2. Hydrodynamic voltammogram at a glassy carbon electrode of 100 ppm of Clotiazepam in 0.05 M sodium hydroxide.

400

The effect of the flow-rate on the signal size at the potential chosen is shown in Fig. 3. A flow-rate of 6.0 ml min^{-1} was used for all subsequent experiments.

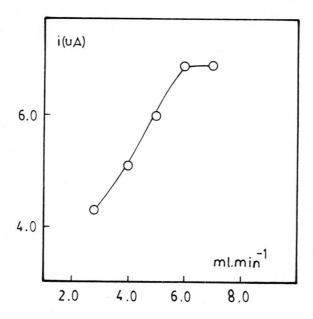

Fig. 3. Influence of flow-rate on the signal size.

Signals were shown to be rectilinear with respect to Clotiazepam concentration between 0.5 and 20 ppm. Typical signals obtained at 0.5 - 2 ppm levels of the drug are shown in Fig. 4. Coefficients of variation on ten determinations at 10 ppm and 0.5 ppm were 1.56 and 1.68 %, respectively.

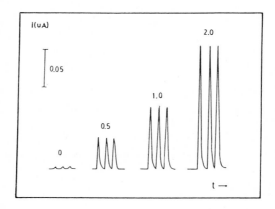

Fig. 4. Flow injection signals obtained for the determination of Clotiazepam.

DISCUSSION

FIA determination of Clotiazepam with amperometric detection, using a static mercury electrode has been studied in this laboratory (ref. 7). The detection at the glassy carbon electrode offers the advantage that it is not sensitive to dissolved oxygen making degassing unnecessary; also the detection limit is greater than that obtained with a static mercury electrode.

The developed method is applicable to the determination of the drug in its pharmaceutical formulations.

REFERENCES

1 H.Schutz, Benzodiazepines,A Handbook, Springer-Verlag, 1982.
2 M.Nakanishi, T.Tsumagari, Y.Takigawa, S.Shuto, T.Kenjo & T.Fukuda, Arzneim-
  Forsch./Drug Res., 22(1972) 1905.
3 J.Ruzicka & E.H.Hansen, Flow Injection Analysis, Willey-Interscience,
  New York, 1981.
4 R.M.Alonso & L.Hernández, Anal. Chim. Acta, in press.
5 A.G.Fogg & A.M.Summan, Analyst, 109 (1984) 1029.
6 R.M.Alonso, Doctoral Thesis, Bilbao (1983).
7 R.M.Alonso, J. Pharm. and Biomed. Anal.,in press.

M.R. Smyth and J.G. Vos
Electrochemistry, Sensors and Analysis
© Elsevier Science Publishers B.V., Amsterdam — Printed in The Netherlands

ELECTROANALYSIS OF CEFAZOLIN, CEFTRIAXONE AND CEFOTAXIME

B.OGOREVC[1], M.R.SMYTH[2], V.HUDNIK[1] and S.GOMIŠČEK[3]

[1]Boris Kidrič Institute of Chemistry, Ljubljana (Yugoslavia)

[2]National Institute for Higher Education, Dublin (Ireland)

[3]Faculty of Natural Sciences and Technology, E.Kardelj University, Ljubljana (Yugoslavia)

SUMMARY

The cephalosporin derivatives cefazolin, ceftriaxone and cefotaxime were studied by direct current (d.c.), sampled d.c., differential pulse polarography (d.p.p.) and cyclic voltammetry. While cefazolin gives rise to one reduction wave, both ceftriaxone and cefotaxime give rise to two reduction waves. The electrode reactions giving rise to these waves are discussed. Optimum pH ranges for the determination of cefazolin, ceftriaxone and cefotaxime are 1.0-3.0, 3.0-5.0 (wave I) and 2.0-3.0 respectively. The linear concentration range for cefazolin is 0.1-100.0 ug/ml (d.p.p.) with a limit of detection (L.O.D.) of 0.01 ug/ml, for ceftriaxone 0.2-20 ug/ml with a L.O.D. of 0.02 ug/ml and for cefotaxime 0.25-15 ug/ml with a L.O.D. of 0.02 ug/ml. For n=10 and at a concentration of 0.5 ug/ml, the relative standard deviation for the polarographic method is 1.0 %, 1.7 % and 1.4 % for cefazolin, ceftriaxone and cefotaxime, respectively.

INTRODUCTION

Cefazolin (CEFZ), ceftriaxone (CTRX) and cefotaxime (CFTM) are cephalosporin derivatives, which are widely used in clinical therapy. Most cephalosporins are electrochemically active. The literature dealing with their electroactivity and resulting analytical applications can be divided into two parts: (i) those papers concerning the direct polarographic activity of cephalosphorins (1-10), and (ii) those papers dealing with the polarography of their degradation products after acidic, neutral or alkaline hydrolysis (11-14).The main electrode reaction responsible for direct polarographic activity of cephalosporins is claimed to be the reduction of the $\overset{3}{\Delta}$ double bond of the cephem nucleus, which is dependent on the presence and nature of the substituent at position 3. However, various authors have different standpoints on the reaction mechanism of this reduction(3-5,7,13) Certain cephalosporins contain additional,or possibly other,reducible groups,the polarographic response of which can be used for analytical purposes (8,9). Only one compound, i.e. cephadroxil, has been reported to undergo oxidation at the glassy carbon electrode (7). No report has yet been published on the polarographic properties of cefazolin.

In this work, cefazolin, ceftriaxone and cefotaxime were studied by direct current (d.c.), sampled d.c., differential pulse polarography (d.p.p.) and cyclic voltammetry (c.v.). The voltammetric behaviour of these compounds is reported and methods described for their analysis.

Cefazolin

Ceftriaxone

Cefotaxime

EXPERIMENTAL

Reagents and apparatus

All reagents and chemicals used in this study were of analytical grade. A Britton-Robinson (B.R.) buffer, 0.04 M in boric, orthophosphoric and acetic acid (with 0.2 M NaOH to adjust pH), or a Clark-Lubs (C.L.) buffer 0.2 M in hydrochloric acid and potassium chloride (with water in different ratios), were used as supporting electrolytes. Polarographic measurements were recorded using a Solea-Tacussel polarograph (PRG 5) operated in conjuction with a slightly modified polarographic stand CPR 3B. The electrode system was constructed with a Pt wire as counter electrode, a saturated Ag/AgCl/KCl as reference electrode and a dropping mercury electrode as a working electrode. Cyclic voltammetric scans were made employing an additional function generator (GSTP 3) and a hanging mercury drop electrode (Metrohm EA 290) as a working electrode. For recording of polarographic waves, scan rates of 2 mV/s or 4 mV/s were used. In the sampled d.c. mode, a free drop time or drop time of 3 s was used; for d.p.p. operation, a drop time of 0.9 s and pulse amplitude of 100 mV was applied. For cyclic voltammetric studies, a scan rate of 80 or 100 mV/s was used. Before and during all measurements, solutions were kept at 25.0 $\pm$0.2°C unless otherwise specified.

Materials

Cefazolin as cefazolin-sodium was supplied by Krka Pharmaceutical Works (Novo mesto,Yugoslavia); ceftriaxone as ceftriaxone-disodium $\cdot$ 3.5 H_2O was supplied by Hoffmann-La Roche, Basle; cefotaxime as cefotaxim-sodium was supplied by Jugoremedija (Hoechst A.G.). The substances in powder form were first dried under a slight vacuum over silica gel. The stock solutions were made by dissolving an appropriate amount of substance in pure water and stored at ca. $4^{o}C$.

Procedures

A 20 ml aliquot of the appropriate supporting electrolyte solution was placed in the measurement cell and purged with argon for about 15 min. After each standard solution addition, the solution was purged with argon for another 30 s before the corresponding polarogram or voltammogram was recorded.

RESULTS AND DISCUSSION
Polarography of cefazolin

Cefazolin exhibits a single, well-defined polarographic wave with a half-wave potential of about -0.63 V in B.R. buffer pH 3.0. Its position and shape is pH - and also concentration - dependent. A characterization of the limiting current was made following studies of the effect of the height of the Hg column, temperature and drop time on this wave. These studies revealed that the wave was diffusion-controlled. The Hg column height test did not exclude, however, that an adsorption process was also involved. This finding was confirmed by studying the pH-dependence of the limiting current. It was found to be independent of pH in the range 1.0-7.0. At higher pH's, a strong maximum (width 170 mV) appeared which could be removed by addition of 0.001 % Triton X-100. Moreover, the limiting current measured at sufficiently negative potentials showed that it was independent of pH up to pH 11.4, where a new wave appeared. This wave at -1.58 V increased with time, clearly indicating an alkaline hydrolysis of cefazolin at this pH. The d.p.p. peak current dependence on pH involves a maximum at pH 2-3 which decreases on increasing pH (Fig.1). From the S-shaped pH-dependence of half-wave or peak potential of cefazolin, it can be concluded that single protonation of two sites are very probable. A logarithmic analysis of the d.c. wave shape implied an irreversible, two-electron reduction process. Its irreversibility was confirmed by a cyclic voltammetric study using different scan rates, pH values and concentrations. This study showed a weak chemisorption of both reactant and product on the Hg surface, which is independent of the potential applied. Because of these complicating adsorption phenomena and without an identification of electrode reaction product(s),no definite conclusion about the reaction mechanism of reduction of cefazolin can be made on the basis of these results. It is likely, however, that reduction occurs at the Δ^3 double bond as proposed for other cephalosporins containing a thioether bridging at the 3-position e.g. cefamandol, ceftriaxone,moxalactam or cefuroxim (5, 9, 15).

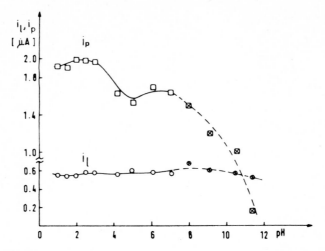

Fig. 1. Dependence of d.c. limiting current and d.p.p. peak current on pH for cefazolin. CEFZ conc.: 1.1×10^{-4} M; medium: C.L. buffer pH 1-2; B.R. buffer pH 2-12; $\otimes$, $\boxtimes$: wave (peak) with maximum.

Polarographic behaviour of ceftriaxone and cefotaxime

Both ceftriaxone and cefotaxime give rise to two polarographic waves, the more positive one of which is assigned as wave I and the more negative as wave II. Limiting currents for both waves at pH 3 (CTRX) and pH 2 (CFTM) were found to be diffusion-controlled, although influenced by adsorption, especially wave I for CTRX. Figure 2 demonstrates the pH-dependence of potentials, slopes and heights of d.c. waves of CFTM. It clearly indicates the splitting of wave I at pH values higher than 4. For ceftriaxone, wave II is distinct up to pH 6, but decreases in height on increasing pH; wave I splits into two waves at pH's higher than 6. Logarithmic analysis for wave II in both cases indicates an irreversible two-electron reduction process. Waves I involve more complexity, especially in the case of CTRX. The "log plot" for CTRX is divided into two portions with different slopes. This fact, together with data from the cyclic voltammetry of wave I of CTRX at low pH values, showing a sharp peak of high current (Fig.3), led us to the assumption that not only chemisorption of CTRX, but also some other surface phenomena must be taking place, causing a rapid change in the capacity of the double layer (e.g. reorientation of CTRX molecule at the electrode surface). On the basis of our results (i.e. logarithmic analysis, dependence of half-wave potentials on pH, heights of limiting current), together with a comparison of literature data for compounds of similar structure (16, 17), a double two-electron reduction mechanism could be proposed, which involves a reduction of the methoxyimino group via a hydroxylamine intermediate to the corresponding amine and methanol.

$$\text{>C=N-OCH}_3 \xrightleftharpoons{\text{H}^+} \text{>C=N-}\overset{+}{\underset{\text{H}}{\text{O}}}\text{CH}_3 \xrightarrow{\text{2e}^-, \text{2H}^+} \text{>CH-NH-}\overset{+}{\underset{\text{H}}{\text{O}}}\text{CH}_3$$

$$\text{>CH-NH-}\overset{+}{\underset{\text{H}}{\text{O}}}\text{CH}_3 \xrightarrow{\text{2e}^-, \text{2H}^+} \text{>CH-}\overset{+}{\text{N}}\text{H}_3 + \text{HOCH}_3$$

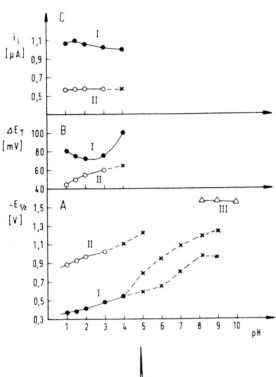

Fig. 2. Dependence of half-wave potentials (A), slopes (B) and heights of limiting currents (C) of d.c. waves of CFTM on pH of medium; medium: as in Fig. 1
CFTM conc: 1.1×10^{-4} M
x: ill-defined waves
ΔE_T: $E_{3/4} - E_{1/4}$

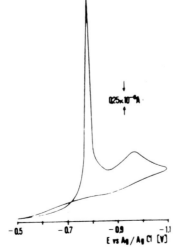

Fig. 3. Cyclic voltammetric behaviour of ceftriaxone.
CTRX conc: 1.1×10^{-4} M
medium: B.R. buffer, pH 2.0
Experimental parameters: scan rate 80 mV/s, starting potential -0.5 V

Analytical applications

On the basis of the electrochemical study presented above, analytical parameters for the polarographic (mainly d.p.p.) determination of the three cephalosporins were investigated as shown in Table 1.

Table 1
Parameters for d.p.p. determination of cefazolin, ceftriaxone and cefotaxime in pure buffer solutions. Optimum apparatus settings: pulse amplitude, 100 mV; drop time, ca. 1 s; scan rate, 2-4 mV/s

Compound	Optimum pH range	$-E_p$ (V vs Ag/AgCl)	pH	Supp. electro- lyte	Linear concentration range ($\mu g/ml$)	Limit of detection ($\mu g/ml$)
CEFZ	1-3	0.64 ± 0.03	3.0	B.R. C.L.	0.1 - 100	0.01
wave I	3-5	0.63 ± 0.00	4.2	B.R.	0.2 - 20 (10 - 100)*	0.02
CTRX						
wave II	1-3	0.82 ± 0.02	3.0	B.R. C.L.	4 - 100	0.5
wave I	2-3	$0.44 \pm 0.07*$	2.0	B.R.	5 - 350*	3.0
CFTM						
wave II	2-3	0.93 ± 0.01	2.0	B.R.	0.25- 15	0.02

*: d.c. sampled polarography

Stability tests proved that stock solutions (10 mg/ml) of all three cephalosporins in water can be used during 3 days, provided they are stored at ca. 4°C. The stability of the polarographic signal itself is good enough for analytical purposes. The precision of d.p.p. determinations for all three cephalosporins is excellent. For n=10 and at a concentration level of $0.5 \mu g/ml$, the relative standard deviation is 1.0 %, 1.7 % and 1.4 % for cefazolin, ceftriaxone and cefotaxime, respectively.

REFERENCES

1 I.F.Jones, J.F.Page and C.T.Rhodes, J.Pharm. Pharmac., 20 (1968) Suppl. 45S-47S.
2 D.A.Hall, J. Pharm. Sci., 62 (1973) 980-983.
3 M.Ochiai, D.Aki, A.Morimoto, T.Okada, K.Shinozaki and Y.Asahi, J. Chem. Soc. Perkin Trans., 1 (1974) 258-262.
4 D.A.Hall, D.M.Berry and C.J.Scheider, J.Electroanal. Chem., 80 (1977) 155-170.
5 E.C.Rickard and G.G.Cooke, J.Pharm. Sci., 66 (1977) 379-384.
6 A.G.Fogg, N.M.Fayad, C.Burgess and A.McGlynn, Anal. Chim. Acta, 108 (1979) 205-211.
7 A. Ivaska and F. Nordström, Anal. Chim. Acta, 146 (1983) 87-95
8 L.Camacho, J.L.Avila, A.M.Heras and F.García-Blanco, Analyst, 109 (1984) 1507-1508.

9 F.I.Şengün, K.Ulas and I.Fedai, J. Pharm. Biomed. Anal., 3 (1985) 191-199.
10 F.I.Şengün, T.Gürkan, I.Fedai and S.Sungur, Analyst, 110 (1985) 1111-1115.
11 J.A.Squella, L.J.Nuñez-Vergara and E.M.Gonzalez, J. Pharm. Sci., 67 (1978) 1466-1467.
12 A.G.Fogg, N.M.Fayad and C.Burgess, Anal. Chim. Acta, 110 (1979) 107-115.
13 A.G.Fogg and M.J.Martin, Analyst, 106 (1981) 1213-1217.
14 L.J.Nuñez-Vergara, J.A.Squella and M.M.Silva, Talanta, 29 (1982) 137-138.
15 B.Ogorevc, V.Hudnik and S.Gomišček, in preparation.
16 P.Zuman, in O.N.Reilley (Ed.), Advances in Analytical Chemistry and Instrumentation, Vol.2, Interscience Publishers, New York, 1963, p.219-253.
17 M.R.Smyth and W.F.Smyth, Analyst, 103 (1978) 529-567.

M.R. Smyth and J.G. Vos
Electrochemistry, Sensors and Analysis
© Elsevier Science Publishers B.V., Amsterdam — Printed in The Netherlands

DETERMINATION OF ANIONS IN PHARMACEUTICAL MATRICES BY AN AUTOMATED

SINGLE-COLUMN ION-CHROMATOGRAPHY SYSTEM WITH INTEGRATED SAMPLE PRETREATMENT

AND ELECTROCHEMICAL AND SPECTROPHOTOMETRIC DETECTION

Etienne Grollimund and Juerg B. Reust

Analytical R+D, SANDOZ LTD., CH-4002 Basle / Switzerland

INTRODUCTION

In recent years, the importance of anion analysis in pharmaceutical matrices
(raw materials, active principles, final formulations) has steadily grown, not
only for technological interest but also due to regulatory affairs. Some of
the requirements for an analytical method suitable for use in the
pharmaceutical industry are:

- true multi-element/multi-ion analysis
- covering a wide concentration range (even within the
  same sample for different analytes from trace to minor
  constituent range)
- on-line sample pretreatment
- usable for a wide range of matrices
- fully automated

The described method meets all of the above requirements and is suitable for
the determination of anions, such as chloride, nitrate, sulphate, sulfide,
etc., from the percentage range down to the ng-level in a wide range of
pharmaceutical matrices

INSTRUMENTATION

The system consists essentially of the following parts: a) preatreatment system, b) separation system, c) detection system, d) controlling system, and e) data handling system.

a) PRETREATMENT SYSTEM

The pretreatment system allows a conventional operation mode with direct injection of the sample into the eluent. The advantage of our design is the possibility to select by software, eventually with minor hardware changes, from a choice of pretreatment methods the optimized form for a specific problem.

In the following schemes, a complete chromatographic analysis in the tandem mode for time efficient analysis of two samples with preconcentration of the analyte and removal of the interfering matrix, is shown in sequences.

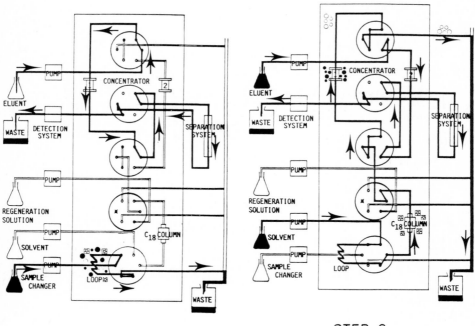

STEP 1   STEP 2

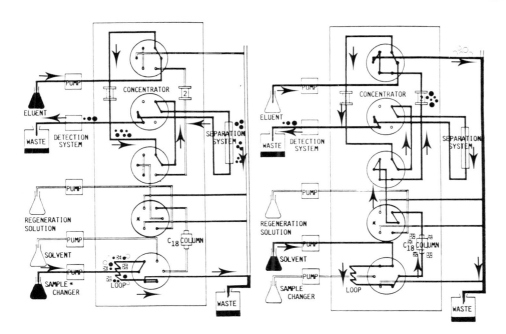

STEP 3   STEP 4

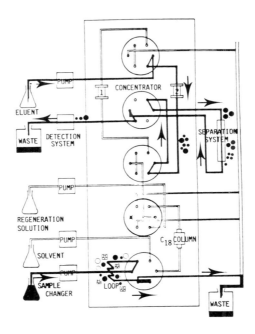

STEP 5

Fig. 1 :   STEP 1

- The system is flushed and equilibrated through concentrator column 1, the analytical column and the detection system by means of electrode flow.

- The injection system is flushed and the loop filled with sample solution containing the analytes (●●●●) and the matrix components (○□▩○).

Fig. 2 :   STEP 2

- The system is flushed and equilibrated through concentrator column 2, the analytical column and the detection system by means of electrolyte flow.

  - The sample solution is transported with solvent (eg. water) through the C_{18}-pretreatment column, where part of the matrix (▩▩▩) is adsorbed. The rest is transported through the concentrator column 1, where the analyte (●●●) is accumulated. The rest of the matrix (○○○○) is swept out of the system.

Fig. 3:   STEP 3

- By means of counterwise flow of the eluent through the concentrator column 1, the analytes (●●●●) are transported to the analytical column where they are separated and eluted into the detection system.

- At the same time, the injection system is flushed and the loop filled with the sample solution 2.

Fig. 4:   STEP 4

- Sample 1 is completely eluted.

- Sample solution 2 is transported through the pretreatment column and the concentrator column 2. Part of the matrix (▩ ▩▩) is adsorbed on the C18-pretreatment column, the analytes (●●● ●) are accumulated on the concentrator column and the rest of the matrix (○○○○) is swept out.

Fig. 5:   STEP 5

- The analytes (●●●●) of sample 2 are eluted from the concentrator column 2, separated on the analytical column and transported to the detection system.
- The injection system is flushed and the loop filled with sample 3.
- Continuation at step 2.

b) SEPARATION SYSTEM

Precolumn: 4.8 mm x 1 cm, Anion Ionconcentrator Inserts,
                      WATERS ASSOCIATES INC., Milford, MA, USA
Analytical Column: 4.6 mm x 5 cm, Anion IC-PAC, WATERS ASSOCIATES
                      INC., Milford, MA, USA
Eluent Pump: Model 2150 HPLC Pump, LKB, Bromma, S
Capillaries: 0.5 mm, stainless steel or Teflon
Sample Changer: Model A40/II, HOOK AND TUCKER INSTRUMENTS LTD.,
                      New Addington, UK
Temperature Control: Foreseen to be controlled within 0.1 degree

c) DETECTION SYSTEM

Conductivity   213A, WESCAN INTRUMENTS INC., Sta. Clara, CA, USA
Detector: Tubular two electrode arrangement
                      2 µl cell volume
Spectrophotometric   LC55, PERKIN ELMER INC., Oakbrook, IL, USA
Detector: 12 µl cell volume
Amperometric   E641/E656, METROHM LTD., Herisau, CH
                      Wall-jet type
                      2 µl cell volume
                      300 mV vs. Ag/AgCl/KCl (3 M)
                      Ag working electrode, in-situ polymerized into
                      PMMA according to [1]

d) CONTROLLING SYSTEM

The system is controlled by a personal computer (HP150, HEWLETT-PACKARD

INC., Palo Alto, CA, USA) and own software programs in basic via an Interface

system developed at SANDOZ [2].

e) DATA HANDLING SYSTEM

The same personal computer is used in its terminal mode as a terminal to the

laboratory automation system LAS 3357 by HEWLETT-PACKARD, which is run on a HP

10000 E-series computer of HEWLETT-PACKARD. This system includes all the

necessary data handling possibilities used for chromatographic   experiments.

The results can be transferred from this system into the analytical data base

of the department automatically.

CHEMICALS

Eluent 1:	0.001 M potassium hydroxide
Eluent 2:	1.48 E-3 M sodium gluconate
	5.82 E-3 M boric acid
	1.30 E-3 M sodium tetraborate x H_2O
	12 % (v/v) acetonitrile
	0.25 % (v/v) glycerin
Quality:	Analytical grade chemicals, MERCK, Darmstadt, FRG
Degassing:	All solvents with high-purity argon
Sample size:	100 to 500 µl
Pretreatment:	4.0 mm x 3 cm Spherisorb ODS, 10 µm, M. KNAUER, Bad Homburg, FRG

RESULTS AND DISCUSSION

Conductivity detector

The conductivity detector offers a wide range of anions to be determined. The simultaneous determination of six different anions are shown in Fig. 6 for two different concentration levels. It is important to note that much lower concentration levels can easily be reached by augmenting the sample size to

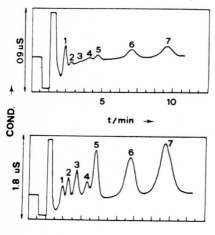

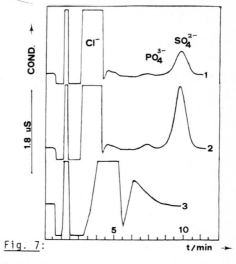

Fig. 6:

Conductometric Determination
of Anions

Preconcentrated from 500 µl
Amounts in ng
1 CO_3^{2-} not det. 2 Cl^- 20/200
3 NO_2^- 50/500 4 Br^- 50/500
5 NO_3^- 50/500 6 PO_4^{3-} 250/2500
7 SO_4^{2-} 250/2500

Fig. 7:

Anion Determination in
Taurocholic Acid

201 mg sample in 100 ml H_2O,
500 µl injected
1 sample (61 µg Cl^-, 0.28 µg
PO_4^{3-}, 1.2 µg SO_4^{2-})
2 first standard addition
3 no pretreatment used

milliliter volumes. Figures 7 and 8 show anion determination in two different
compounds of interest to our department. Both substances contain extremely
high amounts of two anions (chloride and bromide, resp.), where the absolute
amount of analyte is in the ng to μg level. From the figures also the effect
of the pretreatment system is clearly visible and shows that no determination
would be possible by direct injection of the samples.

Further advantages of the pretreatment system are the reduction of cost,
longer column lifetime, and better reproducibility.

Further investigations will include several different modes of operation than
the mentioned tandem mode with analyte accumulation. Additionally the system

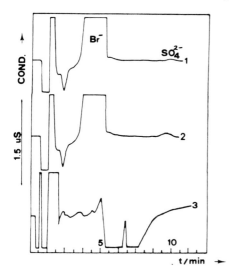

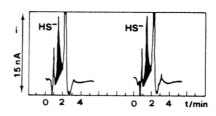

Fig. 8:

SO_4^{2-} Determination in
HBR-Research Compound

100 mg sample in 100 ml H_2O,
500 μl injected
1 sample (25 ng SO_4^{2-}, approx.
  100 000 ng Br⁻)
2 first standard addition
3 no pretreatment used

Fig. 9:

Amperometric Determination
of Sulfide

10 ng in 0.001 M KOH, no
preconcentration used

will also be tested in the two-column operation mode ('DIONEX' operating
principle with separation and suppressor column).

Amperometric detector

The amperometric detector allows the determination of electroactive analytes
such as sulfide, cyanide, etc. Oxidation as well as reduction of the analyte
is possible. Many toxic analytes, such as sulfide, can not be detected by the
conventional supressed ion-chromatography mode, nor by conductivity detection
at low concentrations, therefore making the combination of a single-column
system with amperometric detection very attractive.

All attempts to determine sulfide in the desired concentration range failed
with commercially available electrodes. On the other hand, the silver
electrode design developed at the University of Berne (in-situ polymerization
of PMMA around a silver disk, [1]), allows a determination of sulfide in the
concentration range from 5 to 100 ng/g (cf. fig. 9).

Further investigations will include the evaluation of an electrochemical
pretreatment step and the use of different pulse modes for improving
sensitivity and continous cleaning of the electrode surface. Also experiments
towards the determination of other analytes of interest will be undertaken. Of
additional importance is the possibility for certain analytes (eg. chloride)
to be analyzed by two independent detector systems, adding reliability to the
result.

ACKNOWLEDGEMENTS

The authors like to thank the electrochemistry group (head Prof. E. Schmidt) of the Institute for Inorganic, Analytical and Physical Chemistry of the University of Berne for their cooperation and assistance with respect to the electrochemical aspects of this work. The help of Mr. H.-R. Linder towards hardware design was greatly appreciated.

REFERENCES

[1]   Felix Buechi, Juerg B. Reust, Hans F. Siegenthaler, Automated
        Calibration-Free Determination of Heavy Metals by Anodic Stripping
        Voltammetry in a Thin-Layer Flow-Through Cell, in preparation
        and
        Antonio de Agostini, Diploma Thesis, University of Berne, 1984

[2]   R. Best, Chem. Ing. Tech., 57 (1985) 269-77